Fred Whipple's Empire

The Smithsonian Astrophysical Observatory, 1955–1973

David H. DeVorkin

A SMITHSONIAN CONTRIBUTION TO KNOWLEDGE

Smithsonian Institution
Scholarly Press
WASHINGTON, D.C.
2018

Front cover: Fred Whipple, late 1960s. Smithsonian Institution Archives, SIA Acc.16-263, Box 5, "Whipple 65-72." *Back cover:* The MMT Observatory, on the grounds of the Fred Lawrence Whipple Observatory, Mount Hopkins, Arizona, 1997. Courtesy of the MMT Observatory. Photo by Howard Lester.

Published by SMITHSONIAN INSTITUTION SCHOLARLY PRESS
P.O. Box 37012, MRC 957, Washington, D.C. 20013-7012
https://scholarlypress.si.edu

Library of Congress Cataloging-in-Publication Data
Names: DeVorkin, David H., 1944– author. | Smithsonian Institution Scholarly Press, publisher.
Title: Fred Whipple's empire : the Smithsonian Astrophysical Observatory, 1955–1973 / David H. DeVorkin.
Other titles: Smithsonian Astrophysical Observatory, 1955–1973
Description: Washington, D.C. : Smithsonian Institution Scholarly Press, 2018. | Series: A Smithsonian contribution to knowledge | Includes bibliographical references and index. | Copyright 2018 Smithsonian Institution
Identifiers: LCCN 2018009955 (print) | LCCN 2018011723 (ebook) | ISBN 9781944466176 (ebook) | ISBN 9781944466183 (paperback)
Subjects: LCSH: Smithsonian Astrophysical Observatory—History. | Whipple, Fred L. (Fred Lawrence), 1906–2004. | Astrophysics—United States—History—20th century. | Astronomy—United States—History—20th century. | Science and state—United States—History—20th century. | Cold War—History.
Classification: LCC QB82.U62 (ebook) | LCC QB82.U62 S62 2018 (print) | DDC 522.197444—dc23 | SUDOC SI 1.60:SM 6
LC record available at https://lccn.loc.gov/2018009955

ISBN-13 (online): 978-1-944466-17-6
ISBN-13 (print): 978-1-944466-18-3

♾ The paper used in this publication meets the minimum requirements of the American National Standard for Permanence of Paper for Printed Library Materials Z39.48–1992.

To the Whipple family

CONTENTS

PREFACE

This book explores the forces and drives that brought two astronomical institutions together—the Astrophysical Observatory of the Smithsonian Institution in Washington, D.C., and the Harvard College Observatory in Cambridge, Massachusetts—to become the Harvard–Smithsonian Center for Astrophysics in the early 1970s, one of the largest institutions devoted to astronomy and space science in the world. Astronomer Fred Lawrence Whipple was key in its transformation process, initiated in 1955 by Harvard astronomer Donald Menzel working in concert with Smithsonian secretary Leonard Carmichael and Harvard dean McGeorge Bundy. Initially, Whipple wanted to create an academia-based institutional model for conducting space science in the United States, making his newly minted Smithsonian Astrophysical Observatory a central organizing unit. Instead, after the U.S. government created the National Aeronautics and Space Administration (NASA) to do just that, Whipple deftly adjusted, building highly competitive programs in astrophysics, space astronomy, geophysics, geodesy, and ground-based optical and radio astronomy.

This book examines how Whipple constructed his empire, first through the expansive era preparing for the International Geophysical Year and then through the early NASA years, creating an astronomical enterprise unlike any the world had seen. We follow his continued ambitions through the late 1960s and into the early 1970s, aided and abetted by a new Smithsonian secretary, S. Dillon Ripley, to examine how his continued efforts resulted in both congressional scrutiny of the Smithsonian and a re-evaluation of the relationship of the Smithsonian Astrophysical Observatory to Harvard and, ultimately, to science in America.

ACKNOWLEDGMENTS

This project has occupied my mind for over 15 years, among other duties and diversions. Along the way there have been many who have stimulated, aided, and supported this work. Specific interest started when Patrick McCray asked me to co-sponsor his NSF proposal "Astronomy During the Cold War: The Case of the Smithsonian Astrophysical Observatory, 1955–1975" (NSF Grant Number SES-0323336, 2003–2006) managed by the American Institute of Physics, which led to his monograph on Project Moonwatch, *Keep Watching the Skies!* At about the same time, Teasel Muir-Harmony, a recent graduate of St. John's College, interned at the museum and then stayed on as a researcher assisting in the general archival work in the project. She became interested in SAO's satellite-tracking program setting it in context of international diplomacy, which resulted in her master's thesis from Notre Dame, "Tracking Diplomacy: The International Geophysical Year and American Scientific and Technical Exchange with East Asia, 1955–1973." She and Patrick helped shape the course of my research. At about the same time Louise Thorn (now Devoy) from the University of Leicester interned and worked on the project for an academic year, greatly systematizing records relating to how the SAO grew over time, and assisting in the preparation and execution of many oral histories. A grant from the NASA History Office, "The Space Sciences at 50: An Institutional Case Study" (HSEES06-0023, 2009-2011), shepherded by Steven Dick, Bill Barry, and Steve Garber, made it possible to complete both the oral history and archival work, greatly assisted by interns, fellows, and contract researchers. In particular, Tiffany Hindman, James Neeley, John Hughes, Samantha Thompson, Michelle Krok, Caroline Tung, Chris Hearsey, Sam Zeitlin, and Fernanda Luppani in the first two categories, and Joan Mathys and Audra Wolfe in the third. Mathys provided invaluable expertise and insight performing a broad range of assignments, and Wolfe helped me think afresh about what it was I was trying to do. Archivists, librarians, and colleagues who provided insight, aid, encouragement, and critical commentary are in the dozens; prominent are Pam Henson, Donna Coletti, Ellen Alers, Jane Odom, Steve Maran, Roger Launius, Greg Good, Stefanie Jankowski, Dan Lewis, Dan Brocious, Jennifer Wright, Tad Benicoff, Patti Williams, Martin Collins, Mike Neufeld, Marc Rothenberg, Steven Dick, James David, Zuoyue Wang, Irwin Shapiro, Giovanni Fazio, Jonathan McDowell, Howard Plotkin, Joe Bassi, Andrew Butrica, Gloria Lubkin, and, most deeply, Owen Gingerich. On the management side over the years, Jo Ann Morgan, Jens Feeley, Ahoua Twitty, Steve Garber, and Theodora Blair offered critical guidance and care to the budgets and progress reports. Finally, I want to thank the Department of Space History's Writer's Group bi-weekly workshop nurtured so patiently by Margaret Weitekamp, especially those attending who conscientiously took notes during the meetings. Having motives and intents uttered over a decade ago brought to my attention afresh certainly kept me on my toes.

I am deeply indebted to the many people who submitted to formal oral and video history interviews, to follow-up questions, and finally to reviewing and commenting on how I have used the information and insight they imparted. These activities were generously supported by NSF, NASA, and the Sloan Foundation's Smithsonian Videohistory Program. Most of all, I want to express my deepest appreciation to Babette Whipple and her family. Her warm hospitality, which included several extensive inspection tours of their home and its contents, provided a personal dimension into the life and character of Fred Whipple accessible in no other way.

INTRODUCTION

For better or worse, the Cold War redefined American science.

—Stuart Leslie (1993)[1]

The system of national laboratories is the main American contribution to the transformation of science in the twentieth century.

—Spencer Weart (2003)[2]

On the last day of 1957, as Sputnik's *beep* washed the world in excitement and concern, some saw great opportunity. Fred Whipple, Harvard astronomer and director of the Smithsonian Astrophysical Observatory, was asked to speak to alumni at a Harvard Foundation fundraiser. "Whipple Speaks on Sputnik," the foundation's newsletter reported. But what Whipple did was challenge their world: Sputnik was a "visible sign" that basic research "is the raw material for control of that world."[3] Whipple well knew that a "huge reservoir of scientific and technological talent" was available in the Boston area, ready to be recruited.[4] Sputnik was his catalyst. After another rousing speech, to the Boston Commercial Club on April Fool's Day 1958, Whipple received a passionate letter from a partner in a real estate firm that specialized in industrial parks, offering to help: "I couldn't help wondering if, by providing very adequate facilities on a rather comprehensive basis, it wouldn't be possible to make the Boston area the world center for space research."[5] That was exactly what Whipple had in mind.

Cold War Alignments

Popular opinion links the launch of Sputnik to Whipple's vision of "space research," but its roots are a legacy of World War II and the Cold War.[6] The years following World War II saw both qualitative and quantitative changes in the world of science, such as the emergence of space research, in effect "an enlistment and integration of the bulk of its practitioners and its practice in the nation's pursuit of security through ever more advanced military technologies."[7] Historians have described the many ways science aligned with the government/military complex in the Cold War era.[8] American physicists, for instance, eagerly entered into a "Golden Triangle" of military agencies, high-technology industries, and research universities to create campus-based research facilities on scales previously unimagined. In this new universe, theory and practice, science, engineering, and intellectual curiosity—fueled and guided by government contracts—all became redefined.[9]

Whipple and his cohorts saw this pattern emerging in the late 1940s and early 1950s in departments of physics and engineering at leading universities such as MIT, Berkeley, Stanford, and Michigan. Whipple also saw resistance to the pattern at Harvard,[10] and in the astronomical community at large. Astronomers resisted this new alignment, remaining comfortable and satisfied in the immediate postwar world with traditional sources of funding, primarily private and corporate philanthropy governed tightly by an elite confederation of observatory directors. They opted at first for carefully measured and limited projects and programs of military and civilian support for astronomy, taking care to preserve the institutional status quo and support strategies and methodologies that had been in place for a century.[11]

Whipple was already embedded in, and keenly attuned to, the opportunities presented by Cold War national security interests. In 1947, he headed a working group advising the Joint Research and Development Board of the Joint Chiefs of Staff, which concluded that "all basic upper atmospheric research is fundamentally and inherently of value to the general problems of national defense."[12] Returning to Harvard after the war as a senior professor, Whipple knew that the Harvard College Observatory was still considered a leader in the United States and in the world, though concerns had been raised internally about its future. Its long-time director, Harlow Shapley, though highly successful keeping Harvard astronomy visible, had weakened the Harvard infrastructure through overextending programs and a lack of attention to maintenance, and Harvard's telescopes could no longer compete with the giants in the West.

Harvard was not the only astronomical institution whose health was in doubt at the time. The Astrophysical Observatory (APO) of the Smithsonian Institution, based on the National Mall in Washington, D.C., also faced an uncertain future. Supported since the 1890s mainly by private philanthropy, it maintained a narrow agenda with small solar-observing stations across the United States, in South America, and in Africa. When its highly charismatic director, Charles Greeley Abbot, retired as secretary of the Smithsonian, the APO's future came into question. In the postwar world, the Smithsonian itself faced new challenges. The creation of the General Services Administration in 1949 raised the possibility that the Smithsonian would be subject to its oversight, and in 1953, Secretary Leonard Carmichael sensed that the Smithsonian was vulnerable because its research bureaus could no longer compete with major research universities in America. The Smithsonian had moved from a leader to a backwater of government research, and so Carmichael acted to strengthen its weaker bureaus, searching out collaboration with the academic community, such as Harvard.

Cold War Reorganizations

A few new institutional arrangements emerged at American observatories. The completion of the 200-inch telescope on Palomar Mountain after the war fostered a complex relationship between the Mount Wilson Observatory, the California Institute of Technology, the Carnegie Institution, and the Rockefeller Foundation. The McDonald Observatory in Texas, privately funded and operated shortly before the war by the University of Chicago, sought new alignments as formal astronomy training emerged at the University of Texas.[13] However, these did not involve direct association with a federal or military institution beyond some individual contracts. After World War II, American astronomers, encouraged by the National Science Foundation (NSF), created a national system to allow all astronomers, regardless of institution, competitive access to modern observatory facilities, access that earlier had been a privilege of the elite. Most astronomers favored this system, but there remained strong voices that continued to champion the elite institutional form.[14]

Centralized national institutions are nothing new in science, or in astronomy.[15] They were typically created to coordinate and direct attention to specific problems on scales impracticable or even impossible for single academic institutions. In the late nineteenth century, both the Smithsonian and Harvard had been de facto central institutions for science, providing standardized instrumentation and support for coordinated observations in geophysics and astronomy.[16] Whipple and his former Berkeley mentor and Harvard colleague, Donald Menzel, sought larger opportunities. Menzel, especially, forged new relations with the U.S. Navy, the U.S. Air Force, and the National Bureau of Standards.[17] He also responded strongly to Carmichael's appeal and directed Whipple to build a hybrid of public and private resources that followed, but vastly expanded upon, the model the Smithsonian had set under its first secretaries. It was to be a new form of astronomical institution that was large enough to tackle programs of national scope, yet nimble enough, with multiple sources of funding, to direct significant resources toward solving late-breaking issues in science as well as national security.

This is the background for appreciating how two astronomical institutions changed in response to shifting priorities and new funding opportunities in the Cold War era. These opportunities were made possible by shared interests in new agencies and services supporting and encouraging astronomical research, including the Office of Naval Research (ONR), the Army Ordnance Corps, the Air Force Office of Scientific Research (AFOSR), the Atomic Energy Commission (AEC), NSF, and the National Aeronautics and Space Administration (NASA). Thus, after 1955, the new Smithsonian Astrophysical Observatory (SAO) became the embodiment of a Cold War scientific institution, designed by a small group of scientist entrepreneurs, gathered and led by a person with intimate knowledge of how national security needs and priorities could be translated seamlessly into and made synonymous with scientific research agendas (e.g., satellite tracking and geodetic services), which stretched the limits of astronomical practice.

Here we explore how Fred Whipple, in cooperation with others who shared his style and outlook, utilized Cold War national security concerns as opportunities to create one of the largest institutions devoted to astronomy and space science in the world by the mid-1960s. The question is, how did he do it?

LIST OF ABBREVIATIONS

AAAS	American Association for the Advancement of Science
AAF	Army Air Forces
AAS	American Astronomical Society
ABMA	Army Ballistic Missile Agency
ADS	Astrophysics Data System
AEC	Atomic Energy Commission
AFCRC	Air Force Cambridge Research Center
AFSC	Air Force Systems Command
AMB	Astronomy Missions Board
APO	Astrophysical Observatory of the Smithsonian Institution
ARDC	Air Research and Development Command
ASC	Astronomy Survey Committee
AUI	Associated Universities, Inc.
AURA	Association of Universities for Research in Astronomy
BoB	Bureau of the Budget
BuOrd	Navy Bureau of Ordnance
Caltech	California Institute of Technology
CAMROC	Cambridge Radio Observatory Committee
CfA	Center for Astrophysics
COSPUP	Committee on Science and Public Policy
COSPAR	Committee on Space Research
DIA	Defense Intelligence Agency
DOI	Differential orbit improvement program
EBICON	Electron bombardment induction conductivity
EMR	Electro-Mechanical Research
ERC	Electronics Research Center

ESA	European Space Agency
GEOS	Geodetic Earth Orbiting Satellite
HCO	Harvard College Observatory
HCOC	Harvard College Observatory Council
HEW	Health, Education and Welfare Department
IAF	International Astronautical Federation
IAU	International Astronomical Union
IGY	International Geophysical Year
IQSY	International Year of the Quiet Sun
IUE	International Ultraviolet Explorer
JPL	Jet Propulsion Laboratory
LOR	Large Optical Reflector
LOX	liquid oxygen
LPR	Long Playing Rocket
LRL	Lunar Receiving Laboratory
MIT	Massachusetts Institute of Technology
MMT	Multiple Mirror Telescope
MOI	Major Optical Instrument
MOL	Manned Orbiting Laboratory
MOR	Major Optical Reflector
NACA	National Advisory Committee on Aeronautics
NAS	National Academy of Sciences
NASA	National Aeronautics and Space Administration
NBS	National Bureau of Standards
NCAR	National Center for Atmospheric Research
NEROC	North East Radio Observatory Corporation
NORAD	North American Air Defense Command (prior to 1981)
NRAO	National Radio Astronomy Observatory
NRC	National Research Council
NRL	Naval Research Laboratory
NRO	National Reconnaissance Office
NSF	National Science Foundation
ONR	Office of Naval Research

OAO	Orbiting Astronomical Observatory
OOR	Office of Ordnance Research
OSO	Orbiting Solar Observatory
OSRD	Office of Scientific Research and Development
OST	Office of Science and Technology
PASP	Publications of the Astronomical Society of the Pacific
PSAC	President's Science Advisory Committee
RAND	Project RAND ('Research And Development')
RDB	Research and Development Board
RRL	Radio Research Laboratory
SAMSO	Space and Missile Systems Organization
SAO	Smithsonian Astrophysical Observatory
SCROGE	Smithsonian Computations Relating Orbital Glimpses Everywhere
SEC	Secondary Electron Conduction
SECOR	Sequential Correlation of Range
SIE	Science Information Exchange
SSB	Space Science Board
STAG	Smithsonian Telescope Advisory Group
STRI	Smithsonian Tropical Research Institute
TOS	TIROS Operational Satellite
TPESP	Technical Panel on Earth Satellites
UHR	Ultra-High Resolution
USNC	U. S. National Committee for the IGY
USRA	Universities Space Research Association
VLA	Very Large Array

Chapter 1

The Many Worlds of Fred Whipple: From Meteors and Comets to Artificial Satellites

Born in 1906, Fred Whipple (Figure 1) grew up on a rural Iowa farm his family rented and managed near the small town of Red Oak. He lived through the trauma of seeing a younger sibling die of scarlet fever and then soon after contracted polio himself. Seeking escape, he read everything in reach at a rural library some 10 miles distant, becoming deeply engrossed at first in fairy tales and then in science fiction. Whipple loved his pet chickens and dog (Figure 2), and liked to build things, fix farm machinery, and puzzle over how things work and why they behave as they do. Once, when he regaled his father, Harry Lawrence Whipple, with his own theory of how a whip makes its characteristic cracking noise, he was met with indifference. "I couldn't get any interest out of anybody in my great theory . . . I remember that."[18] His fascination with machines extended to the idea of lofting heavy objects to great heights and distances, a vivid memory from the World War I era, when he read about Germany's long-range siege cannon.[19]

After the death of his brother, and now the only child, Whipple wholly absorbed the religious devotion of his parents, becoming active in the Christian Endeavor Society, the growing evangelical Presbyterian youth organization. Sometime after his brother's death, his mother, Celestia MacFarland Whipple, became obsessed with mortality, having remaining family members suffer elective appendectomies and displaying the remains in bottles on a shelf in the bathroom.

In 1922, the family left the farm and moved to Long Beach, California, where his father opened a small family market. Whipple finished high school there and in 1923 entered Occidental College, then a small, liberal Presbyterian school north of central Los Angeles in Eagle Rock. Now exposed to a wider world, he encountered the Old Testament, which raised more questions than it provided answers.[20] He soon transferred to the new Southern Branch of the University of California (now UCLA) and became attracted to the sciences and to mathematics. For Whipple, mathematics was ". . . an amusing game and a useful tool, but as a life goal I couldn't tolerate it." In his junior year, he took an astronomy course taught by Frederick C. Leonard, a colorful lecturer and later an avid meteoriticist. Leonard sponsored him for a fellowship to Berkeley. Whipple later recalled, "So I sort of drifted along with that. Astronomy was a good thing to do."[21]

At Berkeley, Whipple concentrated on mathematical astronomy: the theory of orbits applied to the motions of Earth and the planets. This was standard fare for astronomers in the United States, and Berkeley was then one of the strongest centers for PhD production.

Figure 1 Fred Lawrence Whipple (age six) and mother, Celestia MacFarland Whipple. *Courtesy Sandra and Laura Whipple.*

The Berkeley director, Armin Otto Leuschner, was an industrious German American astronomer trained at Michigan, Berkeley, and later Berlin. On the Berkeley faculty since 1890, Leuschner established the curriculum that included access to the Lick Observatory some 70 miles south. He provided more than access, however, linking the Lick staff to the Berkeley staff in faculty exchanges and student fellowships.

Figure 2 Whipple and his favorite pets on the farm. *Courtesy Sandra and Laura Whipple.*

Figure 3 (*R to L*) Whipple, Dorothy Woods, and Whipple's parents, ca. 1929. *Courtesy Sandra and Laura Whipple.*

Whipple met Dorothy Cornell Woods at UCLA in a mathematics class. Beyond mathematics, they shared an avid love for tennis. They married in 1928 after Whipple had moved to Berkeley (Figure 3). She joined him, enrolling in mathematics at Berkeley, where she was elected to Phi Beta Kappa in 1929, graduating with highest honors in June of that year. By now, the questions he had been raising about his religious upbringing since his exposure to liberal thought at Occidental remained unanswered, turning his doubts into alienation. In his last attempt to search for the soul, he and Dorothy became fascinated with spiritualism and attended sessions with a psychic medium. However, Fred was soon disillusioned.[22]

From Pluto to Meteors to the Upper Atmosphere

Leuschner created "Leuschner's Method" of orbit computation, which was "the Gospel, the Law and the Prophets in the Students' Observatory."[23] Immediately after the Lowell Observatory announced in March 1930 that it had discovered another planet in the solar system, beyond Neptune, more massive than Earth, Leuschner charged Whipple and another graduate student, George Bower, to calculate its orbit. Whipple dove right in, even though he also now had a two-month-old son, Earle Raymond, who needed attention because Dorothy was recovering from the birth. Still, he and Bower managed to produce the first reliable orbit whose general characteristics stood the test of time and real data.[24] Their published orbit was Whipple's seventh publication; the first six were jointly authored papers with other graduate students giving the elements or ephemerides for newly discovered comets.

In 1930 Whipple was awarded a Lick Fellowship for the summer, mentored by Donald Menzel. Menzel, a 1924 PhD student under Henry Norris Russell and Arthur Holly Compton at Princeton, was firmly grounded in modern theoretical astrophysics. He had been on leave from Lick in residence at Berkeley, and he exposed Whipple to modern stellar astrophysics. Becoming aware of trends in the profession, Whipple knew that an astrophysics thesis would be more competitive at elite West Coast observatories. Once at Lick, he conducted spectroscopic observations of variable stars, completing a thesis in late 1931, "A Spectrophotometric Study of the Cepheid Variables Eta Aquilae and Delta Cephei." He tried to fit the varying behavior of Cepheid spectra to Arthur Stanley Eddington's theoretical pulsation model, but was not satisfied with the result. Still, his thesis was accepted.[25] This would not be the last time he would try to break into astrophysics.

Leuschner's teachings left Whipple with a firm grounding in celestial mechanics and a sense that there were links between geophysics and astronomy—specifically, geodesy. Leuschner also believed strongly in teamwork and organized his Students' Observatory that way. He was an active and effective networker, having been onetime executive secretary of the National Research Council and chair of its Physical Science Division, as well as president of the American Association of University Professors just prior to Whipple's years at Berkeley. He also attended to a wide range of campus administrative activities.[26] Whipple no doubt appreciated these characteristics as he and his family prepared to move East to Harvard in late 1931.

The Harvard College Observatory

As he finished his thesis under Menzel, Whipple pondered what to do next: an observatory staff position or an academic post? He was aware of Menzel's frequent clashes with the highly conservative senior staff at Lick, and similarly sensed that his own personality would not fit there, or even at Mount Wilson. A visit by the charismatic Harvard College Observatory director Harlow Shapley, sometime in the spring, sealed Whipple's fate. Menzel regarded Shapley's visit as a high point for the year. He and Florence Menzel had gathered up all the students to meet the great man from the East. As Whipple recalled after their meeting, Shapley "thought I was an up and coming young man who would add to the luster of the local firmament."[27]

Whipple indeed found the Harvard Department of Astronomy a very different universe; just as he had hoped, it was the "big puddle." Harvard College Observatory (HCO), of which the department was a part, was established in 1839 and by the late nineteenth century was one of the leading observatories in America.[28] The programs E. C. Pickering established at Harvard, including facilities in both the Northern and Southern Hemispheres, defined the fundamental database for what had become the mainstream definition of Samuel Pierpont Langley's "New Astronomy." At Harvard, it was still called "astronomy," but the practice embraced the physical characteristics of the stars, generating and then gathering together in one place vast catalogues recording their brightnesses and spectral characteristics using the newly applied techniques of photography and spectroscopy.

Both the observatory and department were then dominated by Shapley, who succeeded Pickering in 1921 to become one of the most visible names in astronomy of that day. Shapley maintained E. C. Pickering's bureau-based organizational structure, symbolized by the compartmentalized rotating desk he inherited from Pickering (Figure 4), with senior faculty leading large research teams focused on specific problem areas. But unlike Pickering and in tune with the times, Shapley gave more breadth to his staff and strengthened ties with Harvard by creating a new graduate training program.[29] He also strengthened fellowship support for women working toward obtaining a master's degree at Radcliffe College.[30] During Shapley's first years, he saw to it that Harvard offered an AB degree with honors in astronomy, and the doctoral degree was approved soon after.[31]

Figure 4 Harlow Shapley at the compartmentalized desk created by his predecessor E. C. Pickering to manage the observatory's "bureaus." *Ad Astra Per Aspera, Through Rugged Ways to the Stars*, by Harlow Shapley, New York: Charles Scribner's Sons, 1969. *Courtesy AIP Emilio Segrè Visual Archives, Shapley Collection.*

Under Shapley, the Harvard staff grew rapidly. Bart Bok, educated at Groningen and Leiden and hired by Shapley in 1929, was already a leading specialist on galactic structure and a highly popular and inspiring teacher. Cecilia Payne came from England as a graduate student, earned her doctorate in 1925, and was then employed as Shapley's research assistant, earning widespread notice.[32] And soon after Whipple was brought on board as an observer, Shapley tapped Menzel for a senior faculty position. Menzel recalled that when he got his invite in the spring of 1932, "I had no doubt . . . I ran all two blocks from the P.O. Box to my house to inform my wife."[33] He soon began working to build up what would become one of the strongest teams of young astrophysicists on the planet, directing them to studies of theoretical stellar atmospheres and the physics of gaseous nebulae (Figure 5). By 1937, six members of Shapley's senior staff, including Menzel himself, were starred in a national roster of elite scientists, Whipple being added that year.[34]

Shapley never changed Pickering's agenda of cartographic mapping, but did expand it to encompass the whole of the stellar universe, as well as the newly realized universe of galaxies. During Shapley's reign, the campus buildings were expanded (Figure 6), and two observing stations were established, one locally beyond Concord, Massachusetts, 26 miles from Cambridge, and one moved from Arequipa in Peru to South Africa. The latter, the "Boyden Station of the Harvard College Observatory," at Mazelpoort, some 16 miles from

Figure 5 (*L to R*) Cecilia H. Payne, Donald H. Menzel, Annie Jump Cannon, Fred L. Whipple, and Leon Campbell, inspecting a motorized mechanical planetarium ca. 1932. UAV 630.271 (171), olvwork431823. *Courtesy Harvard University Archives.*

Figure 6 Harvard College Observatory in 1947. The large dome houses a 15-inch refractor, the largest in the world in the 1840s. UAV 605.270.1 (SC-57). *Courtesy Harvard University Archives.*

Bloemfontein, opened in November 1933 with some six telescopes ranging up to 60 inches in aperture, then the largest in the Southern Hemisphere.[35] The local station, known as the Oak Ridge Station and then the Agassiz Station after 1952, housed instruments that had been on campus and would soon be augmented by another large reflector.

Shapley gained visibility as a leading spokesperson for science in general and astronomy in particular. He was active in Washington, D.C., promoting "Science Service." But he also made enemies, none more contentious than James Bryant Conant, president of Harvard since the spring of 1933. Shapley had been one of his competitors for the post, promoting the sympathies of civil liberties advocates and liberal-minded lawyers like Felix Frankfurter and his former classmate Grenville Clark, a member of the Harvard Corporation, its core governing board.[36] Shapley also became an outspoken member of the "Committee of Eight" in the late 1930s that led a faculty protest over the nature of Harvard appointments and the apportionment of appointments by department.[37] What lingering acrimony this period created between Conant and his liberal faculty was soon swallowed by the more pressing specter of war, but it persisted.

Whipple's first assignment was to assist, and soon to direct, the Oak Ridge Station, dedicated in September 1932. It included an array of wide-field photographic astrographs, some equipped with thin objective prisms, to produce the vast stores of stellar magnitudes and spectra that Harvard maintained and augmented since the 1880s. Oak Ridge also operated a series of wide-angle patrol cameras and a menagerie of smaller telescopes for a wide variety of projects. But Shapley had bigger plans for Oak Ridge.

Writing to Whipple at Berkeley in June 1931, Shapley assumed he'd be visiting Los Angeles first to see his family before setting out East. If so, he asked him to stop in at Mount Wilson and inspect their 60-inch telescope to "study the problems . . . since we are adopting the fork-type mounting," to which he added coyly, "I think we should be able to make a very much better telescope than the Mount Wilson 60-inch, though it probably will never have such a noble record of accomplishment."[38] In 1934 Oak Ridge would boast a telescope of 61-inch aperture, still the largest on the East Coast in the 21st century.

From all this, Whipple realized that the "puddle" he was entering was not only larger but also turbulent, stirred by projects, plans, and changing priorities. Shapley was endlessly ambitious and no doubt became another role model for Whipple. Although Whipple continued to try to break into astrophysics, showing some promise, he spent most of his time and effort improving and using the equipment at Oak Ridge, discovering asteroids and comets, computing their orbits, and computing asteroid orbits from improved systematic observations by the patrol cameras. In 1933 he discovered a main belt asteroid and named it after his mother, (1252) Celestia. By 1940 he had four comets flying his name and had garnered almost 60 articles on their motions and physical characteristics.

Shapley also directed Whipple to help the visiting Estonian astronomer Ernst Öpik conduct his "Harvard Meteor Expedition" in Arizona. Öpik, a man of firm and particularly bold beliefs, was certain that a large portion of meteors had hyperbolic orbits and therefore came from deep interstellar space, a vast reservoir of comet nuclei, asteroids, and meteors centered on the Sun but at nearly interstellar distances. Adapting Öpik's methods, Whipple applied statistical methods for analyzing families of meteor orbits and from there began studying meteor radiants.[39]

The Arizona Expedition, which was under way when Whipple arrived, used widely spaced visual and photographic observing stations to triangulate meteor trails for their orbits and origins. Whipple improved both the observational techniques and the reduction methods. By 1936 he had introduced synchronous motors to drive precision image choppers for better timing, and tested them with a set of cameras on campus and at Oak Ridge. He soon obtained a sufficient number of useful plate pairs that he expected would confirm Öpik's hyperbolic orbits. However, upon careful analysis, especially using only dual observations that could be accurately timed with a visual record, Whipple realized that all the orbits were elliptical, meaning they originated in the solar system. Öpik was none too happy with Whipple's results, which made for some unpleasantness, but this episode proved formative in Whipple's career. He continued to refine his techniques to increase precision to the point at which he could even discern how meteor velocities changed as they passed through Earth's upper atmosphere.[40] Whipple was by now a recognized mathematical astronomer and an adept instrumentalist. But his personal life was far from stable. It is not clear if or when Dorothy had moved with him to Harvard, but by 1932, Earle Raymond had been sent away to live with the family of a maternal uncle, William Cornell, in Riverside, California. The Whipples divorced in 1935.[41]

But personal setbacks did little to diminish Whipple's energies and enthusiasm for research. He collaborated with Cecilia Payne, now Payne-Gaposchkin after she married Russian astronomer Sergei Gaposchkin, analyzing novae spectra. And with newly minted,

Bok-mentored PhD Jesse Greenstein, he explored enigmatic radio noise from the Milky Way. But he always returned to his meteor studies, which were consistently successful, especially a 1937 study of a Geminid meteor shower, in which he was able to determine velocities, decelerations, radiants, and magnitudes. From these data he created a profile of the densities and temperatures in Earth's upper atmosphere and soon extended this work into a comprehensive review paper for the American Philosophical Society, his largest study to date. It led to his promotion from instructor to lecturer in 1938 and set him on a solid career path at Harvard. By then Whipple had produced a refined picture of Earth's atmospheric density profile between 58 and 99 kilometers. He compared his density distribution with density profiles obtained with other methods and found a seasonal effect as well as good agreement with the density curve of Lindemann and Dobson up to 70 kilometers, "in their remarkable work on this subject."[42] Whipple also reversed his analysis to show how the integrated light from a meteor trail could reveal the mass of the meteor that was evaporating during its violent transit through the atmosphere. Its brightness at any one moment was also related to the amount of mass evaporated at that moment through the frictional forces encountered. Comparing these, Whipple estimated the mass of the meteor at any point, and from its deceleration, as determined by his triangulation techniques, he could derive the resistance of the atmosphere itself at that point and hence the density of the medium.[43]

This analysis solidly linked the nature of Earth's atmosphere to the study of asteroids, comets, and meteors. But it called for more power and precision, such as a new camera system with the widest possible field and the greatest possible sensitivity, together with sufficient photographic scale to make precise positional measurements. In early 1940, Whipple therefore turned to a young observatory staff member, James G. Baker, who was showing promise as an optical instrument designer. Baker had been refining the new Schmidt catadioptric design that had captured astronomers' attention. The Schmidt was the key, but how far could it be stretched? Before Whipple and Baker could find out, life got in the way.

A Breakdown, Followed by War Work

What Whipple experienced and accomplished in the late 1930s and early 1940s shaped many decisions in his future life and career, including the nature of the institution he built at Harvard in the 1950s. But psychological pressures were also building within him, evidently since childhood and no doubt deepened by his divorce and subsequently his anxiety over the welfare of his son, Earle. Earle had returned to live near his father in 1940 at a boarding school in Marlborough, Massachusetts, but it was hardly a stable existence. These factors led to Whipple's complete breakdown during a meeting in Chicago.[44]

In June 1942, Yerkes Observatory Director Otto Struve invited Whipple to a conference to present his meteor studies. Harlow Shapley strongly encouraged attendance because Whipple had been burdened with a heavy teaching load that included year-round courses in navigation for wartime needs and was still struggling to refine his meteor studies. Once in Chicago at the meetings, however, Whipple disappeared. After frantic searching by his colleagues, Jesse Greenstein finally found him in a basement restroom in a stupor. Whipple was rushed to a local hospital for evaluation and was soon transferred to a sanitarium near Milwaukee, where he stayed for several months, undergoing repeated rounds of electroshock. Bart Bok drove out to Milwaukee in September and brought Whipple home. The horrible episode shook observatory staff. The sanitarium doctors had reported to Shapley and Struve that talking sessions revealed insights into a very closed-down life, and Whipple's deep but unfulfilled need to be less of a "lone wolf." There were also moments of crisis; once when his mother visited and he withdrew completely, once when he escaped from the sanitarium and wandered the streets, and once when he had to be revived after submerging himself in a bathtub full of water. Shapley darkly mused at one point that if a person as seemingly self-controlled as Whipple could fall apart, what did that mean for the rest of his Harvard staff, including himself?[45]

Indeed, astronomy at Harvard was then torn by wartime anxieties. The staff had very different views about the growing war and what to do about it. Some were pacifists, like Cecilia Payne-Gaposchkin and Bart Bok. The former felt it was her duty to keep astronomy alive, and somehow aid humanity. Bok agreed to teach navigation, at first, like the others, and both worked to keep lines of communication open between astronomers of all nations. Shapley tried to keep it all together, supporting Bok's efforts, managing the coursework as well as the contracted war work of his staff.[46]

After he returned to Harvard, Whipple still toyed with astrophysical problems, but felt compelled to get involved in war work. He had rejected an invitation from Robert Oppenheimer to join him at his "ranch" in New Mexico to head up a computing center for the Manhattan Project. He wanted to stay close to home, to Earle, and work in a small group, staying in touch with his astronomical interests.[47] But by the end of the year, secure in the knowledge that his son was prospering at Hillside Academy and was well taken care of there, Whipple decided to join Harvard's Radio Research Laboratory (RRL), admitting to a colleague that "my astronomical research period is over for the duration. I expect to be murderous in my research beginning very soon, probably at Harvard though."[48]

Clearly still recovering from his terrible ordeal, Whipple was somewhat ambivalent about it all. Yet he must have known that whatever he did, he had better remain close to home and his support structure. Whipple joined the countermeasures unit to search for ways to confuse enemy radar with reflective "chaff." The RRL, created at MIT in 1942 within its Radiation Laboratory, was transferred to Harvard by the end of the year, and eventually engaged more than 800 staff in secret war research, mainly radar. The countermeasures unit was part of the contribution of the U.S. Army Air Forces to the British project "Window."[49]

Whipple's major contribution to the project was to apply theory developed by an antenna expert at the RRL to optimize the length, width, and weight of chaff shards to maximize radar reflection. Whipple also helped to develop a high-speed mechanical device to make the chaff and release it rapidly from aircraft to confuse S band radar. Throughout his civilian service, Whipple worked directly for F. E. Terman, the director of the RRL, supervising five technicians and two research associates conducting field tests in Florida. Whipple's system was activated after the Eighth Air Force suffered heavy losses at the hands of the Würzburg radar-directed firing batteries, and so he went overseas in January 1944 for seven months to help introduce the so-called American Chaff system, considered to be very successful.[50] His European tour also put him in contact with a wide range of high-level operations research personnel. These included just the types of officers, as high up as Curtis LeMay, who would become program managers in the postwar world and who would look for ways to apply new revolutionary technologies in planning for future wars.[51]

Whipple later testified that his wartime experience was seminal in training him how to operate within a government and military structure. It also provided Whipple with contacts for securing support and insights on how to sell his expertise to the military: "Yes, I think that's what built me up to being willing to take grants and contracts and to understand how the wide world operated with a better understanding of that. Consequently I got into administrative things to push projects that I thought ought to be done."[52]

While at the RRL, Whipple still found time for research and writing, as did other Harvard staff. He finally was able to complete his large work on meteors and the upper atmosphere, based upon the paper he was to have given in Chicago. It appeared the next year as a major contribution that established his meteor trail method as a basis for his future work. At war's end, Whipple would seek support from the Navy Bureau of Ordnance for extensive meteor observations and reductions.[53]

Even before his breakdown and war work, Whipple was aware of the value of his upper atmosphere work—specifically, its relevance to ballistics calculations. During the war, his young colleague Theodore Sterne corresponded frequently about it, first from Fort Bragg and then from Aberdeen Proving Ground, where he led a large computational group after the war. Whipple's cameras had recorded the path of an airplane flying over Boston, and Sterne wanted to know the details of the system and possibly apply it to the "trajectories of projectiles." Sterne regarded "the physics of the flight of meteors—resistance, temperatures, air structure, measured data, reductions, techniques" to be directly relevant to fire control. As early as October 1941, Sterne reported that "The work here is fascinating. Some of it is a great deal like dynamical astronomy. I have the privileges of an army officer, the fun of doing science, and the satisfaction of being of direct use in a time of crisis."[54]

Thus the war provided Whipple with ample evidence that he had finally found his niche. He recalled his wartime experience as "rather satisfying, because one could be involved in a project from the theoretical side, through the experimental, to get production techniques made, selling it to the Army Air Force[s], going to the manufacturers to get them into production and going overseas and getting involved in the tactical use of it. So I was able to be in all aspects of the project, from beginning to end."[55] His wartime experience also placed him in London later in 1944, when the buzz bombs were striking. After September, when he heard of the larger hypersonic missiles, "I realized that we were definitely going to go into space because these rockets were clearly powerful enough to get us out of the gravitational field. So I knew then that this was coming."[56] He also likely realized that aiming these new weapons properly required refined knowledge of the upper atmosphere. He had certainly gained that perspective in his correspondence with Sterne before the war. And there was another benefit of war work that he most certainly knew of through this time: the availability of vastly improved wide-field Schmidt systems. He returned to full-time faculty status at Harvard late in 1944 and was appointed associate professor of astronomy on July 1, 1945.[57]

Postwar Patrons for Research: Meteors, the Upper Atmosphere, and Rockets

Whipple was primed to face the postwar world with a newly found sense of accomplishment and authority. He was also becoming recognized as an important resource for improved knowledge of Earth's upper atmosphere. The subject was still largely accessible in the open scientific literature in the early 1940s, as the military had not yet decided that it bordered on sensitive areas.[58] Thus his papers on meteors and the high atmosphere in 1942 and 1943 established Whipple's name, and he was sought out more and more for his advice, starting as early as February 1944 in an inquiry from the Frankford Arsenal in Philadelphia.[59] By the end of 1945, the second edition of his popular Harvard book *Earth, Moon and Planets* was in press and he was often called upon to speak and advise. Whipple also consulted with Douglas Aircraft's Project RAND about the need to protect future space vehicles, whether they be warheads or artificial moons, from meteor impacts during flight. His report was a supplement to a secret Project RAND study on the feasibility of a world-circling spaceship. Other contributing consultants included Princeton's Lyman Spitzer, who reviewed the nature of Earth's upper atmosphere and explored the characteristics of a space telescope.[60] Whipple estimated the mass range for possible impactors, as well as their expected impact frequency, based upon the number density of meteorites from trail brightness statistics.[61]

In January 1946, aligned with his participation in Washington-based panels and committees for the Research and Development Board and his contracting for the Navy Bureau of Ordnance, Whipple became a charter member of an inter-agency civilian "V-2 Panel." U.S. Army Ordnance was about to launch a series of reconstructed V-2 missiles made from parts scrounged in Germany. They were being shipped to its White Sands Proving Grounds in southern New Mexico, and the Army was then was searching for scientists in military laboratories who would be capable of designing instruments for them to explore the nature of the upper atmosphere and near space, the realms through which the missiles flew. Army Ordnance reached out to Navy and Signal Corps laboratories, and a meeting was arranged in Washington that brought them together to establish the panel. Scientists from Princeton and Harvard were also invited, including Shapley, who brought along Whipple. Although Shapley was enthusiastic, he made it clear to Whipple that they were there to advise, but not to participate in any active way. So Whipple encouraged the use of the missiles to study the upper atmosphere in the ways he suggested in his RAND report. His direct involvement, feeding directly into his own plans, would be limited to providing a camera network that could record both artificial and natural meteor trails at magnitude levels faint enough to produce realistic statistics on the frequency of impactors large enough to concern missile and, eventually, satellite performance. This knowledge was also of value in assessing the ability of an artificial mass, like a warhead, to survive reentry into the atmosphere.[62]

Called upon frequently to serve the needs of the nation as a member of various oversight panels for military research and development, Whipple exploited the leverage he enjoyed as an elite academic. He was the only academic member of the V-2 Panel, and he was among very few who were not working at a laboratory sponsored by one of the services. Whipple was also tapped to take the chairmanship of the Panel on the Upper Atmosphere within the Geophysics Committee of the Research and Development Board. The former chairman, Thomas Johnson of the Naval Research Laboratory, had recommended Whipple not only as the best qualified to succeed him, but also because he was a non-government scientist associated with an elite academic institution.[63]

By this time, Whipple was also being courted by other observatories and military laboratories, such as Lowell, and the Naval Research Laboratory, where in 1949 he was asked to succeed E. O. Hulburt as Superintendent of the Optics Division, a position he had formerly applied for. By then, however, he was assured of being promoted in June 1951 to a full Harvard professorship.[64]

Indeed, Whipple was quickly establishing new networks and teams to track meteors and then to analyze the data rapidly and efficiently. By the end of 1945, with Shapley's guidance and the concurrence of Harvard's administration, Whipple finalized a $60,000 BuOrd contract to create a more powerful system of cameras and operate them for two years, called the "Harvard Meteor Project."[65] He was bringing in enough funding to offer support to astronomers such as Zdeněk Kopal, then at MIT in its Center of Analysis. Kopal had expertise in, and access to, high-speed computational tools such as the Bush Differential Analyzer. Whipple also recruited Luigi Jacchia, who had been on the Harvard staff for some time and was contributing his mathematical expertise to a wide array of problems. Jacchia, an Italian astronomer, had arrived at Harvard in 1939 and had made himself very useful there and at MIT. In 1955 he would be one of the first people Whipple tapped as a charter member of the new Smithsonian Astrophysical Observatory (SAO) to focus on upper atmosphere profiles.[66]

Whipple also turned to James G. Baker to vastly improve the field of view and space penetration of his meteor cameras. Baker had established a vigorous wartime optical laboratory, but at war's end Harvard abruptly closed it, divesting itself of wartime assets. Baker "felt if both my arms had been cut off" and for several years remained deeply unsure of his future. After a leave working to complete a massive report on wartime work of the Office of Scientific Research and Development (OSRD) Optics Division of the National Defense Research Council, he returned to Harvard in the late 1940s still troubled.[67] Following Baker's early years at Harvard helps to capture the atmospherics and infrastructure of the Shapley era that Whipple eventually exploited.

Baker devoured coursework and quickly distinguished himself, becoming a junior fellow of the Harvard Society of Fellows, which ensured his support from that point on. Menzel was the first to recognize his talents for building new instruments and put him under the guidance of instrumentalists such as Joseph Boyce and George Harrison at MIT, which crystallized his focus on optics. By 1939 Baker had taken courses on practical photography and the theory of optics from Whipple, focusing on optical aberrations and adapting the Schmidt design to provide large, flat, and accessible photographic fields.[68]

The war accelerated Baker's immersion into optical design. Through Shapley, Baker interned in January 1941 at Wright Field and then returned to Harvard to establish an Army Air Corps contract optical program.[69] By August, Baker's Optical Research Laboratory was housed on HCO grounds and attracted experienced opticians from the Amateur Telescope Makers of Boston. After delivering excellent camera lenses in the summer of 1942, the OSRD funded an expanded program and the facility moved to larger quarters. By April 1943 there were some 29 employees, although the turnover was high.[70]

In addition to traditional lens systems, Baker applied the Schmidt concentric catadioptric optical system to aerial camera design. An early design was capable of 120-degree coverage, but it had a curved field, "which was not very popular among the military people."[71] The photographic emulsion had to be deposited on a shell-shaped surface, so an additional reverse optical system was needed to flatten them out for aerial mapping on 40 inch–square prints. After the war, Kodak found a way to mold their film to fit the curved focal plane of Baker's system and so, as we shall see, this design ultimately became Whipple's Super-Schmidt meteor camera.

Baker jumped at the chance to work for Whipple. He knew it would be relatively straightforward to extend his wartime designs into a modified Schmidt system to provide a 120-degree field with sufficient scale and aperture to capture faint meteors as well as shaped charges shot from V-2 missiles.[72] Although Baker completed the design rapidly, it took several years before the system was fully operational at observing stations on Sacramento Peak and Organ Pass.[73] The new systems created a flood of data. Two of Whipple's observers, Richard E. ("Mac") McCrosky (Figure 7) and F. W. Wright, recorded some 12,000 trails between 1952 and 1955 averaging 400 trails per 100 hours of observing time from the dark New Mexico sites. McCrosky, supported on Whipple's Office of Naval Research grant, developed a graphical method that could produce radiants, velocities, and heights very quickly. This allowed Whipple's team to perform meaningful statistical calculations that greatly refined knowledge of the density variations found between sporadic meteors and meteor streams of different origins. McCrosky developed the technique for his Harvard thesis.[74]

These density variations had also been inferred from the trajectories of V-2 flights from White Sands. With several years of in situ data gathering, the rocket experimenters, including Whipple, realized that his profiles of upper atmospheric properties from meteor data were far different from those derived by the rocket trajectory analyses, which themselves varied widely. Further, and most vexing, all their atmospheric profiles differed from the one adopted by the National Advisory Committee on Aeronautics (NACA), the body responsible for the establishment of what was called the "Standard Atmosphere." The V-2 experimenters and the NACA looked to Whipple to sort out the problem.[75]

By late 1951, with data from his first Super-Schmidt showing how poor his earlier camera data had been, Whipple decided, from a coordinated series of balloon and rocket

Figure 7 Richard McCrosky field testing a Super-Schmidt meteor camera, ca. 1952. *Courtesy AIP Emilio Segrè Visual Archives.*

flights and indirect measures, that the new rocket data provided a more reliable profile than did either his old (pre-Super-Schmidt) data or the NACA Standard. Using the rocket data, Whipple rederived the meteoric density ranges by calculating the altitudes where their trails ended, and found to his fascination that their densities were low, between 0.3 and 1.0 grams per cubic centimeter, less than one-third that derived from terrestrial meteorite samples. What Whipple's cameras had been tracking all along were not meteoroids, he realized, but cometary debris. And this evidence also neatly fit a new model he had created for comets. They were not solid meteoric bodies but an admixture of rock particles and ices, a view that has stood the test of time and is commonly referred to as Whipple's "dirty snowball" model. It transformed the role of comets in solar system history.[76]

State of Mind

Whipple's life stabilized with his second marriage to Babette Samuelson in the summer of 1946. They met during the 1945 Christmas holidays at a party thrown by the Menzels, where she sensed a certain strength of character in the man, a man with a mission, and saw in him "somebody I could help to achieve his goal by being the good little wife." Indeed, she was more than that as a professional practicing psychologist, and she knew it: "He married me because I was a psychologist and I could get rid of his nightmares"[77] (Figure 8). Whipple's son, Earle, eventually moved back to Cambridge and lived with them briefly just before he entered Harvard. But just after he arrived, Earle's mother, Dorothy, who had remarried and had moved to San Francisco, committed suicide. These were certainly difficult years, but they were also years promising new life: by the summer of 1951 the Whipples had two daughters, Sandra and Laura.[78]

Despite his successes, Whipple never fully accepted that he would remain in meteor research. By the early 1950s the rocket work had demonstrated to him, as well as to others, the potential of scientific rocketry and eventually of satellites. In the mid-1940s, he claimed repeatedly that his meteor programs were just a "five-year plan to see how much can be found out about these little particles of the solar system."[79] He said much the same thing in 1947 and 1949, keeping to a five-year plan, but slipping it each time.[80] By then, Whipple had become chairman of the Department of Astronomy at Harvard, worrying about students

Figure 8 Fred and Babette Samuelson Whipple, October 1957. *Courtesy Sandra and Laura Whipple.*

and other campus-related obligations. He accepted this responsibility as a necessary distraction; as he admitted to his colleague Jesse Greenstein, "Both Bok and Shapley love to run things anyhow so the responsibility does not sit too heavily." Still, the responsibility took him away from what was really exciting, such as developing his new and revolutionary theory of comets.[81]

Indeed, his meteor network responsibilities also kept him away from analysis and interpretation, and at the time meteoritics also had a somewhat checkered reputation owing to its popularity with amateur meteorite hunters and prospectors.[82] But his frustration could also have been because Baker's Super-Schmidt network was still not complete. Still, he hoped that once it was operational, it would "throw considerable light on the evolution of the solar system." He was excited about this prospect, but lamented that he was "becoming more and more snowed under by administrative work—which I frankly do not like."[83]

The Super-Schmidt network became fully operational in 1951 and 1952, and the results were spectacular, capturing far more photographic trails than ever before.[84] By spring 1953, Whipple was looking forward to the completion of the project, corresponding with the Bureau of Ordnance and then with the Office of Naval Research about the future use of the cameras. He was also using the same tracking studies to better understand how to maximize the survival rate of ultra-speed ballistic projectiles through the atmosphere, and promoted his operational Super-Schmidt network as the "ultimate [in] present-day capabilities" made possible under contracts with a wide range of military bureaus. By the summer of 1953, the Air Force had completed its second Super-Schmidt installation in New Mexico, following the Harvard system, and regarded Whipple as a central player.[85]

Why Conquer Space?

Whipple was always available to the media, eager to spread the word linking astronomy to space travel dreams. "I keep seeing your name in the news," one of his former colleagues from the Radio Research Laboratory, more recently at Stanford, wrote him in March 1947, "[talking] about rocket ships and asteroids and cosmogony."[86] Whipple's access to the media was no doubt a gift of association with Harlow Shapley, who always encouraged his staff to be ready to present their work to the public whenever sky phenomena were likely to grab its imagination. No one was more willing than Whipple, who had something to say whenever a comet was in the sky or when meteor showers were expected or when aurorae were reported. Rarely if ever sensationalist, and usually quite prudent, Whipple had a certain talent for quotable quotes. After the war, Whipple's appearances in the media continued to center on meteors and comets, but soon included space travel.

Historians have examined how disparate communities and individual interests within military services and between its branches negotiated with the intelligence community and the scientific establishment to build various political arguments for spaceflight. Most important was to establish a visible capability of global overflight, but to do it in a way that was peaceful, embodying "articles of faith, creeds, acolytes, ceremonies."[87]

Whipple navigated the turbulent waters of spaceflight advocacy as one of the few academics in this nested set of special interests, and eventually it factored into his decision to take on the leadership of the SAO. He thus joined a small band of passionate spaceflight visionaries, centered on Wernher von Braun but propelled by the efforts of Willy Ley, Cornelius Ryan, and others who lived in an overlapping set of communities ranging from aerospace companies and spaceflight advocacy groups to military and national security institutions.

Just as he had always claimed that all science was of interest to the military, Whipple envisioned that access to space would open up the universe for the astronomer. Virtually every problem astronomers might address would be enabled by access to space, he promised at the First Annual Symposium on Space Travel held at the Hayden Planetarium in October 1951, including searching for life on other planets where "the best way to discover the answer to these questions is to make a space trip to the planet itself and investigate on the spot."[88] Whipple also foresaw that instruments would be carried into space on both robotic and piloted space stations to provide continuous surveillance for practical ends such as weather prediction, climate studies, solar radio disturbances, and improving knowledge of the shape of Earth. It would be as big a technological challenge to handle the data flow, but in 1952 Whipple expressed confidence that the "huge 'electronic brain' calculating machines that we have today" were up to the job. This was all destined as man's "conquest of nature."[89]

There were always security classification barriers when proselytizing for spaceflight. Whipple admitted, ". . . we were all handicapped, not assisted, by knowing quite a lot about military plans for large ballistic missiles. . . We were under top security wraps in discussing satellites or space travel." So he and his cohorts never disclosed anything like "the actual steps that probably would occur but had to jump over the realistic near future that we expected."[90] The key was to appeal to all sides and to stay out of the line of fire between the promoters and critics of spaceflight. His talks and essays reflected von Braun's, Arthur C. Clarke's, and Fred Singer's enthusiasms, though he added cautionary remarks.[91] In the fall of 1952, he co-authored with von Braun "The Exploration" to dramatize for *Collier's* readers what an expedition to the Moon would be like, writing it as a trip diary set amidst the realistic landscapes Chesley Bonestell crafted depicting lunar scenes and humans living and working there.[92] But he also made clear to the popular but technically savvy readership of *Sky & Telescope*, in "Astronomy from the Space Station," just how challenging it would be to perform astronomical observations in space.[93] Possibly his most evangelic expression came in a series of talks in 1953. Asking the question "Why Conquer Space?" Whipple gave four personal reasons "for wishing to see man in space." First came "pure curiosity," which needed no justification. Then, there would be an "enormous impetus to science," which also went without justification. But then, as if to justify the first two, there would be the real political rewards in Cold War context: "I feel that the level of a culture from the historical point is measured largely by its technological achievements." Those who do it first "will be marked for all time in the history of the world as a technological leader." His deepest justification was epistemological: "It represents man's conquest over nature." And as an astronomer who had thought deeply about collisions of meteors, comets, and asteroids and their dire consequences, Whipple warned that conquering space was "one more step toward insuring immortality for the race, if not for the individual."[94]

By the time "Why Conquer Space?" appeared in print in the fall of 1954, in the first issue of the new journal *Astronautics*, Whipple was actively engaged in planning for a scientific satellite within von Braun's circle. He was also discussing the possibility with his colleagues on the V-2 Panel, by now renamed the Upper Air Rocket Research Panel. And he was already deeply involved in planning for the International Geophysical Year (IGY) of 1957–1958, having been designated chair of the IGY Technical Panel on Rocketry. It was this panel that would eventually form a subpanel to consider an Earth satellite, and Whipple was on that one too.

S. Fred Singer, former scientific liaison officer for the Office of Naval Research (ONR) in London and now at the University of Maryland, made sure that Whipple was invited to a quiet but high-level meeting on 25 June 1954 at the ONR in Washington, led by a Lieutenant Commander Hoover. Von Braun would be there, as would Fred Durant, a CIA officer and president of the International Astronautical Federation. Singer felt he needed Whipple there to help "in applying the brakes (you'll see what I mean)." Singer warned Whipple that he needed clearance if he wanted to attend.[95] Whipple had the right clearances and did attend to hear von Braun propose using an Army Redstone missile with three upper stages of clustered Loki anti-aircraft rockets to loft an inert 5 pound body that would orbit for several weeks or more.[96] This seemed too hasty and wasteful to Whipple and Singer, who wanted something both instrumented and visible from the ground for tracking so they could use its motions to determine the properties of the uppermost portions of the atmosphere as well as the shape of Earth.[97] By the end of September, von Braun submitted a formal proposal for an instrumented artificial satellite dubbed "Project Orbiter" and presented it at the end of the International Union of Geodesy and Geophysics General Assembly in Rome.

Just as he advised von Braun to launch something useful, Whipple also advised the Scientific Advisory Board of the U.S. Air Force not to wait to launch something overly complex. In a September 1954 report he encouraged them to start immediately with smaller payloads and learn along the way: "Are we to fly before we can walk?" He envisioned four classes of payloads, ranging from von Braun's inert slug to a fully instrumented, manned physical laboratory. He immediately dismissed the first as not useful or informative. The second class, an "observable object," could be very useful, for political as well as scientific goals. Something weighing 10 to 30 pounds that might "contain a simple mechanism to increase its surface area after reaching its orbit" would make it an easily seen and impressive naked eye object, maybe 5 feet in radius. It could carry corner reflectors for microwave radar visibility as well. The most complex practical level would be a "minimum unmanned physical laboratory" similar to present sounding rocket payloads and on the order of Singer's own idea for what he called MOUSE, some 50 to 100 pounds. Above all, Whipple urged quick action: "I seriously question the wisdom of postponing the possible experience and research that we could gain from the simpler, cheaper types of satellites, until we can plunge into the extremely difficult and expensive later stages of the unmanned physical laboratory."[98]

Whipple listed eight objectives, military and scientific, that should be the drivers for the program. At the top was pure Cold War rhetoric: "World-wide demonstration of our power to send a missile to any part of the globe." But immediately beyond this fact lay a number of scientific observations, all of which could be pursued by the tracking systems he was envisioning based upon his Super-Schmidt network that he knew the Air Force had bought into. First, optically visible satellites could provide precise data for improving knowledge of the gravitational field of Earth and, consequently, predict the trajectories of "extremely long-range missiles" and better test the "re-entry problem." Most of all, Whipple advised that highly precise knowledge of the density, temperature, and pressure profiles for the upper atmosphere were critical for effective ballistics. Leaving for others the question of why precision was needed when bombs were so powerful, he pointed out that recent sounding rocket measurements of the high atmosphere confirmed that trajectories were extremely sensitive to atmospheric properties.[99]

In his report to the board, Whipple expressed no qualms about the value of precise data on the upper atmosphere for military application. In another presentation to the board at about the same time, however, "On the Security Classification of Upper Atmospheric Techniques and Data," representing the Geophysical Panel on the Upper Atmosphere, he was just as passionate that such data remain unclassified. "Science can progress rapidly only through the exchange of ideas, techniques, and data," he opened, adding that "the effects of security classification are to dull scientific initiative and inspiration." These "unhappy effects," he suggested, could "be alleviated by the establishment of large research organizations in which the basic scientific research activities are not compartmentalized." This is exactly what he had in mind a few months hence when faced with the opportunity of bringing the Smithsonian to Harvard.[100]

By early 1955, space crusaders faced two questions: Would the United States decide to launch a satellite during the IGY? If so, who or what organizations would be given the lead?[101] To answer these questions, the U.S. National Committee for the IGY formed a special study group creatively dubbed the "Long Playing Rocket (LPR) Committee" to consider and issue a draft report on feasibility, scientific value, and budget. Whipple chaired the committee, which labored over how to balance what was admittedly a sensitive military capability with the IGY's mandate for openness.

The politics of overflight and how this issue might have influenced the ultimate decision by the United States has been discussed and continues to be debated.[102] With an IGY imprimatur, the LPR argued, it would be politically easier to gain access to tracking sites around the globe, and, overall, as others argued, reflecting Whipple's views, it would make international cooperation much smoother. Overall, IGY involvement would facilitate worldwide tracking and data collection. As Whipple argued, the "IGY association was necessary to 'ease permission to go over other countries.'"[103]

Throughout these years, Whipple's expansionist enthusiasms could not have been greater. But his home base at the Harvard College Observatory was then in turmoil in the wake of Shapley's retirement. It was far from clear just what Harvard wanted to do with astronomy.

Chapter 2

Harvard Astronomy Post–World War II

By late 1946, most of the senior staff members were back at Harvard, recovering and regrouping. But it was not a warm reunion. "[T]hink where Shapley is and tremble," Henry Norris Russell advised a promising student in 1946, at a time when Shapley's political and social activities, as well as his constant effort to enlarge Harvard's facilities, had diluted his effectiveness as observatory director. Shapley's senior staff worried that Harvard astronomy was operating in the red, and many repairs to the buildings and instruments were long overdue.[104] But they did not agree on how to rectify the situation. Adding to the tension, President Conant imposed a new review mechanism on Shapley in 1948, creating a Harvard College Observatory Council (HCOC) composed of senior faculty to advise Shapley.[105]

Despite Shapley's initial successes with the Rockefeller Foundation and other funding sources, there had been no effective growth of the endowment since the mid-1930s. Yet by World War II, the observatory had grown considerably; in 1927 the Boyden Station was moved from Peru to South Africa and greatly expanded, and in 1940 Menzel had created the Climax station in the Colorado Rockies for solar work. There were also enlarged facilities at nearby Oak Ridge including the new 61-inch reflector.[106] Unlike in the late 1930s, there were now relatively few Harvard-trained PhD astronomers starred by the American Association for the Advancement of Science (AAAS).[107] The infrastructures available for training and instrument development were stretched beyond tolerable levels.

As one of its first actions, the HCOC tried to mount a small fund-raising campaign, but the Harvard administration blocked it, claiming it conflicted with its own general campaign. Conant's imposition of the HCOC and then his rejection of their initiative underscored the poor relations between him and Shapley that had occurred since the 1930s. Shapley's politics and blatant disdain for some of Harvard's loyal patrons were at the root of the tension; because of Shapley, Harvard felt the "pressures of McCarthyism [and demonstrated how they] could shatter personal relationships and academic camaraderie," even in a field as remote as astronomy. Conant never openly criticized Shapley, but he did dislike the man, finding him "personally arrogant, politically naïve, and an inept administrator."[108] Both Whipple and Menzel therefore had to navigate between Conant and Shapley as they sought out funding for their projects. They both were members of a growing network of committees and panels organized first by the military services and then by Vannevar Bush's Joint Research and Development Board and its successors under the Joint Chiefs of Staff.[109] And both succeeded in raising project support. Their successes, however, created only more tension for their staff, because of Shapley's failure to prioritize. For example, as Whipple was pressing Baker for new meteor cameras, Menzel wanted an improved coronagraph, and Bok wanted a new large Schmidt camera for South Africa. Baker therefore found himself overwhelmed and without institutional support: "All these things happened simultaneously."[110]

Shapley's Pending Retirement

Shapley was scheduled to retire in late 1952 and Conant was determined to hold him to it, despite the objections of more than one member of the Harvard Observatory Visiting Committee.[111] Conant was then pushing to normalize the observatory as an academic department, under a department chairperson and subject to the rights and limitations of Harvard's academic departments. But Shapley and the HCOC hotly rejected this directive.[112] When Conant remained adamant, the HCOC appealed to dean of the faculty, Paul Buck, to select a new director before Shapley retired in September, and Bok was their choice.[113] From this resistance, Buck and Conant knew that to affect any changes in Harvard astronomy they could not rely on present staff or on the traditionally friendly Visiting Committee. Harvard's only recourse, under the rubric of collegiality, was to appoint an independent blue-ribbon committee to deliberate over the future of Harvard astronomy.

The Oppenheimer Committee

In February 1952, Conant asked J. Robert Oppenheimer to assess the observatory and department. Oppenheimer was then a member of Harvard's Board of Overseers and he and Conant had known each other for years.[114] Oppenheimer created a committee, turning first to Bengt Strömgren, then director of the Yerkes Observatory, whom Oppenheimer also had in mind as a candidate. He also recruited his former Manhattan Project business manager at Los Alamos, C. Donald Shane, director of the Lick Observatory. Ira S. Bowen, a leading physicist and the new director of the combined Mount Wilson and Palomar observatories, and Robert Bacher, a Caltech experimental physicist and Los Alamos alumnus, completed the committee. Oppenheimer had originally wanted Harvard physicist Ed Purcell to join the committee but Purcell, who would be awarded the 1952 Nobel Prize in Physics the following week, felt conflicted because he had been closely involved with Harvard astronomers in a radio project.[115] In picking Oppenheimer, Conant succeeded in placing the fate of Harvard astronomy in the hands of physicists and astrophysicists far removed from Shapley's world.[116]

Dean Buck assured Oppenheimer that he had a free hand and that he could ignore appeals from Harvard astronomers.[117] Still, there had to be hearings with the staff, which were held in June, and then the committee made individual reports. Strömgren was the first and most forceful:

> Senior staff consists of people who were very good when appointed. Although they are good now, the group is by no means strong. If left to itself, it would not have the strength to build up a first class institution in observational astronomy, even

> if the money were available, nor would it have the strength needed for the development of a good theoretical center. Present rate of scientific accomplishment not satisfactory.[118]

Strömgren's indictment extended to the dilapidated facilities that he claimed were "inadequate to the extent that a continued flow of Harvard PhD's, trained as at present, would be a liability, rather than an asset, to American Astronomy."[119] He also felt that none of the present staff should have been tapped to take over. Some, like Whipple, were certainly marketable elsewhere. Oppenheimer agreed: "Whipple is a man of real distinction. He would be the easiest one to find a place for elsewhere, although he is tired and discouraged, but he has good taste." Beyond Whipple, they thought none of the senior staff was worth saving and it would be better to start from scratch. Reflecting directly on Shapley, they agreed that the real problems were systemic, that the "set up . . . has been bad." As Oppenheimer expressed it, "With good direction and the kind of supervision which Physics has had, the astronomers who have been here for 10 or 15 years might have been very different people."[120]

Throughout this period, Whipple kept his distance from the debate, concentrating on research and projects. At one point, he asked Buck for a sabbatical at an unnamed observatory "to renew my scientific contacts there and to lay plans for my scientific work in the future."[121] Buck granted his request, but it was never acted upon.

The committee clearly valued Whipple and likely saw him as an emotional casualty of Shapley's mishandling. The committee also was not uniformly negative; C. D. Shane, especially, whom Menzel also had mentored at Berkeley, felt that Conant's and Oppenheimer's indictments would create "humiliation and loss of morale" and no end of lasting friction between Harvard College Observatory (HCO) staff and members of the committee, which "would be unfortunate for American astronomy."[122] Conant therefore eliminated ad hominem statements and concentrated on the deteriorated infrastructure, illustrating dramatic losses by comparing the state of Harvard astronomy in 1932 with what it was in 1952.[123] Shapley's empire had to be trimmed: save what was worth saving. But Conant's intent was always, as he confided to Oppenheimer, to "hit the astronomers over the head" and he was not hesitant to consider abandoning observational astronomy altogether.[124] Conant saw three options: (1) Hire a new outside director to build up the observatory infrastructure; (2) with an outside director, scrap the southern station and maintain the Oak Ridge site as a training facility; or (3) close both stations and concentrate on theoretical astrophysics within the physics department.[125]

The Harvard Corporation approved option 2 and immediately provided some $300,000 for upgrades sooner than Conant had expected or even requested, even before a new director was to be appointed. Conant rationalized this decision to Oppenheimer as the need for expediency and relayed the corporation's hope that Oppenheimer would accept the additional charge of finding the new director as soon as possible.[126] There was reason for haste, as Conant was about to step down from the Harvard presidency to become the new U.S. High Commissioner for Germany. "Sorry to raise a lot of trouble," Conant added, "by leaving this astronomy matter in your hands. Please take it on!"[127]

Menzel as Acting Director

Shapley had advised Conant that Menzel was "definitely the most unpopular member of the major Observatory staff." Accordingly, Conant selected Menzel as acting director. When he got the news, Bok noisily resigned in protest as associate director in October 1952, which only further strained emotions at the observatory.[128]

As acting director, Menzel quickly set about to do what the administration requested. A number of the buildings had to be either wholly reconstructed or abandoned and destroyed. Displaced staff had to be somehow accommodated and vestiges of the Pickering/Shapley years erased. Those vestiges, offices containing staff of the highly regarded amateur American Association of Variable Star Observers (AAVSO), staff of the popular astronomy magazine *Sky & Telescope*, and even the remnants of James Baker's World War II optics unit, all had to move their operations off campus. The many nooks and crannies that had specialist library niches were closed down and consolidated with the main library, and the library itself sent its most valuable volumes to the Houghton Library on campus. The library also had to regularize relations and policies with the main campus facility, Widener Library.[129]

If culling the library was painful, weeding the famed Harvard Plate Collection was, to the more traditional members of the senior staff, considered a sacrilege, most ardently expressed by variable star and spectroscopic parallax expert Dorrit Hoffleit.[130] The collection of photographic plates was an almost continuous record of the night sky that had been taken over a half century and was cherished as a "natural history of the heavens."[131] Both actions took months and there were protests all along the way (Figure 9). Matters turned darker and hotter when Menzel took steps to meet the corporation's vote for option 2, which meant divestiture of the Boyden Station. Bok went ballistic and created a public relations furor in South Africa and elsewhere. Twenty-six Harvard graduate students and staff wrote a protest letter to Dean Buck, with copies to the Harvard Corporation and the Harvard Board of Visitors.[132]

Menzel, however, also worked hard to correct staffing issues, paying far more attention to the problem than had Shapley. He could not control Bok, Hoffleit, nor especially Shapley. However, he was on good terms with Whipple and suggested that Payne-Gaposchkin be promoted to full professor, a move long overdue.[133] Whipple generally agreed with Menzel, and after it became evident to Payne-Gaposchkin that Menzel really did support her, she too began to warm to his leadership. Another of Menzel's top priorities was to rectify the dreadfully low pay and status of many Harvard staff, culling those who were not performing and recognizing those who were. Inflamed by Bok, Hoffleit, and others, however, the eviction of the AAVSO, and the decision to make Boyden a cooperative venture in collaboration with European and South African universities all combined to keep the staff in an uproar.[134]

Menzel defended his actions. His "Interim Report on Observatory Problems," issued in February 1953, prompted Buck to report to Oppenheimer that "Menzel is doing a surprisingly good job and deserves commendation." A few days later, Lawrence Terry, chair of the Overseers Committee on Astronomy, endorsed Menzel.[135] But an outside candidate

HARVARD OBSERVATARY MEMORANDUM

From: D. H. Menzel Date: 12/8/53

To: A. A. V. S. O. , Att. Mrs. Mayall

Subject: Moving

We hope to start tearing down Building A as soon after the first of January, 1954, as the contractor can undertake it. We should appreciate it, therefore, if you would plan to make the move by that date.

Figure 9 Menzel's eviction notice was posted in the AAVSO offices with an added cartoon from the *Saturday Evening Post*, personalized to suit the situation. *Dorrit Hoffleit Papers held by the Schlesinger Library, Radcliffe Institute, published with the permission of the Harvard University Archives. Cartoon © SEPS licensed by Curtis Licensing Indianapolis, IN. All rights reserved.*

still had to be found. In September 1953, the new dean of faculty, McGeorge Bundy, made a personal appearance at a meeting of all observatory staff "to make clear the finality and authority of this decision." Bundy added privately to Oppenheimer, "This I shall do with the vigor of ignorance."[136]

Shapley's Successor

Shapley's concerted attempts to undermine Menzel, and Menzel's complaints about Shapley's taking credit for everything, convinced Oppenheimer and his committee in 1952 that what Harvard needed was a truly powerful person from the outside who could stand up to Shapley and overcome his liabilities. But through 1952 none of their targets, among the most powerful names in astronomy, would bite.[137] No one "sufficiently eminent, qualified and appropriate to be brought to Harvard" would even negotiate. So in spring 1953 the committee agreed that Menzel should be appointed as director, but for a limited five-year term. The committee also wanted additional appointments—at least one new senior astronomer and member of the HCOC—to occur simultaneously. Menzel's appointment, therefore, was only a "contingent recommendation" because they did not believe that his nomination alone, without this supplementary position of new blood at a very high level, would be sufficient to fix Harvard's problems.[138]

Menzel asked Payne-Gaposchkin to endorse him to Dean Buck, and she gave him flying colors; under his six-month acting period, the HCOC had been "working in a spirit of growing solidarity; the atmosphere has been one of cohesion rather than division." Under Menzel's tutelage, more graduate students were seeking broader astrophysical problems. Most critically, she felt that during his tenure they now had strengthened ties with Harvard physics and chemistry and MIT applied science, with an "increased emphasis on cross-field research . . . in tune with the whole trend of astrophysical research."[139] Payne-Gaposchkin had moved completely into Menzel's camp. With endorsements like this, and the fact that no external candidate wanted the job, Bundy late in 1953 called Menzel into his office and asked him, "What would be your reply, Donald, if I asked you to become Director instead of Acting Director?" Menzel accepted on the spot, asking Bundy to come to HCO to make the announcement himself.[140] The five-year probation was still in place, but it seamlessly became permanent within two years.

Menzel certainly planned for change. One of his first actions as acting director had been to identify some 22 issues he and the council had to address, from the divestiture of Boyden and a major shift from traditional catalog work to what were more fruitful, problem-based projects. He knew that Conant and his successor as president, Nathan Pusey, as well as Buck and Bundy, had convinced the Harvard Corporation what the Harvard astronomy had to look like, and it was not business as usual with the present Harvard staff. Menzel singled out only a few names to retain, such as James Baker, but others might do better elsewhere.[141]

Menzel's actions convinced the corporation that the observatory was aligning with Harvard goals and would eventually regain its distinction. Even so, with Conant now gone, replaced quickly by a new president, Menzel faced very much the same challenges, albeit in a calmer atmosphere. Pusey, a humanist, put less pressure on the sciences, and replaced Conant's "assertive, meritocratic spirit of FDR's New Deal" with "the affluent, confident mood of the Eisenhower and Kennedy years."[142] Even so, Menzel would have to demonstrate innovation and positive growth, strengthening his infrastructure and making a strong distinctive mark in order to restore Harvard's faith in the observatory. Indeed, within a very short time, serendipity helped him achieve both these goals.

Chapter 3

The Creation of the SAO

In May 1955, the *Harvard Crimson* heralded the arrival of the Smithsonian Institution on the Harvard campus. The article opened grandly, quoting Whipple, director designate: "The interests of both observatories are very similar, and the change will enable us to launch a broad attack on the area of astronomical geophysics."[143] Whipple continued:

> The problems before us are not of the sort that can be solved by individuals . . . and we will now have many more people working here, and thus have a greater opportunity to exchange ideas. I should say that the potentialities of both institutions have been greatly increased by the shift of the Smithsonian here.[144]

The *Crimson* reported that "high-level Smithsonian astronomers" would both teach and do research, which would aid training at the university. The Smithsonian's outstations, in Chile and California, would remain in operation, and the combined focus of both observatories would now be "astronomical geophysics," which "deals with the sun and its effect on the earth, especially through the impact of light on the earth and its atmosphere."[145]

The *Crimson*'s coverage reflects Whipple and Menzel's goals rather than the present state of affairs. First, Whipple's explicit use of the term *astronomical geophysics* is a reflection of his priorities. Menzel used the term as well, but it was also an entrepreneurial statement. They both knew that the Office of Naval Research (ONR) was particularly interested in the promise of such a crossover, and, to be sure, the International Geophysical Year was rapidly being organized.[146] It also reflected Shapley's penchant for promoting interdisciplinarity since the 1930s.[147] Yet the term rarely if ever appeared in astronomical literature either before or after this time. Astronomy and geophysics had long been linked institutionally and intellectually, but not in this adjectival form before 1960.[148]

The assertion that there were "high-level Smithsonian astronomers" was definitely premature. Whipple would be the first, and there was no physical transfer of staff from Washington to Cambridge. Nor was there concordance of interests between the two institutions. His prediction that the institutions would grow certainly came true, and from his words here, as reported by the *Crimson*, Whipple was sensitive to the expressed concerns of the Smithsonian that the work of the Astrophysical Observatory (APO) of the Smithsonian Institution would continue. What Whipple apparently did not say, however, was how he would pursue the APO's singular goal.

The Astrophysical Observatory of the Smithsonian Institution

The APO, founded by Samuel Pierpont Langley and then led by Charles Greeley Abbot for some 50 years, never varied in its mission, from its inception in 1890 to its death in 1955. Although this might be considered unusual, unimaginative, or even unacceptable behavior today by modern professional standards, it was not uncommon, nor was it seriously questioned, during that time.

The idea of establishing a physical observatory to study the heavens from the grounds of the Smithsonian goes back to the first secretary, Joseph Henry, who in the 1850s secured congressional authorization to establish a magnetic observatory,[149] and to close the loop called as well for a solar-monitoring observatory.[150] In 1887, his call was answered by the second secretary, Spencer Baird, who invited Langley to leave Pittsburgh's Allegheny Observatory and create a true astrophysical observatory, the first of its kind in the New World, to pursue his "New Astronomy," recording the amount and character of the Sun's heat.[151] The so-called solar constant was a critical factor in meteorology, because the Sun, then as now, drove the general structure of Earth's atmosphere. Langley had turned to this eminently practical problem in the mid-1870s and quickly garnered national attention from astronomers.[152]

Soon after his arrival in 1887, Langley also became the third secretary when Baird died.[153] He found that money was not a problem for building his observatory, but finding a suitable site was. He looked for "seclusion and quiet" but found so many restrictions around Washington that he finally settled on a temporary site immediately south of the main building[154] (Figure 10).

Although Langley built the APO, and set its agenda, he did not tend to it daily as Smithsonian secretary, becoming preoccupied with aviation and other pursuits. The funds allocated by Congress and by Smithsonian patrons, however, were sufficient to hire staff. Very much along the lines established by Henry, these funds also served to support research elsewhere, and Langley gained great equity within American scientific circles by funding emerging and established scientists.[155] But he left daily operations of the APO to one of his first hires, his chief assistant, Charles Greeley Abbot.

Charles Greeley Abbot

Abbot arrived at the APO in 1895 armed with training in chemical physics from MIT. His task was to refine the value and character of the solar constant by taking daily observations of the solar intensity in different parts of the spectrum as observed at sea level and at varying altitudes above sea level. His goal was to extrapolate its value "as it would be observed at mean solar distance outside the atmosphere altogether."[156] Like Langley an avid experimentalist, Abbot was also a dutiful data collector, employing devices such as their silver disk pyrheliometer for measuring the light and heat of the Sun. He was perfect for the job (Figure 11).

Figure 10 The offices, shops, and solar-observing facilities of the Astrophysical Observatory, located in the South Yard of the Smithsonian Institution during the Langley and Abbot years, ca. 1899. *Smithsonian Institution Archives, Image # 80-12955.*

Abbot's attack on the solar constant as a mission of the Smithsonian Institution had two phases. First, he worked to refine its value, but then he campaigned for the rest of his career to determine its variations, linking them to variations in Earth's climate. He accomplished the first phase by designing and executing a series of stratospheric balloonsonde ascents in California between 1913 and 1915.[157] His second phase was more elusive.

Over the decades, he constantly improved his instrumentation and established widely spaced observing stations with trained staff operating carefully systematized and calibrated instruments. Through the 1920s, Abbot parlayed his findings into a creed telling of firm connections between solar cycles and weather patterns. Seeking popular recognition, he delivered stump speeches on how his work would lead to true long-range weather prediction.[158] Abbot's claims, however, created problems for C. F. Marvin, chief of the Weather Bureau, because his office was being deluged with hundreds of letters asking for predictions and for advice about everything from when to schedule sporting events and weddings to

Figure 11 Charles Greeley Abbot demonstrating a silver disk "pyrheliometer" sometime between 1913 and 1917. *Courtesy Library of Congress.*

when to plant crops. Such raised expectations, from what Marvin felt was flimsy evidence, deeply concerned the conservative meteorologist. Marvin, however, was a Washington colleague and ally and knew how to play the game. Accordingly, he called for more observations, justifying the continued growth of the APO.[159] Abbot parlayed this support to gain patronage from a wide range of benefactors that made it possible to build new stations in the American West, in Chile, and in southwest Africa.[160]

Abbot succeeded Charles Walcott as sixth secretary of the Smithsonian in 1928, but his focus remained on the Astrophysical Observatory and its programs. With support from the Research Corporation of New York, he established a new division of Radiation and Organisms in 1929. During the 1930s, when the Smithsonian, along with other governmental bodies such as the National Bureau of Standards, suffered retrenchment, Abbot's APO prospered. Its staff grew steadily, split between the observing stations and instrument development in Washington. His instruments were used at stations around the world, both by Smithsonian staff and by a growing host of cooperating observatories (Figure 12). In Langley's tradition, these devices were reliable, standardized, and wholly interchangeable, and made the Smithsonian's reputation.[161]

Abbot exploited a close interlocking network of elite friends and colleagues within scientific Washington and in the astronomical, geological, and meteorological communities.[162]

Figure 12 Andrew Kramer, APO instrument maker from 1892 to 1953, photographed in the early 1930s working on typical solar-monitoring instruments. *Smithsonian Institution Archives, Image # SIA2012-6495.*

Within this circle, especially starting in the 1920s, what Abbot believed he was finding was a phenomenon very much in vogue at the time—cyclic behavior—that the Sun's energy varied in measurable and ultimately predictable cycles. In the 1920s, cycles were a fascination. The Carnegie Institution of Washington created a "Committee on Periodicity" and hosted "cycles conferences" looking for correlations in all natural phenomena. There were many voices in support, therefore, of Abbot's continued work, holding out the hope that his conclusions would be vindicated.[163]

Dodging Criticism

Abbot faced his harshest criticism in 1939 with bluster. A young, mathematically gifted Harvard astronomer, Theodore Sterne, followed up the work of an Indian meteorologist in London who showed that most of Abbot's periodicities fitted known atmospheric anomalies and so were not solar in origin.[164] Abbot had summarily dismissed these findings, which stirred Sterne into action. Adept in Fourier and least squares techniques, Sterne scrutinized the periodicities and became concerned about their reality. Some did have statistical significance, but they could well be not only of the solar constant but also of the "solar constant, plus random errors, plus possible systematic errors," none of which had been or could be accounted for.[165] Sterne and two Harvard colleagues, Walter Orr Roberts and Karl Guthe, concluded that the periodicities Abbot reported were not found through harmonic analysis, "but only from a partly subjective examination of a rather loose sort."[166]

Abbot blithely brushed away this blunt rejection by the Harvard astronomers. When Shapley invited Abbot to Harvard to respond in May 1940, Abbot haughtily claimed this was clear evidence of his veracity. One of his quietest critics was, however, very close at hand. In 1947, his long-time assistant at the APO and now his successor, Loyal B. Aldrich, admitted privately that, although he had the greatest affection and respect for his old boss, "Mr. Abbot's papers on this subject in my judgment tend to emphasize the favorable factors and to neglect the unfavorable, rather than to weigh impartially all factors. As a result, one is left with a feeling of doubt."[167]

The APO After Abbot

Abbot retired both as secretary and APO director in 1944. Aldrich carried on, defending mission and staff, but he now reported to a new secretary, ornithologist Alexander Wetmore, who was less than sympathetic.[168] Aldrich was near retirement himself and lacked Abbot's engaging charisma and skill moving through scientific Washington. Over the next several years he echoed Abbot's appeals for the APO's mission, but that was literally all he did, without the alliances that Abbot was always able to marshal. Aldrich was simply being true to his word; he made no promises about solar variations and none about weather influence.

In his first annual report, celebrating the Smithsonian's centennial, Wetmore stated bluntly that that "its position is strong in some respects and weak in others."[169] He was most concerned about inadequate facilities and lack of support staff to properly serve the some 2 million annual visitors to the National Museum. But he also feared that the Smithsonian was no longer competitive in a new world dominated by research institutions and laboratories flourishing in universities and industry.[170] By 1950, his concern deepened over "increasing pressure on all sides" to regiment science "in certain countries." He cautioned that the Smithsonian should not overstep its bounds, such as adding "functions unjustified by normal demands or necessities." Yet, he defensively added, it had to continue to foster "small and perhaps unpopular projects."[171]

These words were no doubt a response to calls then in Washington for more accountability and oversight of government agencies.[172] One of his chief advisors, John L. Keddy, a politics and government specialist who had been the Smithsonian's Bureau of the Budget examiner and was now assistant secretary, knew that Congress and the executive branch were considering some form of general government reorganization, which became more ominous when a Senate committee on "Expenditures in the Executive Department" convened and started asking questions. The newly created General Services Administration was really shaking things up in Washington late in 1949, so Keddy strongly advised Wetmore to alert Senate and House Regents of the grave danger afoot, sure that lurking on the Hill was a bill that "proposed subordination of the Smithsonian to the General Services Administration."[173]

Another member of the secretary's staff, John E. Graf, agreed. He had been Associate Director of the Smithsonian's United States National Museum since 1931, responsible for administrative matters, and now assistant secretary as a specialist in federal budgetary procedures. Their charge was to professionalize the Smithsonian administration in the eyes of Congress. The steps that led to the APO's demise on the Mall need to be understood in this context.

Aldrich reported to Keddy and Graf, who at first were not concerned with his leadership, but always urged him to fill any open positions quickly, since an unfilled position in the government "indicates the agency does not know how to manage its funds."[174]

Real problems arose in 1951, however, when the House Appropriations Committee threatened to halve APO funding. Even though it was overtly a reflection of general government cost cutting amidst the growing Korean War, there were also concerns because some in Congress were impatient that the Smithsonian's decades-old promise of providing weather prediction still had not happened. Even though Wetmore rapidly restored the APO's support from general funds, the Smithsonian had been warned.[175]

In 1952, to help prepare for his own successor, Wetmore initiated a general reappraisal of Smithsonian research and curation. It was a watershed in revealing how the research staff viewed itself, and for what it was ready to fight. Graf chaired the Smithsonian Institution Planning Committee and Keddy co-chaired it. The committee created subcommittees to look at the research infrastructure and also suggest ways to improve outreach without taxing the researchers. Aldrich was on the central committee as a research head, but played little part in the proceedings, or in deliberations that were dominated by concerns from the National Museum of Natural History, mainly on how to improve or regularize the use of a nature reserve in a lake along the Panama Canal called Barro Colorado Island.[176] But there was also a very active and influential Subcommittee on New Work Programs that deliberated over the next several years on the future of research at the Smithsonian. By summer 1954 their overall statement was to continue the course, but to foster more collaboration with American institutions with complementary resources, "seeking either private or governmental funds."[177]

Throughout this time, Aldrich's annual report to the secretary remained unchanged from the previous nine years: "Our goal has continued to be to improve if possible the standard set by previous work."[178] Wetmore took these reports in stride, but Leonard Carmichael, his successor, scrutinized his directors' reports in light of the findings of the deliberating committees. And as had Wetmore, Carmichael found there was more cleanup necessary.

Enter Carmichael

Leonard Carmichael, the seventh secretary of the Smithsonian Institution, took office in January 1953. With a PhD in psychology from Harvard awarded to him in 1924, and professorships at Princeton, Brown, and the University of Rochester, he had been president of Tufts University since 1938 and was the Smithsonian's first secretary from academia. Carmichael was also a colleague and confidant of several members of the Smithsonian Board of Regents, including Vannevar Bush, who had been a Smithsonian regent since 1940 and a leader of scientific Washington during World War II.

Carmichael moved quickly to revitalize the institution. As had Wetmore and Abbot, he pushed to raise visibility by improving exhibits and expanding media outreach. But his deeper concern was for the health and future of the Smithsonian's research viability after the Smithsonian Institution Planning Committee found that "a need exists for a critical appraisal of the relative importance of fundamental research in relation to curatorial and service functions."[179] Carmichael also knew of the ongoing debates over the scope of the emerging new National Science Foundation and its role in the federal support for science. Would it be a funding agency, or an agent for coordination and control of those funds not only in academe but also in other federal agencies and bureaus? He certainly did not want the latter option, and most strongly reacted to a portion of the proposed directive that called for bureaus and agencies to "confine themselves to 'those research and development projects *which meet the criterion of necessity*'" as defined by their authorized missions.[180] Even so, thinking of how to protect the Smithsonian's various investments in research, from exploration biology to the monitoring of solar radiation, he worried less about their value to humanity than to their value to the institution's health and competitiveness.

The "criterion of necessity," Carmichael realized, as much as he might have objected to it, meant evermore that the work of the APO had to be defensible. Abbot, still very much present as an Associate of the Institution, claimed it was, urging only that more public outreach was critical to survival.[181] But Aldrich confined his attention to managing the APO and its Division of Radiation and Organisms. In September 1953, his second in command and planned successor died unexpectedly. And the third senior member, A. Kramer, the instrument maker, was about to retire after some 40 years of service. There were about a dozen other employees at the time, from field station directors to meteorologists to clerical assistants and computers, but none of them had standing or reputation as successor to Aldrich.

This state of play worried Keddy, who was annoyed with Aldrich's dry reports and appeals for "the need for extreme accuracy" and for the continued improvement of instruments and reduction methods. In late 1953, Keddy scribbled, ". . . same thing again. . . ,"[182] some four times in the margins against each of Aldrich's claims. At one point, he boldly scrawled, "no vision." Keddy knew that Carmichael had been talking with his academic colleagues about the future of the APO. His Tufts colleagues recommended he contact astronomers for advice, such as Leo Goldberg, then at Michigan, Charles D. Shane at Berkeley, and Donald Menzel at Harvard, who was then the best known and most influential solar astronomer in the United States.[183] One trusted colleague at Tufts darkly advised that he and his colleagues "have been unanimous in their feeling that it is high time that a more modern research program was undertaken—in addition of course to continuing the routine measurements of solar radiation." Among several prominent scientists critical of the APO, he named Fred Whipple, who knew of Sterne's analysis.[184]

With Keddy's continuing urging, Carmichael contacted Menzel and, finally Shapley, more than confirming his concerns.[185] There was never any question in Carmichael's mind concerning the scientific and social importance of the problem, as delineated by Langley, Abbot, and then Aldrich, and the importance of continuing the search for correlations with weather. The question was, how and where best to do it?

Whither the APO?

Anticipating Aldrich's retirement, Carmichael quickly found it unlikely that a strong candidate could be attracted to Washington to direct the APO's far-flung stations. So he started looking further afield for a new base of operations. At first he was attracted to a facility in Freemont Pass in the Colorado Rockies at Climax called the High Altitude Observatory (HAO). Created by Donald Menzel, and managed at first from Harvard as a remote station, the HAO was rapidly growing into a significant solar observatory funded by the U.S. Air Force and linked to the University of Colorado in Boulder. It was directed by Menzel protégé Walter Orr Roberts, who more than anyone built the observatory and maintained it through World War II, and collaborated with Sterne.

Roberts had been recommended by Shapley and seconded by Menzel to continue the APO's solar agenda, but reconstitute it along modern astrophysical lines.[186] Carmichael visited Climax in August 1954 and was much impressed with Roberts's creation, "a unique and really brilliantly effective group of co-workers." They also met with the president of the University of Colorado about a possible university connection.[187] Roberts was already building stronger ties with the University, since his Harvard ties were weakened due to, in the words of a HAO Board member, an "irreconcilable conflict between Roberts and Menzel."[188]

Indeed, as historians have clearly shown, Carmichael's search for a new home for the APO came at a time of many tensions, from parochial East-West differences among astronomers, to the role of the federal government, the Smithsonian in particular, in the support of science, to the need to continue a line of research activity that was no longer in vogue, and, to be sure, to personality clashes.[189] As with the deliberations by Oppenheimer's committee at Harvard, the process of coming to the decision included many of the same names, mainly directors of major observatories.

Through autumn 1954, Carmichael consulted with astronomers and meteorologists in Washington and Boston about Roberts, finding both positive and negative reactions.[190] Complicating the matter was that observatory directors on the East and West Coasts differed over the need for, and organization of, a national optical observatory that would foster new technologies, such as photoelectric photometry. Most of them wondered whether the Smithsonian's move should not be deliberated in the context of a national facility.[191]

Indeed, Carmichael's inquiries came at a particularly sensitive time when national observatories for both optical and radio astronomy were being debated by national panels and members of those panels perceived Carmichael's inquiries in parochial ways. R. R. McMath of Michigan felt strongly that Roberts's observatory "was not the right place" for the Smithsonian to land.[192] And others advised that the APO "should stay where it is at the moment, but later be moved to the National Observatory site."[193] Through this effort Carmichael found factions within the astronomical community, but he was also hearing some names come up frequently as possible candidates to revitalize the APO. Roberts was suggested repeatedly, but also were Fred Whipple and Richard N. Thomas at Harvard and George Field at Princeton. Thomas, a bright and promising Harvard PhD and junior member of the staff since his graduation in 1946, had broad interests in stellar atmospheres, working with Payne-Gaposchkin, Menzel, and Roberts. He also had strong interests in hyperballistics and meteor studies, and at times worked with Whipple, Sterne, and others. Menzel strongly supported Thomas.[194]

Carmichael was far from sure what advice to take, but one thing was clear. Given the volatility of the profession, "the Smithsonian Astrophysical Observatory should retain its identity, no matter what new association is established."[195] Menzel continued to push for Harvard, and promoted Thomas as a key player. Thomas even visited Washington to outline an ambitious network they called "The Solar Associates." He argued that it would combine the forces, and the resources, of Harvard and the APO as well as the Bureau of Standards and Roberts's HAO, all fueled by the "U S Government."[196] Carmichael listened carefully, as he wanted a strong connection with a university. But it had to be with the right institution, willing and able to continue the legacy programs of the APO but with modern competitive techniques. Ira Bowen, who had been on Oppenheimer's committee, felt that the APO legacy of routine observation was best left in government hands, but if it were linked to Harvard and Menzel, something new and innovative might emerge.[197] And giving it further thought in December 1954, Bowen admitted that a link to Harvard had advantages.[198] Goldberg agreed: A Smithsonian association with Harvard would no doubt be positively received in the highest circles.[199]

The highest circles, for Carmichael, were the Smithsonian regents. Jerome Hunsaker, regent and pioneer aeronautical engineering professor at MIT, had a high opinion of Roberts's abilities but worried that he was a "one man show" unless there was a strong university affiliation.[200] Hunsaker confirmed that a multi-institutional arrangement was important, but the Smithsonian had to retain its identity. As Carmichael considered his options, thinking of Roberts and Colorado or Thomas at Harvard and the solar associates scheme, evidently he, Menzel, and Bundy were also encouraging Whipple to consider the job. In one of his many phone calls to Carmichael, Menzel reported that "Whipple, as he thinks about the matter, is becoming more interested in the Smithsonian opening." Menzel added that Dean Bundy "expressed a real enthusiasm for an association with the Smithsonian."[201]

By the end of the year, Carmichael decided to move the APO to Harvard, and after more meetings with Menzel and Whipple in mid-January, they refined the scope of the "solar associates" wherein the APO would move to Harvard, with connections to

Roberts's Climax, the U.S. Bureau of Standards, and the Air Force's Sacramento Peak Solar Observatory, virtually all major American institutions concerned with solar-terrestrial relations within Menzel's domain. All this would be fueled by "U.S. Gov. contracts."[202] But Roberts's description of what he wanted, a "private non-profit corporation" with an independent board of directors that answered to the University of Colorado, made Graf and Keddy suspicious because, they advised Carmichael, it meant exposing Smithsonian's "hard core of money" and that "both Harvard & Climax are expecting to lay profane hands on that money."[203] In the end, Carmichael told Roberts that the Smithsonian wanted ties "with some other centers that are in a sense in the Harvard orbit."[204] This killed Roberts's hope for a Colorado-based empire.[205] With Roberts now sidelined, Graf counseled Carmichael, "Our first move is to catch a Director and then listen to his ideas on various and sundry subjects affecting the Astrophysical Observatory."[206]

The choice was now between Thomas and Whipple. Menzel argued that a top man was necessary. Thomas was good, but young and a bit brash. However, a top man at Harvard would require a salary well beyond the present federal pay scale for the Smithsonian, which "is not sufficient to attract a top man in the field, like Whipple." Menzel envisioned that Whipple would have a large chunk of the combined observatory to worry about, and that the Smithsonian would require a full floor of one of Harvard's bigger buildings. But it could also fund the construction of a major new building, with a brass plaque over the door identifying the facility as "a matter of practical application as well as of pride." The Smithsonian, Menzel added to Bundy, would be "in a position to make further grants for the work beyond the basic budget, and in addition, accept the various government contracts."[207]

Whipple, of course, was not a solar astronomer. But Menzel knew that his meteor work had "already brought us into problems of the upper atmosphere and the broad field of astronomical geophysics." He stressed that "this new cross-field study is important and exciting." "Cross-field" research was then a passion at Harvard, and Menzel felt this was a most effective strategy. Harvard was well placed to take advantage of the situation, taking the lead in training new scientists to work in the hybrid field: "The demand for such scientists is great, in the government, in universities, and in industry." He reminded Dean Bundy of Harvard's strong efforts in radio astronomy and emphasized that his own and Whipple's theoretical studies in aerodynamics had "multiple application in fields ranging from pure astrophysics to the design of airplanes or the calculation of explosion [*sic*] in shockwaves." Finally, demonstrating his determination to clean out the old, he applauded the parallel initiatives at Harvard and MIT to modernize meteorology: Harvard would do well to move boldly into the meteorology of the upper atmosphere "rather than in the relatively non-productive field of synoptic meteorology. . . . An increasing number of meteorologists believe that the clue to climatology lies in the upper atmosphere."[208]

Menzel's vision of what the Harvard-Smithsonian relationship represented and might someday produce, the mere scale of it, required Whipple's seniority and his extensive experience in Washington—his numerous memberships on boards and advisory panels. Accordingly, Bundy advised Nathan Pusey that they were working on a joint government appointment for Whipple. Bundy, though enthusiastic about Whipple, remained cautious "because of the number of painful disengagements which I have had to work on in the astronomical area in the last year." But now, he felt, it was worth it: "The Smithsonian is a very different institution from those which we have been dealing. It has income of its own, first-rate standards, and we think a disposition to make its major appointments by criteria that mesh with ours."[209] Bundy therefore endorsed the plan as long as all parties could agree "on the man who might be the principal link." That man was Whipple, for whom Bundy had developed a particular sympathy:

> One of the things which is of great interest to Whipple, the man who might well be director for the Smithsonian if the deal comes off, is the possibility of researches leading toward the establishment of a space station. I always laugh when he says this, but he is dead serious. The research work for this sort of undertaking will be on government contracts, and furthermore, some of it will have to be classified. So Whipple is hopeful that if we can work out a pattern with the Smithsonian, it may be possible for him to have a classified research project through the Smithsonian, without breaking our own Harvard rules against classified work. I have warned him that this would require Corporation approval, but I have also said that I do not think it impossible that an arrangement can be made which would safeguard Harvard's position, while permitting him to proceed with this dear dream.[210]

When asked after the fact, Whipple recalled no hesitancy about accepting the offer, or having to be convinced by Menzel. His intentions for accepting the directorship were clear, definitely along the lines Bundy expressed: "I took the job . . . so that I could then operate this photographic satellite observing program under the aegis of the Smithsonian, rather than Harvard. That's why I went along with that. And that's how it worked out."[211] More in the moment, however, in 1964 he admitted to a colleague that "I went into the Smithsonian Astrophysical Observatory area because it was quite clear that the space plans would be greater than Harvard University would be interested in pursuing with me."[212]

In 1955, it is not difficult to imagine Whipple's vision extending seamlessly from the practical necessity of satellite tracking to von Braun's "Project Orbiter" and even to a space station via access to government and military funding.[213] This is the world Whipple needed access to, with an institutional base that was willing to participate. And it may well have been an appreciation of this vision that attracted Bundy and Pusey to him, rather than to a younger and more impetuous talent such as Thomas, a brilliant but junior astronomer at the time, and one Bundy even turned down for a Harvard faculty position in late 1956.[214] Menzel, Bundy, and definitely Carmichael knew that Whipple was of sufficient breadth and stature, and well experienced in government and military patronage, to meet their collective vision. As wild as advocacy for a space station may have seemed in 1955, somehow Whipple's "dear dream" was of the level that met Bundy's vision for Harvard's future. Indeed, for Whipple, as we shall see, bringing the Smithsonian to Harvard allowed for wider freedom of action to redefine just what astronomy at Harvard would encompass. It would be a larger venue in many ways, but one that retained the fundamental ingredients

Whipple knew he needed. Whipple in fact had the option of moving to Washington at the time, and his decision not to do so indicates that he was more than comfortable bringing the Smithsonian to Cambridge where he could pursue his goals "without losing contact with the scientists in this area."[215]

Closing the APO in Washington

In mid-March, Keddy recorded the state of play. Aldrich would be "terminated" as of 1 July 1955 and Whipple would take over as director. The APO headquarters would be transferred to Cambridge on 1 July as well. All save one of the observing stations would close on 30 June and there would be transfers of staff as a result to Washington or to the remaining station at Table Mountain, and one staff member, possibly, to Cambridge. Table Mountain in California would continue through 1956, and then be re-evaluated. The instrument shop in Washington would be maintained through the year. If a pending Air Force contract materialized, all APO data were to be card punched for future use in meteorological studies. There would definitely be support to "employ young scientists part-time at Cambridge" who were pursuing PhD degrees at Harvard, and, finally, the APO Division of Radiation and Organisms was to be transferred to the Office of the Secretary for the time being.[216]

Keddy's clinical report ignored the anger and frustration of the remaining Washington staff. Aldrich had pleaded with Carmichael since mid-1953 and, getting nowhere, finally resigned in January 1955 in failing health.[217] Graf, closing the loop, informed Aldrich simply that the Washington headquarters would be closed down and all subsequent analyses would be carried on at Cambridge.[218]

Carmichael drafted a letter of appointment for Whipple as an exercise to "think on paper." Whipple would be responsible for receiving the data and managing staff in Cambridge for its timely analysis. Whipple had to be prepared to manage the analysis, keeping "the incoming calculations of the field stations up to date" and to do so the Smithsonian would provide a mathematically adept astronomer as Whipple's assistant who would "make a critical study of the whole subject and attempt to provide some evaluation of the program." Carmichael assumed that this man would eventually become a civil servant as assistant director, and hoped that the sole remaining professional on the APO staff might be considered a candidate for the position.[219] But it was up to the Smithsonian to determine how Whipple's joint appointment fit government guidelines, for facilities, staff, and present assets.

Meanwhile, Menzel urged Bundy to gain Pusey's approval of the overall plan. The Harvard College Observatory Council had approved the move and the conditions of Whipple's joint appointment, and Menzel made it clear that this infusion of Smithsonian support would "provide a research potential approximately doubling the number of appointments on permanent tenure now existing in the Department of Astronomy." These new appointments, moreover, would "greatly increase our ability to train young scientists for research and teaching."[220]

Building and Housing a New Staff

While Carmichael's staff and Menzel drafted versions of the "Memorandum of Understanding," Whipple focused on housing details, determining that his operation required some 10,000 square feet and that a new building would cost some $150,000. Any new building would take time, so Whipple opted for space in Harvard College Observatory's already crowded complex. But soon, the Harvard administration decided to build the building and amortize it by charging Smithsonian rent. By late June, with a signed agreement in hand, Menzel reported that Harvard was making headway on a new building, hoping for a groundbreaking in the fall.[221]

Coordinated press releases from the Smithsonian and from Harvard in early May produced wide media response, which, beyond the *Crimson*, included *Sky & Telescope*, the popular magazine that had been closely associated with Harvard for decades, and numerous newspapers, both local and national. Whipple and Menzel continued to define the boundaries of the new relationship and plan for the new building. But their most contentious issue was approval for appointments to the new combined staff. Carmichael and Menzel at first agreed that appointments to what was now called the Smithsonian Astrophysical Observatory (SAO) were to be executed with some Harvard oversight, including the director of the Harvard College Observatory, and at least one member of the Harvard Observatory Council, designated by the dean of the faculty. The SAO director, however, would have "full responsibility" for the "direction of its scientific programs," under advisement by the Smithsonian secretary but would be expected to coordinate the research "of the two organizations to the mutual benefit of both for the furtherance of research and education in astrophysics."[222] There was little description, however, of any balance of power between the two directors—apparently not an issue for Menzel and Whipple. Moreover, what little oversight Harvard retained in the initial wording disappeared in the final memorandum of June 1956:

> [A]ppointees to career positions of higher scientific standing in the Smithsonian Astrophysical Observatory shall be selected with the advice and consent of the Director of the Smithsonian Astrophysical Observatory. It is recognized that he may avail himself of such advice at Harvard University . . . as he considers appropriate, especially for those appointees who shall also serve at Harvard University.[223]

In effect, this wording gave the Smithsonian director considerable latitude in hiring, but did not give him authority to make Harvard appointments. Nor did it give the Harvard director veto power over Smithsonian appointments. How this balance would play out in future years would be one of the issues framing the relationship between Harvard and the Smithsonian. And it did not take long to become painfully apparent.

Whipple's highest priorities, as he told the Harvard College Observatory Council in March, were for a "statistician—then an atmospheric physicist" who would competently analyze the APO data, as Carmichael requested.[224] On a trip to Washington, following up Menzel's advice, Whipple stopped at the Aberdeen Proving Ground to court Theodore

Sterne.[225] Sterne was one of the first to analyze Abbot's claims and to find them wanting. From the start, however, Sterne expressed concerns about the nature of the appointment. He did not expect another analysis to bring anything new to light, and though he had been assured by Menzel that he would be given a Harvard appointment, became annoyed when Whipple said they would make an attempt, but could not yet guarantee it. Sterne initially turned them down. "It would mean the abandonment of a status of considerable distinction in Ordnance and the Army, in which my work has been a source of much satisfaction to me, for an even lower status than I had at Harvard in 1941." But after Thomas "worked on me to such an extent," Sterne decided to reconsider.[226]

Sterne's reaction was a warning. He was valuable to both Menzel and Whipple because he was highly capable in data analysis and was experienced on many fronts, including high-speed computers. He had been chief of the ENIAC electronic computer team in the late 1940s and in 1953 was designated "Scientific Advisor to the Director of the Ballistic Research Laboratories" at Aberdeen. Even though he had had no unclassified publications since 1946, those before the war were solid contributions to stellar interiors, celestial mechanics, binary star analysis, and developing numerical methods in astronomy. Menzel used Sterne's initial letter of refusal as "concrete, unsolicited evidence" that a solid Harvard appointment would be necessary to secure "the quality of people we could get at the Smithsonian."[227]

Menzel pushed Bundy to expedite the appointment process for at least four new appointments. He suggested that Bundy create an ad hoc selection committee consisting of Norman Ramsay and George Kistiakowsky, Harvard faculty who knew and could appreciate Sterne's classified work. Other friendly faculty included Edward Purcell and J. H. Van Vleck. But not Strömgren: "We simply do not trust his judgment."[228]

The next name on Whipple's list was John L. Rinehart, "whose work in ballistics from the experimental standpoint has brought him scientific distinction." Rinehart had worked at the Naval Ordnance Test Station, China Lake, California, had contributed to hyperballistic studies of high-speed impacts, and was already on Whipple's SAO staff as assistant director but in need of a Harvard appointment.

Third came Max Krook, a mathematical physicist from Birmingham University, who had been appointed a Harvard lecturer in the early 1950s. Krook had performed distinguished work in stellar atmospheres and had contributed to a study of the possible influence of interstellar matter on the solar constant, and hence on climate change.[229] Krook was also well liked as a valued Harvard colleague, collaborating with Menzel, Thomas, and graduate students.[230]

And fourth was J. Allen Hynek, then professor and chair of astronomy at Ohio State University. Hynek had an eclectic portfolio; a specialist in observational stellar atmospheres and variable stars, he was then secretary of the American Astronomical Society. In the 1940s he had worked with James Van Allen, first on the Applied Physics Laboratory proximity fuse team and then with his rocketsonde group as project manager, astrophysical consultant, and public interpreter. Hynek was interested in a one-year appointment at Harvard to assist Whipple in a new project to examine meteors by high-powered radar in association with MIT's Project Lincoln, which would provide a million-watt transmitter.[231] Hynek would also assist in developing a popular astronomy series for WGBH-TV. If "this trial period should prove satisfactory on both sides," Menzel and Whipple proposed he be continued with a joint appointment as a full professor.[232]

What was clear to everyone was Whipple's top candidates were not mainstream astrophysicists. They had all engaged in non-academic fields that aligned more with Whipple's project interests than with independent research.[233] Charles Whitney, who was then in his last year at Harvard as a PhD student working on stellar atmospheres, recalled the Smithsonian initiative as very much in line with the Harvard tradition of organizing work into large projects. Since the time of Pickering and Shapley, these projects had been called "bureaus," named for patrons, such as the so-called Milton Bureau that gathered knowledge of variable stars.[234] Whipple hired Whitney on a Smithsonian appointment initially because he knew Whitney could be useful in various capacities, such as extending the meteor calculations to satellite tracking. And with a Harvard PhD, Whitney was also appointed lecturer to continue his work in stellar atmospheres.[235]

Harvard understood that all costs for appointing Sterne, Krook, Rinehart, and Hynek would be borne by the Smithsonian. Even so, dual appointments beyond non–tenure track lecturers or "research associates," especially to those at professorial rank, were not lightly bestowed by the Harvard Corporation, even though under Pusey the faculty was slowly gaining power, partly abetted by Bundy.[236] Menzel saw the appointments as Harvard's contribution; he argued that Sterne's case was

> an excellent example of the way in which Harvard, even though it is costing us nothing since the Smithsonian will pay one hundred per cent of Sterne's salary, can contribute to the effectiveness of the Smithsonian Astrophysical Observatory, which was one of the reasons for their coming here in the first place.[237]

The Smithsonian appointments came through quickly but the Harvard appointments languished until late in the spring, when Pusey finally, following Menzel's suggestion, formed a new committee to deliberate over them. All save for Sterne had responded positively to the Smithsonian offers; he gave Whipple a "conditional acceptance." Sterne also wrote directly to Carmichael, saying he needed assurance that it would include at least a half-time teaching professorship at Harvard. He was happy where he was; only the prospect of teaching again at Harvard made all the difference.[238] Hynek, however, was willing to take the risk, with a leave from his chairmanship at Ohio. Even before his appointment was approved, in February 1956 Whipple reported that the meteor work with Project Lincoln was stalled so he was redirecting Hynek to the "satellite program."[239] Apparently Hynek was wholly content with this, as it included the potential of public service and popular exposure.

Bundy, however, was aware of the dangers of retaining non-tenured staff, especially during the Conant years.[240] Would they dilute the overall quality of Harvard teaching and research? Whipple promised that this sort of dilution would not happen, nor would there be lingering financial responsibility. Smithsonian appointments would not be a dumping ground for "second-rate men who have proved themselves useful or well liked around the Harvard College Observatory and for whom there is no tenure position available." Conversely, Whipple wanted Bundy to appreciate that there were men around who could be "extremely valuable in their specialties, more valuable, possibly, than Harvard Professors." So in effect there should be no "general principle upon which one can view with alarm" hiring men useful to the Smithsonian who were not of Harvard caliber.[241]

Whipple saw value in men with "peculiar special talents" who may otherwise have weaknesses that would disqualify them for the tenured ranks. He also recognized, as he knew Bundy did, that there were weaknesses among Harvard faculty, especially in teaching, for which some of his hoped-for staff, such as Krook, Thomas, and Sterne, showed high promise. The key argument for Whipple was flexibility: Could these people do the job, no matter what the job turned out to be?[242]

The Special Ad Hoc Committee on Hiring

Pusey charged the new hiring committee to seek applicants who would move the department in the directions suggested by Oppenheimer's original committee. But since these new positions were to be paid by the Smithsonian, Pusey asked Carmichael to be on it too, ex officio. The committee (Bowen from Mount Wilson, Struve, now at Berkeley, Kistiakowsky, Purcell, and Ramsey) met in March 1956 to hear Menzel and Whipple make their case. To placate Bowen and Struve, at the outset Menzel agreed that "Harvard should not try to compete in studies which depend upon the use of large telescopes for making observations, but rather concentrate on theoretical investigations calculated to supplement the efforts of other observatories." Harvard would concentrate on theoretical astrophysics, and continue to concentrate on studies of the Sun, meteors, and Earth's upper atmosphere.[243]

Struve, however, felt that "the Department is still trying to do too many things and so suffers from dispersed interest." They needed to cut down, and if they had to choose one observational strength, it should be radio astronomy. Overall, the members held to an outline Menzel provided, resisting astronomers who clearly would flourish best with big telescopes, and suggesting younger names who were specialists in radio astronomy. They also bandied around some names of geophysicists and meteorologists. The meeting turned into a referendum on both the state and the breadth of the profession, as dozens of names were raised, tossed about, and mostly discarded. The committee wanted superstars if they could be captured, and once again spoke of the hottest names in astronomy of the day. There would be a new "Distinguished Professorship," a Menzel-created designation that caused Bundy and Pusey to quip "all professorships at Harvard are distinguished professorships." They agreed to a "Professorship in the Department of Astronomy" that the HCOC wanted to give to Thomas. Even though there were better people elsewhere, Struve felt, he ended up supporting Thomas as "a gamble well worth taking."[244]

The committee, however, was unanimous in rejecting Harvard professorships for all those Whipple wanted from the outside: Sterne, Krook, Hynek, and Rinehart. There was some lukewarm support for Sterne, but no "clear-cut endorsement." Appreciating that these men still filled the needs of the Smithsonian Institution, however, the committee voiced no opposition to designating them "Research Associate." In sum, the committee stated, and Pusey endorsed, its top recommendation to replace Bok with a new permanent appointment of a radio astronomer, not a series of temporary appointments. And as for the Smithsonian appointments, Pusey pushed them back, directing Bundy to "attack the problems raised by the proposed joint appointments."[245]

Menzel and Whipple kept pushing with Bundy.[246] They were bluntly apologetic about the dual appointments, wishing they had made a stronger case for Krook, "whom we consider to be one of the most outstanding astrophysicists in the country, but who has not yet been generally recognized as such." He was setting "new standards" and contributed immensely to the collegiality of the group. However, they admitted that "we erred in our consideration" for Hynek, but appreciated having the committee's responsibility for "this decision rather than to take it ourselves." Hynek was still coming to the Smithsonian and they believed he would "do a very useful job for us." They did not apologize for Rinehart, saying only that as a scientist who had made "outstanding contributions . . . in a field between astronomy and ballistics," he was "little appreciated by either discipline." They reserved more extensive arguments for Sterne and, later for Payne-Gaposchkin whom Menzel had included on the promotion list.[247] Menzel and Whipple fought hardest for Sterne, suggesting at one point that he be appointed the James Smithson Professor of Astrophysics, which would highlight where the money was coming from.[248] Carmichael did not like this idea, replying bluntly that there were "too many potential complications of a fiscal and personal management nature to permit our approval." Moreover, he was quite miffed at their appropriation of the Smithsonian founder's name.[249]

By late June, Whipple and Menzel were still looking for a leading radio astronomer to replace Bok and retain Harvard's reputation in the field. Menzel was also at the time pushing for a cooperative venture to organize a national radio facility that would be managed by Associated Universities, Incorporated, a regional consortium of universities, and naturally saw Harvard as a leader.[250] They knew they had the backing of Harold I. Ewen and his Ewen-Knight Corporation, good friends and patrons of Harvard, so opted to find an astronomer who could work with the radio engineers. Passing over former Harvard graduates and Bok students Dave Heeschen and Ed Lilley, they recommended T. Kochu Menon. Menon had been a student under Ewen, so together they would make a strong team, the HCOC argued. They asked for a five-year appointment for Menon as Agassiz Radio Astronomer, at the level of research fellow and lecturer, once Bok left. This would not be a Smithsonian appointment because an anonymous gift paid all the bills.[251]

By the end of 1956, Hynek, Krook, Rinehart, and Thomas had all accepted Harvard appointments at lower rank; Sterne came as associate professor. The additional appointments the committee had debated, were made possible by the staff positions opened up by the departure of Smithsonian APO staff at the observing stations and in Washington. Most prominent was Thomas Gold, who came from England to work with Menon and Ewen after a particularly complex negotiation process. Struve championed Gold over all others to make Harvard radio astronomy world class.[252]

Bureaucratic details complicated dual appointments. Since they were to teach, and be supported wholly by the Smithsonian, it was not possible to use direct federal appropriations, since, technically, Smithsonian staff did not teach. They would be paid appropriately out of the Smithsonian's private, or "trust," funds, for the specific amount of coursework they directed. The flexibility the Smithsonian long enjoyed by having a core of privately endowed funds as well as funds from federal revenues would now be accessible to Harvard. Menzel still hoped that to avoid possible ill feelings from the Smithsonian side, at the least Harvard should provide "a token contribution" for each lecturer, not to exceed $1,000.[253]

Along with team building, Menzel and Whipple secured laboratory and office space, leveraging support beginning to flow from other institutions eager to establish collaborative relations with the new SAO. In October 1956, Harry Vestine, in charge of the Data Coordination Office of the International Geophysical Year [IGY] at the National Academy, wrote to Pusey asking whether Harvard might agree to be one of three "IGY World Data Centers." Vestine traveled to Harvard with his associate, atmospheric physicist Gerhard F. Schilling, to meet with Pusey, Menzel, and Whipple, carrying a statement from the Senate Appropriations Committee outlining the functions of the IGY Data Center and the "scope of the problem." By early November, Schilling informed Menzel that the Smithsonian "will house archives on satellite, aurora, and airglow."[254] Within a year, Schilling would become another Smithsonian staff member.

This new association made the matter of housing all the more immediate and pressing. The combined staffs already threatened to overflow Harvard College Observatory's Garden Street compound of dilapidated buildings, and the promised new construction was still years away. The Smithsonian could not appeal to Congress until Harvard committed land for the facility. Menzel assured Harvard that the Smithsonian's eventual appropriation would allow for a new building large enough to include a badly needed machine shop that could be filled by in-kind gifts from Harvard patrons. Menzel also felt that if Harvard deeded over some of its neglected telescopes, languishing in Shapley-era buildings that had become structurally unsound, the Smithsonian would build the domes to house them for use primarily by students.[255]

Beyond gaining additional real estate, the new core of the Harvard Observatory Council, Menzel, Whipple, Payne-Gaposchkin, and Bok, continued through 1956 to develop procedures for how the two institutions would work together. They had to coordinate observatory and departmental budgets, and staff now had to juggle new complexities caused by multiple lines of support. Who would pay for what, what had to be charged to whose account, how much Harvard should charge the Smithsonian for office space, and where should that money go, all were up for discussion.[256]

Keeping all the administrative details straight was a daunting task. A sense of the tension at the working level was revealed when Velma A. Adams, a key administrative staff assistant, quit because of the lack of a "close working arrangement" between the HCOC and its own administrative infrastructure. The council had agreed to admit her to HCOC meetings only when it was time to take notes on council resolutions, and not to be present during the deliberations. "Certainly you must have sufficient confidence in your assistants to be willing to discuss frankly with them your thoughts about the overall operating problems of the Observatory; if not, you have the wrong assistants." Adams hoped that her successor—it would require two or more people to cover her tasks—would be so trusted.[257]

Adams feared managing endless expansion, and her fear was justified. In March 1956, with the Smithsonian now firmly in place, the HCOC unanimously agreed that Harvard should petition to become a charter member of a new collaboration of universities that would operate the new "National Observatory," funded by the National Science Foundation (NSF), somewhere in the Southwest. NSF would bear all costs, and it was very important to the HCOC how the board of directors of this new entity would be constituted.[258] No matter what Bowen had advised, Menzel knew that access to big telescopes under clear skies was absolutely essential to the competitiveness of the Harvard College Observatory. But signing up to develop and maintain a national observatory in the Southwest, while a prudent step, would not be sufficient for Harvard.[259]

Multi-institutional observatory associations were a new idea and much the talk of the day. The coming of a national facility was one manifestation, but there were other arrangements, as Menzel well knew, that promised world-class capabilities, and there was heated debate over how representative it would be.[260] Writing to Roger Lowell Putnam in November 1956, Menzel spoke of the patterns set by the "Yerkes-McDonald or the Cal Tec-Mt. Wilson-Palomar" [*sic*] observatories. Putnam, the sole trustee of the Lowell Observatory in Flagstaff, Arizona, had asked Menzel for advice about finding a new director, and Menzel had also learned from "Mac" Bundy at a cocktail party that Putnam was seeking his advice on various matters, one being a possible three-way association between Lowell, the Smithsonian, and Harvard.[261]

Menzel liked the idea. First, the Smithsonian's station at Table Mountain, California, was likely to close, and they would want another western site with similar characteristics. Further, both Whipple and Menzel were very interested in planetary research, Lowell's forte, and if Lowell agreed to expand into stellar and galactic areas as well, then Menzel thought it quite possible, pending the approval of the Harvard Corporation, that they could move the 61-inch reflector to Flagstaff. For years, it had been used as a test bench for new instrumentation, but its value would be greatly enhanced by Flagstaff's high altitude and dark site, where, with Smithsonian funding, a new spectrograph could be added "for studies of planets and planetary atmospheres."[262]

Menzel's pitch to Lowell illuminates his expansive and opportunistic view of the Harvard-Smithsonian relationship. But it reveals too that he was concerned about governance. Unlike the Yerkes and Caltech associations, there would not be a single director; the association of Harvard and the Smithsonian "was in no sense a 'merger,'" Menzel explained to Putnam; each would retain its director and infrastructure, combining elements when it would be to the advantage of both. The director at Flagstaff, for instance, would be resident there, yet also be a member of the Harvard Observatory "Administrative Council."[263] Save for the U.S. Naval Observatory, Menzel continued, "all major observatories in this country have a direct link with a university." Lowell was the outlier: "Isolation is one of the greatest drawbacks of the Lowell Observatory," Menzel asserted, and it needed "the stimulation or association with minds that have a fresh outlook, the opportunity to commune with scholars."[264] Isolated astronomers using mechanical calculators at Flagstaff would now have the benefit of high-speed electronic computing machinery at Harvard. Lowell astronomers could visit Harvard on a rotating basis and utilize all the resources there as research associates, becoming lecturers if they wanted to teach. The other facilities Harvard had fostered, such as Sacramento Peak or Whipple's meteor stations, all would be part of the association, as would the growing radio telescope facility at Oak Ridge, Massachusetts, which was best complemented by an optical facility, such as moving the 61 inch to Flagstaff. Menzel even cited as assets Harvard's lingering association with the Boyden Station and the possibility that they all might join an international cooperative arrangement with five European observatories.[265]

Menzel's description of his burgeoning Harvard empire for the Lowell trustee Putnam nicely encapsulates his vision. It demonstrates once again that the Smithsonian was but a part of a much larger plan, a highly opportunistic plan to be sure. Menzel implied only indirectly that there was a Smithsonian research agenda to maintain, in the guise of closing Table Mountain and somehow moving it to Flagstaff. And nowhere did Menzel mention synoptic solar monitoring as a program, or even Whipple's growing preoccupation with satellite tracking. These were not the higher activities of a mainstream astrophysical observatory, even though they may well be the bread and butter facilitating it. Such was the nature of Menzel's pragmatic opportunism. On the eve of the 1 July 1955 announcement of the move to Harvard, Menzel jubilantly wrote to Carmichael, "I think we ought to break a bottle of champagne over Fred, or something, to say the least. . . . I can't tell you how enthusiastic we all are, and I feel that we are developing a new program that will have far-reaching effects on science."[266]

Not everyone at Harvard shared the joy. Bok would soon be leaving, and Dorrit Hoffleit had been pressured to leave as well, having just lost her office and seeing her style of research end at Harvard. Her mother, supporting her decision, admired her courage to rise "up against them without any right or power," sharing her daughter's view that "what is sorely needed at the Observatory is the encouragement of pure astronomical research and moral integrity."[267]

The SAO's First Years: A Preamble

The choices Whipple and Menzel made reflect goals to pursue problems in astrophysics, geophysics, and meteorology on scales unprecedented in astronomical history. They did it by creating an infrastructure that would support new hybrid ventures, including those conducted in space. Those they hired could perform cutting-edge basic research, but they were also the types of people who would respond positively to building service programs in creative and effective ways.

Whipple organized his growing infrastructure along explicit funding lines and called them *divisions* starting in 1956. The SAO started that year with three divisions: solar astrophysics, meteor studies, and a "satellite program" with about 12 people devoted to tracking what was thought would be the first satellite, the Navy's Vanguard.[268] The first category, the Smithsonian's legacy, was not yet formally established with a division director, since Sterne held out until the end of the year. The latter two were staffed and actively engaged in grant-supported projects by the NSF, the Army Ballistic Missile Agency, and the Air Force. In 1957, the same divisions remained, but the staff reached, by March, some 57 people.[269] By now Sterne had been recruited to head the solar astrophysics division, the meteor group was performing ablation studies for the Air Force, and the satellite program, by far the largest in staffing and funding, was split into two divisions: the development of a worldwide photographic network using a vastly enhanced extension of Baker's Super-Schmidt system, and a volunteer visual tracking network, Project Moonwatch, that was to spawn a huge data collecting and public relations entity that, for all intents and purposes, became the way that local communities and facilities became part of the Space Age. Coordinating all of this, and benefiting from it, was a burgeoning computational analysis group.[270]

By 1958 Whipple was also contracting to provide orbital analysis services for the U.S. Army, and beginning to propose for preliminary studies leading to launching scientific Earth satellites for geodesy and astronomy. The satellite-tracking program now dominated the landscape: Project Moonwatch, variously watched over by Hynek assisted by Armand Spitz and others, was rapidly gaining media attention because of interest in Sputnik, Explorer, and Vanguard. It was a visible program that recruited volunteer observer groups worldwide, lined up with little telescopes set to cover a large swath of the local meridian (Figure 13). The computation and analysis group now included satellite tracking, orbit prediction and analysis, and studies of Earth's albedo. It was organized around large mainframe computer equipment requiring new buildings on the compound, and soon in rented space throughout Cambridge. A new "Upper Atmosphere" division pursued air density studies,

Figure 13 Margie Whidden describes a Moonwatch team structure: how they are lined up along their local meridian, exactly north-south. Each observer looks though a small telescope that is pointing to a different part of the sky along the local meridian. When a satellite passes, only the observer who sees it in sight calls out or taps a button for recording. *Smithsonian Institution Archives, Acc. 16-263, Box 3.*

stellar scintillation, lunar dust studies, and planning for x-ray and ultraviolet space telescopes. Through 1959 and into the early 1960s there was nothing less than explosive growth of all these divisions; from an original staff of fewer than a dozen personnel in 1955 (none who actually physically transferred from Washington), Whipple commanded more than 300 in 1961.

How did he do it? With the Smithsonian as his institutional base, Whipple proposed the satellite-tracking network in the latter part of 1955, arguing that his experience with distributed photographic meteor-tracking systems since the 1940s, vastly improved with the Super-Schmidts, positioned him as the best person to create a new optical tracker four times more sensitive than his meteor cameras. His satellite-tracking network would mimic the global character of Abbot's solar stations. Carmichael quickly accepted the concept because it served national needs mandated by the legislation that made the United States a part of the International Geophysical Year (IGY), and was paid for by the NSF, which managed funds allocated specifically by Congress.[271] It made the Smithsonian a major player.

Staking Out Territory

To establish the SAO's credentials, Whipple organized a series of symposia that drew in leaders of American astrophysics, geophysics, and atmospheric physics. The first symposium approved by Carmichael and led by Thomas explored the interface between meteors, the upper atmosphere, ablation issues, and hyperballistic studies and resulted in a publication on cosmic aerodynamics. It was supported largely by U.S. Air Force contracts for studying Earth's meteoritic accretion, one of Whipple's and Thomas's interests, and centered on employing the hyperballistics of atmospheric meteors as probes of the atmosphere.[272]

Thomas highlighted the ballistic connection and Whipple connected it institutionally to Langley's, and hence, the Smithsonian's, pioneering legacy in aerodynamics. Meteor astronomy in fact was now the "common property" of astronomers and "postwar aerodynamicists and ballisticians," Thomas claimed, because "the level of working velocities [had risen] into the ultra-speed and astroballistic ranges."[273] Thomas organized the symposium with like-minded U.S. Air Force specialists, strengthening connections in that sphere.[274]

In the months leading up to the announcement of the move and the creation of the SAO in July 1955, Whipple, through Washington networking, chaired an ad hoc committee on the "Needs of Astronomy." It had been created to improve communication between members of the 1954–1955 Panel on Astronomy Advisory to the National Science Foundation, leaders of the American Astronomical Society, and directors of major observatories. Eight of the 16 members of this committee met at Princeton on 3 April 1955, with the mandate to provide "guidance" for the NSF in various areas not traditionally considered. These areas included education and the development of new departments of astronomy; ways to improve financial support for astronomy through a combination of NSF and private-endowment sources; means to stimulate the development of new instrumentation; the "Encouragement and Development of Collateral Sciences," such as new computing devices and techniques; and specialized laboratory and theoretical methods such as "astrochemistry, astroballistics, etc." The committee report *New Horizons in Astronomy*, another Smithsonian publication fostered by Whipple, was nothing less than an attempt to reshape not only patronage patterns for astronomy but also the profile of astronomical institutions.[275]

The ad hoc committee endorsed most of the NSF's initiatives, which by then included the creation of national facilities for observational astronomy, both optical and radio. Carefully guided by Whipple, however, the committee called for a collective statement on the needs of astronomy, "concerned primarily with anticipated developments during the next few years, particularly where new equipment, new techniques, cross-field endeavors, neglected areas of research, or special research efforts [that] should lead to distinct progress in astronomical research" were involved.[276] Less clear was how those needs would be organized: multi-institutional shared facilities or nationally directed facilities were both on the table. In effect, Whipple leveraged his position to place his blueprint for the SAO at the same level as the committee's endorsement of the NSF's national observatory scheme.

Whipple secured both NSF and Smithsonian support to prepare *New Horizons in Astronomy* for publication, using it as a vehicle to create a series he hoped would become a new publication of record, *Smithsonian Contributions to Astrophysics*.[277] This new series would "expand [new] avenues of publication" for research in modern areas of astrophysics—namely, hybrid studies not emphasized by traditional journals such as non-optical techniques and instrumentation as well as the "related sciences" of rocketry, computing machines, turbulence, astroballistics, electrodynamics of fluids, and hydromagnetic and plasma problems. It would be a new forum that would correct the narrow focus taken by previous advisory panels that had been dominated by traditional observatory directors. It also legitimized a new institutional platform that could take advantage of a variety of resources that provided maximum flexibility to take advantage of opportunity, and, on occasion, to construct that opportunity. In many ways, Whipple's plan echoed Joseph Henry's original philosophy for the Smithsonian that it "only undertake programs that cannot be adequately carried out by existing United States institutions" and, furthermore, that it provide a forum for publishing the products of that research, in what he called *Smithsonian Contributions to Knowledge*.[278] If Whipple had a personal agenda, it was to legitimize his own specialty within mainstream astronomy. Suggesting that most astronomers were attracted to the largest and most extreme realms, such as nuclear astrophysics and cosmology, Whipple wanted to remind them, and their patrons, of the critical value of "meteoric bodies" that "ionize the high atmosphere and contribute to its chemical composition, scatter sunlight in the zodiacal light, and provide us with a cosmic ballistics laboratory of ultra-velocity particles."[279]

Public Relations and Infrastructure

Whipple quickly established a public relations office at the SAO, one of the first astronomical institutions to take such a step. He and his staff enlisted local writers, editors, and designers to create a lavish brochure celebrating the new association of the Smithsonian and Harvard. Prepared and published in collaboration with *Sky & Telescope*, the brochure graphically illustrated the "new" Harvard College Observatory "and its branches." These consisted of the home base on "Observatory Hill," containing the SAO in three office buildings, the Sears Tower, and numerous outbuildings. Also illustrated were the Agassiz Station near Harvard, Massachusetts; the Boyden Station; the Air Force Sacramento Peak Observatory at Sunspot, New Mexico; an Air Force–sponsored solar radio observatory near Fort Davis, Texas; Whipple's meteor-radio network, still under planning with Project Lincoln; and the High Altitude Observatory, now administratively divided between Boulder and Climax, Colorado. There was also an architect's conception of the new SAO building, a three-story office building containing shops in the basement and student observing facilities on the roof.

The brochure heralded how the new duties of the Smithsonian offices would serve the IGY and "the widely publicized program of the artificial satellite." Whipple and his SAO, the brochure announced, have "recently been assigned responsibility for tracking the satellite and predicting its path as it circles the earth in its 1½-hour period." The brochure made it clear that this new building had to be built soon "to harvest the greatest benefits from this relationship."[280]

The public relations staff Whipple fostered articulated the major programs and projects that gave expression to his vision, his "dear dream." They also provided a mechanism to create a local identity and establish an engaging atmosphere for the place, giving its burgeoning staff a sense of identity and belonging. After all, any clerical, technical, and administrative infrastructure that was capable of juggling and reconciling disparate sources of funding that came with very different constraints and expectations, working within a superstructure consisting of federal rules and Harvard traditions, had to be able to face friction, from securing dual appointments to maintaining harmonious relationships. In a hybrid institution as complex as the SAO and HCO association, it is not surprising that friction continued, and in one critical relationship, grew to the point of rupture.

But what were the Smithsonian's expectations for the new SAO? What did Carmichael hope and expect would be upgraded and modernized and would flourish under new management in Cambridge? This question sets the stage for understanding the complexities and the dynamics of growth of a multi-institutional hybrid body that expanded the definition of what it meant to do astronomy.

Chapter 4

Whither the Solar Constant: A Referendum on Basic Research

As he negotiated the institutional and staff relationships between Harvard and the Smithsonian, Leonard Carmichael pondered the research legacy of the old Astrophysical Observatory (APO). Should he maintain the direction that Langley had established and Abbot had so loyally pursued for so many years? Were there new and more effective ways to pursue the solar constant, or was the search for the solar constant and its variations a problem of the past and far from the frontier problems in astrophysics that warranted the Smithsonian's attention? As the leader of the Institution and guardian of its legacy, Carmichael had to defend its investments, placate its critics, and, most assuredly, maintain the confidence of the U.S. Congress.[281] He turned to Whipple for guidance on options for how the solar constant was to be studied, but wanted to be assured that the study itself was worth continued support. He did not receive that assurance.

Whipple Articulates His Vision for the New SAO

The International Geophysical Year (IGY) was closely tied to synoptic solar observations—it was timed to coincide with a solar maximum. Whipple could have chosen, one would assume, to join a new international working group that convened in 1953 to construct a worldwide solar monitoring network.[282] Yet there is no evidence that he made any attempt in this direction. Even though solar monitoring could be related to his interest in Earth's upper atmosphere, neither he nor Menzel got involved. What Whipple did decide to do, and how he articulated his vision, therefore, set the stage for how he eventually managed Carmichael's expectations for continuing the legacy program of the APO.

From the start, it was clear to Richard Thomas that Whipple was not taking steps to mount and sustain a solid solar research agenda. In March 1955, Thomas wrote bluntly to Whipple that there was no visible effort at Harvard to secure "a person whose primary interest is the sun as such." Base staffing for a responsible program required at least three professional staff, Thomas claimed, to teach and to coordinate research as well as to engage in problems of stellar atmospheres.[283] Whipple would eventually assign solar physics responsibilities to new staff such as Charles Whitney, but it was not happening fast enough for Thomas.

Menzel and Thomas and their students were all, of course, far more versed in solar and stellar astrophysics than Whipple was, or cared to be.[284] After all, Whipple was wholly engaged in developing his satellite-tracking network and all the activities relating to it, which included upper atmosphere research and geodesy.[285] He had expected Sterne to be on board as well, but that matter was only one of many he was struggling with to articulate his vision of the scope of Smithsonian Astrophysical Observatory (SAO) research: "I must say that I am so rushed these days that I find it very difficult to put anything into a very coherent statement."[286] His challenge was to chart a course between his new Smithsonian obligations and what he had already committed to the IGY, and, of course, what he felt would be the most productive area in which to concentrate. He needed someone with a clear perspective and considerable capabilities in all these related areas. Sterne was his man, and his campaign for Sterne's appointment continued well into 1956: "The major problem, as I see it," Whipple wrote in September, "is to be sure that one avoid the backwaters of unnecessary data accumulation and study."[287] Whipple had just returned from International Astronomical Union meetings in Dublin, where the president, Otto Struve, railed against regimentation, the nationalization of observatories, and the need to preserve individual initiative and creativity. Struve appreciated that planning was important: Organization and control had to be fostered, but not to the point of regimentation or the suppression of individual creativity to the collective will. And most poignantly, Struve charged that past legacies were still strangling the discipline, for example, inordinate attention given over to "arranging, classifying, methodizing, [and] simplifying" the heavens.[288] This was advice that doubtlessly impressed Whipple as he pondered his new responsibilities directing the SAO.

Whipple really needed someone to talk to. He needed Sterne, someone who could make critical and sound judgments "untinted by irrelevant emotional factors." Unusually frank in a letter to Sterne just before he had to travel to Washington to present his vision for SAO to senior management, Whipple admitted that:

> I have never been more in the need of sound counseling. My major problem is how to strike a proper balance [between two factors:]
>
> a) my love of doing science myself and getting my hands deeply into problem[s];
> b) utilizing to the full extent the potentialities of my administrative position and contacts without becoming purely an "operator."
>
> For some time I must sacrifice heavily under "a" to attain the obvious goals under "b". However, I do want to get completely enmeshed in this approach to science.[289]

Whipple's testimony that October in Washington reflects his anxiety over achieving not only personal balance but also, more to the point, autonomy. He knew that he had to deal with Carmichael's expectation that Abbot's program would continue, but he did not want to be saddled with that legacy. So he lectured Smithsonian bureau directors, facilities managers, and assistant secretaries Keddy and Graf on the relative worth of individual versus institutional research, trying to convince them that the SAO's programming had to preserve individual initiative within a systematic institutional support structure.[290]

While he campaigned for autonomous research with Smithsonian leaders, Whipple assured Sterne that the Smithsonian would provide it: "Why you have such a desire to become an official active part of such a financially destitute organization as Harvard, when you have the opportunity to conduct pure research with the great opportunities of the Smithsonian Astrophysical Observatory, is quite a mystery to me."[291] A Harvard professorship, Whipple inferred, was no longer worth much. Money would flow, but not from Harvard astronomy programs, which were "essentially as 'broke' as they were when you left here," Whipple claimed. In contrast, the government contracts that he would soon be bringing in "far exceed the total budget of the Observatory and the Department combined. The factor is more than twice."[292]

Sterne finally agreed, and was on board in the summer of 1956 to become second in command in "the new order of authority."[293] Sterne's appointment was even noted by the *New York Times* in July 1956, which observed that he was to become associate director of SAO and, oddly, Simon Newcomb Professor of Astrophysics at Harvard.[294]

Continuing Debate over Solar Radiation Monitoring

With Sterne safely on board, Whipple finally admitted that the central Smithsonian was applying "rather heavy pressure" on the SAO to continue existing solar observing programs. The Smithsonian had just issued a news release touting Abbot's latest findings and conclusions, more ardent than ever. Whipple asked Sterne to comment privately, and Sterne scribbled back, "This place is too staid for the profane comments called up by these enormous interpretations."[295]

Whipple had agreed to keep Table Mountain open for two to three years with a staff of three observers, but he wanted to transfer Montezuma station to the government of Chile as a "friendly international gesture" but was overruled by Keddy and Graf. Whipple got less resistance to preserve the data itself and possibly "summarize the data of past years on punch cards, to make them more easily usable." Aldrich in particular liked this idea, saying he would ask an assistant to explore the possibilities of IBM or Remington Rand support.[296] Keddy and Graf concurred, but once the photographs were analyzed, they would keep only the best of the collection and "the remainder can be sold for window glass."[297]

Table Mountain was still very much alive and had a loyal constituency.[298] One was Preston Butler, who was well known to the Smithsonian administration as a former, and much valued, member of Abbot's staff. Butler pleaded with Keddy and Graff to continue the work of the APO. Whether or not it would result in verifying the variations in the Sun's output, there were still "many problems which can only be solved by long precise observations in the field."[299] Butler's petition convinced Carmichael in early 1957 to call for a review from a larger sphere of science than astronomy alone.

By then, although he had pressing tasks relating to satellite orbit determination, Sterne had made progress evaluating the solar data.[300] Directing Radcliffe graduate student Nannielou Dieter, Sterne compared APO's primary station data from Montezuma and Table Mountain, looking to see if they were recording the same variations. Both of the statistical methods they applied, covariance and serial, led to the same conclusion. They found very little discordance, which attested to the precision of the instruments and their operation. But as to finding real variations, none could be reported: "If the solar constant really changed at all between 1926 and 1955, it changed by amounts having a smaller root-mean-square value than 0.17 per cent of the solar constant itself."[301] In May 1957, they reported no evidence for "periodicities . . . common to the two stations." And in a more detailed report, strengthened by the services of the Littauer Statistical Laboratory of Harvard University, they were now convinced that if any periodicities existed, they would have been found.[302] They were ready to answer Carmichael's call.

Whipple and Sterne invited an eclectic but powerful band of military and government laboratory scientists to examine the state of APO findings and to suggest a future course. Attendees represented Naval Research Laboratory (NRL), the Geophysics Research Directorate of the Air Force Cambridge Research Center, the Naval Radiological Defense Laboratory, and the U.S. Weather Bureau. After little debate, committee members "uniformly agree that the measurements of solar radiation, using present equipment and the present site, are not worthwhile."[303] Even so, the committee felt that with better equipment and better sites, the analysis could be refined, at greater cost, if Carmichael wanted to make that investment.

Carmichael indeed was willing to consider this, since, he suspected, "the general opinion at the Cambridge headquarters of the Smithsonian Astrophysical Observatory seems to be unsympathetic to the continuation of solar radiation measurements."[304] Indeed, Abbot was telling him just that, as well as reminding Carmichael that there was a community of workers dependent on a continued flow of Smithsonian data. "If we finally decide to discontinue it," Carmichael felt, the Smithsonian had to "broadcast widely and in plenty of time" that fact to allow others to fill in the gap, if they wished to do so.[305]

Indeed, Abbot took every opportunity to state his case, sending endorsements to Carmichael from his dwindling circle of supporters.[306] Passionate as ever, his letters were often colored by poems, biblical quotes, and aphorisms, as well as lengthy passages from Langley's original writings on the mission of the observatory. But there were also inquiries from other observatories, asking about the instruments distributed by the old APO, especially the standard silver disk pyrheliometer, which had been distributed worldwide.[307] This, Carmichael came to believe, was the true legacy of the old APO: instrument standardization. Accordingly, he continued to provide staff support as Abbot requested, and did all he could to keep the man contented, consenting to the commissioning of a bust of Abbot and to the display of APO instruments in the Arts and Industries building[308] (Figure 14).

Carmichael's Real Concerns Regarding the Nature of Smithsonian Research

Beyond Abbot's presence and pressure, Carmichael, coached by his assistant secretaries Graf and Keddy, harbored a far darker concern, one that would come back to haunt the SAO in the next decade. They were keenly aware that the SAO's very existence depended

Figure 14 Charles Greeley Abbot, in his 95th year, poignantly sharing his evidence for solar variability with anyone who might listen. *Smithsonian Institution Archives, Image # 94-4039.*

on congressional appropriations, which had been specifically targeted for measuring solar radiation for decades. If Whipple now diverted those funds to other purposes, then the Bureau of the Budget had to be informed. Would the annual $80,000 budget remain, even if the goals were different? "I cannot hazard a guess as to what their attitude might be," Carmichael advised Whipple.[309]

Whipple appreciated Carmichael's sincerity, but he was also wary of Keddy. They had already clashed over travel and relocation funds from various sources not designated originally for travel. Keddy constantly reminded him that there were formalities concerning compensation when on official travel, constraints on procuring consultants, and compensation when not doing the work assigned, such as teaching at Harvard.[310]

Through the spring and summer of 1957, Sterne responded to Carmichael's questions in a series of longer and longer essays about proper practice and the authority to do pure research in a government agency. Sterne questioned the value of continued solar monitoring because he felt it unlikely to produce anything useful. But Carmichael continued to hope that "out of the many fundamental researches on the sun now being conducted some new knowledge is bound to be acquired which can later be utilized for the benefit of mankind."[311] They were concerned that whatever the Smithsonian chose to do, it had to be justified in the eyes of Congress. Indeed, Carmichael was constantly rehearsing what he would say before an appropriations committee.

In August 1957, Keddy forced a showdown. Whipple was summoned to a meeting in Washington with Carmichael, Keddy, Graf, and E. E. Eisenhart, and there faced outright hostility. But it was not only related to the solar work. The meeting had been called to rectify development delays and serious oversight problems with the SAO's key satellite-tracking hardware, the Baker-Nunn system.[312]

By now, Carmichael and his staff had become concerned about the huge undertaking Whipple had successfully initiated for tracking the first satellites. As we will see in the next chapter, the Baker-Nunn was a very ambitious system and its development was taking longer than expected. None had been delivered yet, or was close to being operational, and, most embarrassing, an example would not be available to display for an important meeting upcoming in Washington. Carmichael wanted to be assured that the system would be in place when the first American Vanguard was expected to launch in November. Only after they bore down on Whipple about the cameras did the subject turn to solar radiation.

Carmichael's congenial bearing in his correspondence with Sterne and Whipple was absent in the bulleted stenographic record of the meeting, where Keddy attacked Whipple's project management failures. They were also displeased with Sterne's negative reports on the APO legacy program and the review by conferees from the February meeting. Then, abruptly, Carmichael turned dark: "Let's not have Sterne do anything with this program as he is antagonistic to it," and with this, Keddy took his cue: "He is out to sink it, and you [Whipple] are not interested." Whipple then changed the subject, asking, "Are we ever going to have a building of our own?" to which Carmichael replied that after "careful study . . . we feel, for many reasons, that we cannot ask for more money in '59."[313]

Whipple returned to Cambridge after the meeting and briefed Sterne. One can only speculate on Whipple's state of mind after the confrontation in Washington. The promised new building was stalled, and this meant Whipple had to scramble for space to accommodate his burgeoning staff. He left it to Sterne to answer Carmichael, hoping they could, as Whipple frequently stated, "avoid the backwaters of unnecessary data accumulation." Sterne wrote Carmichael, saying he "felt deeply" his responsibility to provide only the best advice to Whipple and to the secretary that would make the Astrophysical Observatory "into a first-rate scientific institution." His 11-page position paper, meant to give Carmichael insights in any future testimony, expanded greatly on his original thoughts about the "scientific sterility of solar constant work" and the fact that it would be daunting to improve on the already "high level" of Abbot's data.[314]

Sterne called Carmichael's bluff, agreeing that as a government bureau "we have the duty first being morally sure ourselves that the project or purpose is worthwhile, and that the money can be advantageously spent." They should ask Congress only for research they truly believed was useful and not of "debatable worth." He advised "complete candor" in any reports to the Hill:

> Explain if necessary to the Bureau of the Budget and Congress that the old solar constant measurements appear to have continued long enough to have served a scientifically valuable purpose, but that their longer continuation in the old form now appears to be of debatable and doubtful merit, and they have become difficult to supervise effectively.[315]

Sterne then cited a recent essay in *Science* that identified the Smithsonian as one of only two federal agencies, among some 38, that perform research that did "not organize them about practical problems" and was, moreover, "the only Federal agency that conducts its own pure research." After all, Sterne advised, Congress had established the Smithsonian with the sole mission "of increasing and diffusing knowledge," which to his mind was "general approval for the Smithsonian to do basic research for its own sake, without appeal to practicality."[316]

Further demonstrations that the solar constant was indeed constant would not be useful, he argued. Given the high accuracy of the Abbot-era measurements themselves, there was no real hope "to overcome the sterility that I have mentioned unless the instruments are above the atmosphere."[317] This point, Sterne hastened to add, was emphasized by the NRL conferees. Sterne thus deftly countered Abbot's own arguments to show how difficult any improvement would be, and went so far as to state that Abbot's resistance to criticism soured the enterprise, and, to be frank, the reputation of the Smithsonian. The Smithsonian had to realize that any ambitious young scientist would be reluctant to "incur the isolation likely to result from entering a no longer fashionable field." Sterne offered various options that allowed for some work to continue at Table Mountain, not by careerists, but by contractors. This option would relieve the Smithsonian from "violating the principle of letting key men choose their projects."[318]

Carmichael was mulling over Sterne's lengthy position paper when he received a copy of Sterne's and Dieter's second paper on "The Constancy of the Solar 'Constant'" and felt uneasy about both. Carmichael even withheld approval to publish the article, at least until he was assured that Abbot's reputation would not be tarnished. The solution was to cite his observations and not his conclusions. Carmichael heartily agreed with Sterne that "the publication of counter rebuttals and counter-counter rebuttals is unnecessary."[319] Sterne also sent Abbot the manuscript, and Abbot responded predictably with an extensive restatement. But it was clear that his energies were spent. He still argued that his correlations revealed "the larger features of weather." And once again, he predicted that when these correlations became well enough known to be exploited: "[A]ll over the world, the annual production of wealth thereby will many times exceed the whole cost of the Smithsonian Astrophysical Observatory, and its scores of researches, from its beginning in 1890 until its entire modification in 1953."[320]

However much Abbot's presence affected Carmichael, it is clear that both Graf and Keddy held the ancient scientist in great respect, not the least as the embodiment of institutional legacy.[321] And aside from Abbot's claims, Sterne and SAO staff, Carmichael responded, had to "face the facts" regarding the Smithsonian's ability to share in independence of action. He warned that this applied only to "the core of the Institution, our establishment under which the statute of organization was founded for 'the increase and diffusion of knowledge among men.'" This core body of activity, he added, "operates on trust funds." The Smithsonian's charter did indeed give it "wider authority than that possessed by any Government agency," but in practical action, the research activities of the Smithsonian in the intervening century had far outstripped what the trust fund could support, making it "impossible to take advantage of that wide authority." The research bureaus of the Smithsonian, including the SAO, operate by federal appropriations, and therefore "do *not* have the wide research authority of the institution."[322]

Here was the crux of the matter for Carmichael: Any Smithsonian activity supported by appropriated federal funding "must justify their estimates in some detail" and, further, "we cannot employ scientists on Federal funds to [quoting Sterne] 'work with little interference on projects of their own choosing within a general broad field like solar astrophysics.'"[323] In practical matters, what this translated to at the moment, and what it applied to, was who to hire to "supervise the Solar Radiation program." Sterne sensed this and let Whipple know in no uncertain terms that Carmichael's "peremptory request" had to be answered, and, moreover, it was Whipple's problem, not his.[324] So Whipple solved the immediate problem by asking one of the remaining field station directors, Alfred G. Froiland, to keep Table Mountain open and actively engaged in solar monitoring through 1961.[325] His retention did not cost Whipple a position, nor did it prevent him from acquiring significant expertise in solar and stellar atmospheres.

Dénouement

By 1961, SAO's existence no longer depended on the continuation of solar radiation monitoring. Its large-scale satellite optical-tracking program had brought worldwide positive attention to the Smithsonian. And the many areas of scientific investigation SAO staff entered, mostly but not entirely by Whipple's instigation as he responded to the opportunities of the Space Age, brought added visibility. Thus Carmichael ended Smithsonian's solar monitoring in the spring, after Whipple assured him that the National Bureau of Standards would take up the slack.[326] Just two weeks prior to this decision, Carmichael had sent Whipple yet another manuscript by Abbot. Doubtful that there was anything new in it, still Carmichael wanted it published, "in view of Dr. Abbot's age," unless there was anything Whipple deemed to be "positively erroneous" in it. Whipple responded the next day, agreeing that there was nothing new in the paper, save for the fact that Abbot now admitted that the future of solar monitoring lay in satellite-borne observations.[327]

During this era, to help foster communication and hopefully a working relationship between SAO and the Smithsonian's central administration, formal periodic conferences were held alternatively in Washington and in Cambridge. The topics for discussion, and the discussants, changed over time. By November 1962, welcoming a new Smithsonian assistant secretary, James Bradley, and a host of management consultants to Cambridge, Whipple introduced his new assistant director for science, Charles A. Lundquist. Lundquist, from NASA's Marshall Space Flight Center and with a long record in tracking and orbit determination (Figure 15), outlined the research performed at the SAO, emphasizing the use of satellite data to study Earth's atmosphere and geodetic properties, the continuing studies of meteor samples to prepare for the analysis of lunar materials, and plans for orbiting observatories and theoretical astrophysics. Missing from his accounting was any mention of solar constant work.

Figure 15 Charles Lundquist at the Marshall Space Flight Center, sharing insights on orbital mechanics with Wernher von Braun and Herrmann Oberth. *Smithsonian Institution Archives, Acc. 16-263, Box 7.*

During Whipple's years as director, no effort was made by Smithsonian staff to perform solar constant observations from satellites.[328] And no effort was made until the mid-1970s by Smithsonian staff to re-examine the half century of accumulated Smithsonian data in light of new knowledge collected from space until P. V. Foukal and J. E. Vernazza, together with Harvard undergraduate Pamela E. Mack, re-examined the APO data, comparing it with radiometer data from Mariner 6 and Mariner 7. They ruled out any relative change of the solar constant from space greater than two orders of magnitude smaller than the sensitivity limits of the Smithsonian's ground measurements made between 1923 and 1952. Their analysis also confirmed the covariance analysis of Sterne and Dieter, and completely discounted Abbot's arguments, based most recently on his 1958 re-analysis of the effect of a large sunspot group crossing the solar disk, which they attributed to chance.[329]

This result probably would not have stopped Abbot, who until his death in 1973 kept his vision alive in his never-ending defense of the APO's legacy.[330] But he and his institution outlasted the times that created them. Carmichael faced a very new relationship with Harvard astronomers in his continued effort to rationalize that legacy and uphold the Smithsonian's honor. In the first years of this new relationship, he was not assured that Whipple would deliver; the lengthy negotiations over staffing, freedom of action, and accountability naturally left him wary, especially when there was also concern that Whipple would deliver on satellite tracking.

The choices Carmichael made suggest that the program ended institutionally with the closing of Table Mountain. But in effect it ended in 1955 when Carmichael chose Harvard and Whipple over Walter Orr Roberts's High Altitude Observatory, for reasons far more complex than Abbot's pressure or evaluating the intrinsic merit of the effort. This chapter has addressed the institutional reasons why they lingered at all past 1955. But surely, as later commentators attested in light of the spaceborne observations, the covariance analysis of Sterne and Dieter had settled the issue. But just as surely, this required confidence in, and universal acceptance of, the information gained from the spaceborne observations. That is what effectively ended the legacy.[331]

Indeed, none of the satellite projects the Smithsonian proposed within the next few years, ranging from Beacon to Celescope, included any instruments capable of addressing the question of solar radiation. To appreciate why, we need now to turn back to Fred Whipple and his goals and passions, and how they propelled him from satellite tracking into space.

Chapter 5

Optical Satellite Tracking: Establishing the Smithsonian's Central Role

> Unconventional from the start. Adapted a meteor camera using Jim Baker optics . . . and Joe Nunn, a mechanical engineer, both more at home in industry or military than academic world. Hired a staff of non-scientists, Leon Campbell, Jr., ex-journalist with a talent for PR, to organize volunteers of Moonwatch, Chief Peterson, ex-Navy petty officer and communications specialist to set up . . . international network . . . J. Allen Hynek . . . most unconventional . . . Karl Henize, a young, untried astronomer to run operations, a host of wild men, ex-sailors, Marines, cowboys. . . . Somehow it worked. On October 7, 1957, nearly thirty-nine years and one month ago tonight, SAO stood ready to track Sputnik 1. From that point, SAO grew and the Space Age, reaching some 500 staff in the mid-sixties, stations around the world . . . always FLW pursued the unconventional, the risky, the revolutionary, SAO first orbiting observatory, OAO [Orbiting Astronomical Observatory], Celescope, early application of computers to astro problems, modeling atmospheres of stars, analysis of meteorites, which would lead to lunar rock analysis, and, of course, the riskiest of all, the MMT.
>
> —*James Cornell, bullets for Irwin Shapiro, circa 1996*[332]

Unconventional indeed. Whipple's editorial and public information arm first managed *Smithsonian Contributions to Astrophysics* but then provided editorial services for the burgeoning array of weekly newsletters and occasional flyers issued to manage the central facility in Cambridge as well as its worldwide tracking stations, both amateur and professional. This office also had the responsibility, explosively enlarged by Sputnik, to handle media inquiries and to prepare proactive press releases. As a quasi-governmental facility, it also published both technical and popular histories, which have been deepened more recently by historical attention.[333] Here we flesh out Jim Cornell's colorful bullets, especially how the Smithsonian Astrophysical Observatory (SAO) acquired the responsibility for satellite tracking. If there is a gap in this well-covered history at the moment, it is the very question of "why the Smithsonian?" At the time of Whipple's assumption of the directorship, in July 1955, it was not clear that the Smithsonian would be chosen as the lead institution for optical tracking. That decision would not come for another year after the National Science Foundation (NSF) provided seed funds.[334] Nor was it known how the technology and the infrastructure to handle the data flow would be managed. Recounting what is known about both challenges reveals Whipple's deft gamesmanship as he worked to build his empire.

Positioning

On July 29, 1955, President Eisenhower announced that the United States would launch a satellite. The next day, Leland Cunningham, one of Whipple's orbit specialist colleagues from Berkeley and a kindred spirit over the decades, wrote him excited by the news, offering his assistance in orbit computing: "It would seem that there are likely to be spots in this project where I could be of use."[335] For Cunningham and others, no matter who would ultimately instrument the satellite and launch it, they figured it most likely that Whipple would be tracking it. Indeed, Whipple would soon chair a new working group within the U.S. National Committee's (IGY's) Technical Panel on Earth Satellites (TPESP) that would worry about optical tracking.[336]

Whipple assumed that von Braun's Orbiter would be chosen and that he would be tasked with planning Orbiter's "optical observing program." But when the Navy's Vanguard system was picked, he recalled that "I had to start over from scratch with a new satellite observing program."[337] Numerous commentators have discussed how this decision left von Braun and his Army team, extending to the Jet Propulsion Laboratory, out in the cold.[338] But for Whipple, this only meant he had to fight harder for the tracking nod. By fall 1955, he was chair of the "Long Playing Rocket" (LPR) committee and was a member of the TPESP, so he was in a strong position. Vanguard would be larger and more visible than the redesigned Orbiter, so it better fit the satellite geometry Whipple always preferred, because its polished sphere would be more "observable from ground."[339]

The NRL Vanguard team, however, promoted its radio-based "Minitrack" system aggressively. Some of its leaders openly questioned Whipple's calculations for the optical visibility of a satellite, and there were other options beyond Whipple's meteor network cameras known to both Army and Navy groups that were equally competitive.[340] All participants agreed that some form of optical tracking was critical, not only for operational reasons but as well for obvious political and scientific value. Radio tracking, of course, was independent from local weather conditions and provided instantaneous Doppler data.

Even while he was negotiating with Carmichael to define the SAO and with Harvard to acquire staff, Whipple answered virtually every call to speak on his vision for spaceflight. His speeches were no longer long-term forecasts or fanciful portraits. Now they had immediate goals, or were updates on the state of play, to "promote public understanding of the project."[341] At a meeting of the American Astronomical Society in Troy, New York, that November, he addressed the assembly with a fervor reminiscent of wartime. Congratulating the NRL on taking the lead for America, he proclaimed that "the artificial satellite symbolizes, along with artificial nuclear fission, the greatest achievements of man's conquest over nature. First matter and energy, and now space will have become the slaves of man."[342]

The SAO directorship became Whipple's bully pulpit. But it was not the only platform he could have taken. Ironically, before the Vanguard decision, he had been repeatedly approached to take over the Naval Research Laboratory's space science divisions, which would have potentially offered far more support.[343] The NRL would have been the obvious place to put in action the disparate plans he had been making to assume the responsibility for tracking the artificial satellites that would be flying during the International Geophysical Year. That he did not drop the Smithsonian invitation and head to Washington, and he preferred to remain loyal to von Braun's Orbiter indicates that although Whipple was certainly a risk taker and an opportunist, he had his limits. Thus Whipple's decision to fly the Smithsonian banner and push for optical networking, not only utilizing an extrapolation of the powerful astronomical cameras he had developed for meteor tracking, but also the creation of a worldwide popular network of amateur satellite watchers, can be understood best by his penchant for institutional flexibility. He was committed to a technology and kept his institutional constraints as loose as possible; his loyalty to the powerful national governing panels was unquestioned. Unlike individualists like S. Fred Singer, who evidently broke with the pack, Whipple rarely if ever claimed to be anything but a loyal member, someone who could be trusted to do the right thing for the greatest good.[344]

All the tracking methods had serious limitations given what little could be launched into orbit quickly, and though everyone agreed that optical tracking was necessary, there was no *a priori* agreement as to who would provide it. But Whipple did expect that the Army would be the choice. This fact comes clear from the record of a meeting called by the Army's Office of Ordnance Research (OOR) western section, in Los Angeles on 16 July 1955, which also reveals the degree to which Whipple dominated the field. A note taker described how Whipple called the meeting "as head of the ONR scientific use and observation program for the proposed Project Orbiter."[345] Whipple indeed spoke as if he represented the ONR, but he was really acting as consultant for Varo Research of Garland, Texas, a contractor for "designing the 'Warhead'—the satellite itself." Austin Stanton, the entrepreneurial president of Varo, attended with von Braun and E P. Martz Jr., from White Sands, von Braun's official advisor for tracking. Martz brought along the astronomer Clyde Tombaugh from White Sands and others from the Redstone Arsenal. This group met to discuss with OOR staff the characteristics the satellite needed to facilitate tracking. Tombaugh, legendary as the discoverer of Pluto and now working for Martz at White Sands, presented his "programmed rapid-scan satellite search techniques," which he had developed to try and discover tiny natural satellites in orbits closer to Earth than the Moon.

Tombaugh had been developing optical-tracking systems at White Sands Proving Ground and at the Lowell Observatory in northern Arizona for years. At White Sands, his telescopic systems followed and recorded the gyrations of spinning missiles.[346] From 1950 to 1955, working in the Systems Engineering Branch at White Sands, Tombaugh also developed a variety of cameras, including fast Schmidt telescopes mounted on a series of ramshackle but functional multiple-axis, motorized tracking mounts. They never found a natural satellite, but his system seemed to be a viable option for tracking artificial satellites.[347]

Tombaugh's technique could pick up faint fast-moving bodies if the camera was tracking the object closely enough. Whipple immediately appreciated this, suggesting that the system would be even more powerful with one of his Super-Schmidt cameras. Tombaugh eagerly concurred, as long as the Super-Schmidt optical system was mounted on Tombaugh's tracking system. There was much talk of using "super bright" flares and possibly the light from Loki exhaust jets as light sources, but this was deemed too chancy. In conclusion, Whipple felt that the only solution was to "take his Super-Schmidt meteor cameras to the launching site and equip them with variable drives to be used in Tombaugh's program."[348] Whipple's advocacy carried the day, even though von Braun and Stuhlinger remained wedded to some sort of flashing light system.[349]

Whipple secured even greater control of the process when, in early October 1955, Joseph Kaplan, who chaired the Executive Committee of the U.S. National Committee for the IGY, gave Whipple the task to prepare a requirements study that would describe in some detail what was needed to optically track a satellite. How he took advantage of it reveals how his core scientific interests in upper atmospheric research and especially in meteoritic studies influenced his choice. We now look at how those interests played out at the SAO.[350]

Meteor Research as Prelude

On the eve of the announcement that he would direct the new SAO, Whipple made sure that key staff on his now decade-old Harvard Meteor Project knew that his heart remained in meteor research. He had worried in past months that Navy funding would not be supplemented, but by April 1955 had secured at least another year of funding from the Air Force, and more was promised. Therefore, throughout the years he was building the visual and optical satellite-tracking networks, he was also dovetailing these with his meteor network efforts. He leveraged his staff as well: Astronomers such as McCrosky were active in both arenas. McCrosky always knew, though, where Whipple's heart lay. The upper atmosphere, hyperballistic ablation studies ("astroballistics"), satellite tracking and geodesy: All were vehicles to study the nature of meteors.[351]

Insight into Whipple's personal priorities comes not from how much time he spent on any one facet of his rapidly growing empire, but on how little. Generally, he wanted to secure key people for critical roles in his major areas of interest, and then turn them loose with a considerable degree of autonomy. The more competent these people were, the more latitude he gave them. Of course, he had great difficulty and frustration trying to get good staff because most wanted Harvard appointments with tenure or at least the possibility of tenure. He tried first to place his senior staff among the professorial ranks but finally had to settle for the non-tenure-track lower ranks of research associate or lecturer. Both Sterne and Hynek were quite frustrated with second-class status at Harvard, but John Rinehart did not seem to mind at first, adapting to what he recalled was the "Harvard milieu."[352]

With degrees from Caltech and the University of Iowa, Rinehart had contributed to proximity fuse development and pursued hyperballistic studies at the Navy's Inyokern facility at China Lake, California. He was fascinated with fragmentation and deformation problems in terminal ballistics and wanted to "simulate the flight of meteorites," so he pushed for ways to achieve higher and higher velocities to study impacting and crater shapes.[353] Distinguished in the military world, he was well known to R. N. Thomas and, through Thomas, Whipple. Like Whipple's other targets, Sterne and Hynek, he enjoyed a considerable degree of autonomy, and Whipple assured him that this would continue as he became assistant director of the SAO. Rinehart was Whipple's first professional SAO appointment, heading up the meteoritics group and searching for patrons beyond the Navy.[354]

Whipple did not fully anticipate the strictures of Smithsonian governance when he wrote in July 1955 to Rinehart, describing the position. Whipple assured Rinehart that his position came with the authority to "take official action while I am away in Europe this summer." This meant he could hire and fire and that "you will have a real sense of freedom of action." Whipple expected some prudence, lightly joking that he was not expected to "fire me or anything of that sort" during his absence. But overall, Whipple expressed great relief and comfort that Rinehart would be joining him as a "co-partner in the Smithsonian Institution operation" and would be setting up the meteoritics and astroballistics programs. In reflection, he blithely admitted: "Gosh, it does feel good to think that you can just go ahead and do these things now and I won't have to worry about them."[355]

Rinehart's responsibilities included, as one might expect given his prior employment, contact with the Navy offices funding the program. But Whipple also gave him budget authority within limits. Whipple had already worked out a preliminary Smithsonian appropriation for FY1957 with Keddy, allowing for two additional hires and a small motor pool as well as funding for Rinehart's own research. But he left the details to Rinehart. When the Air Force picked up the program, Whipple left it to Rinehart to work out the details.

Whipple also wanted to leverage his staff, sharing or moving them between funded research lines. He advised his staff repeatedly that no matter where the funding was coming from, their work would continue: "So far, my experience has been that we keep on operating the stations because of the extremely valuable results that have been coming out. . . . I hope you can ride through this current period of uncertainty with optimism, holding the assurance that there are other interesting jobs that will be available even if this one should finally fold up."[356]

Juggling funding lines, and especially delegating this task to others, led to confusion and annoyance in Washington. Rinehart, acting with the experience of a project manager in a military facility, where support was fungible, worked diligently through the winter of 1956 with various agencies interested in or capable of supporting his, in Whipple's terms, "astroballistic studies."[357] In June, however, funding lines got seriously crossed between the NSF and the Air Force, which annoyed Keddy. Apparently Paul Klopsteg, associate director at NSF, contacted Carmichael about a grant application from the SAO for meteoritic studies, but, as Keddy related the matter to Whipple and Rinehart in June, "they were wondering whether such a grant would duplicate funds contained in the Smithsonian 1957 appropriation." Carmichael, according to Keddy, was annoyed by the surprise call from Klopsteg, as he was not informed of the proposal: "Such a lack of coordination and failure to maintain full communications respecting critical administrative matters is inexcusable."[358]

Keddy demanded that, henceforth, "no such administrative decisions shall be made without the approval of this office."[359] Whipple, however, was not about to back down, reminding Carmichael just how important this line of research was to the military, and thereby implying how important it was that the lines of communication that had to remain open were really between the primary players, without the Smithsonian administration interfering. Yet another new USAF contract for studying the accretion of meteoric material by artificial satellites was in fact the continuation of what was already at least a $1 million and decade-long investment by Harvard (from Navy funds) that was still a long way from solution, and was "of extremely great interest to the military in these present days of extra-atmospheric missiles."[360]

Rinehart's program was the first to develop under Whipple's new relationship with the Smithsonian. As a result, Rinehart experienced most of the growing pains for learning how to deal with it, including relations with other Smithsonian bureaus. Edward Porter Henderson, the venerable curator of meteoritics at the Smithsonian's U.S. National Museum in Washington, had been listed since 1957 in Whipple's annual reports as a consultant and collaborator to Rinehart's division. Although their contact seemed cordial enough soon after Rinehart was hired as assistant director in 1955, their very different styles led to tensions. Keddy made matters worse when he asked Henderson to comment on Rinehart's papers for clearance; one on meteor ballistics and flight markings in March 1958 led Henderson to claim that Rinehart's work was derivative to the point of plagiarism. Henderson asserted that Rinehart missed "an important fact" about the flight markings he was analyzing, but he was not about to reveal it to him "because I don't particularly want him to steal it."[361] Rinehart was equally negative about Henderson's work, feeling that a report in May 1958 lacked "scientific value" and could not be printed as a product of SAO's Air Force contract. "It is incoherent, verbose, loosely tied together, poor in grammatical construction, and needs much reorganization," Rinehart concluded, an opinion quietly shared by Henderson's supervisor, head curator G. Arthur Cooper.[362]

This friction may have contributed to Rinehart's departure in August 1958, but it was probably not determinative. Even though he enjoyed comparative freedom, Rinehart found Whipple distant and secretive: He "was not about to share his plans and embryonic ideas with anyone," which made his job representing SAO's administration in Washington frustrating.[363] Rinehart left to take up a teaching post at the Colorado School of Mines, later becoming director of research for the Coast and Geodetic Survey.[364]

Whipple's relations with Henderson remained delicate. They eventually clashed over the question of which bureau would be the official repository for meteorites recovered from recent falls tracked by SAO's camera networks.[365] What Henderson may have lacked in analytical methods, he more than made up for by his generosity to laboratories, lending samples far and wide, which gave him considerable political weight.[366] Henderson was therefore not about to share this legacy with Whipple.

Whipple's meteorite division continued to grow and prosper under Edward Fireman, who had worked for Rinehart and by 1959 built a new analytical laboratory boasting equipment like an electron probe microanalyzer. Fireman (Figure 16) was a nuclear physicist hired away from Brookhaven where he had been measuring trace elements in metals and

Figure 16 Edward L. Fireman added analytical laboratory capability to the SAO's facilities. *Smithsonian Institution Archives, Acc. 16-263, Box 3.*

was using his techniques to age date meteorites from their exposure to cosmic rays.[367] Through the 1960s, Fireman's staff engaged in isotopic meteoritic age studies, micrometeoritic interplanetary dust analyses and general chemical and physical analysis of silicate-rich meteorites. The meteor work was in many ways complementary to the development of the satellite-tracking program. It not only provided the original infrastructure for expertise in tracking, but it also acted as a trial balloon: The missteps that were made building up the first program became useful lessons for navigating Smithsonian waters to build the second program.

A "Tentative" Optical Observing Program for Earth Satellites

When Joseph Kaplan asked Whipple to prepare a requirements study for optical tracking, he added Princeton's Lyman Spitzer to help out. Whipple immediately set out to produce two documents: One, a report with Spitzer, would describe in outline the structure of an ideal optical-tracking program that would be issued as a "request for proposal" (RFP) for any qualified agency or organization to bid on. The other was a proposal from the Smithsonian to do just that. Both proposals were submitted within days of each other to Hugh Odishaw in time for the second meeting of the TPESP in late November.

Spitzer was, like Whipple, among the few ardent promoters of spaceflight within the astronomical profession and had provided technical briefings on Earth satellites and their ability to determine upper atmosphere densities for RAND since the late 1940s. He was also prominent and highly respected within the various IGY panels associated with rocketry and satellites as an expert on the high atmosphere. But he gave the lead to Whipple. "I assume that you are preparing your recommendations on optical tracking for the meeting of our panel," Spitzer wrote, implying that he understood that Kaplan's invitation was symbolic. Whipple and Spitzer met in Washington the night before the second meeting of the TPESP on 21 November 1955 to go over the report, and Whipple made the presentation to the panel the next day.

The RFP described an ideal overall administrative structure to manage, execute, and analyze the products of a worldwide optical-tracking system for the Vanguard program. It would include a "central organization for satellite position measurement, prediction and analysis" within an "Administrative Center" that was responsible for the installation and operation of all the stations, as well as for communications to and from the optical and radio stations, and for communication with special sections for orbit prediction and analysis. A separate but connected "communications center" would provide direct contact by radio between other rapid communications links with the optical observing stations, NRL's Minitrack radio position stations, and other groups that might be engaged to make ground observations of the satellite. The communications center would send all information to an "analysis center" and in return would receive predictions. The analysis center would consist of an "interpretation group," a "photographic reduction group," and a "computation group" with high-speed computing machines. There would also be a "theoretical analysis group."[368]

There had to be considerable overlap to facilitate communication: "Theoretical analysis," for instance, would be responsible for reporting any changes in the orbits of the objects, upper atmosphere density fluctuations and distributions, the distribution of mass inside Earth, and the precise positions of the stations, all in a consistent fundamental coordinate system. "Its research would extend far beyond the observational program, but we need not worry about the long range aspects at the moment if the short range current operations are clearly indicated." Next, optical tracking and Minitrack had to work together to achieve rapid response. Minitrack would provide an orbit "giving positions to the order of five minutes of arc or better" with time determinations to a few hundredths of a second. These data, the instant of meridian passage and altitude of meridian passage, would be radioed "immediately to the central communications center," which would in turn relay them to the computing bureau for the immediate calculation of orbital elements. These functions had to act as a single unit within one institutional framework, but it did not have to be "physically in close juxtaposition to the Vanguard program."[369] Whipple thus created an operational space for SAO.

Beyond rapid response, there also had to be highly flexible contingency planning. "In case the electronic tracking system fails completely, the best initial orbital elements will be taken from the observed data near the point of launching," which would come from traditional tracking stations near the launch point to track the earlier rocket stages "to the point that the satellite is actually free in its orbit." To prepare for any contingency, all observing and tracking stations had to be equipped with large column-mounted binoculars "and other useful visual optical equipment, in so far as it can be made available without unusual expense." This last contingency soon became Project Moonwatch, a worldwide net of visual observers that could provide preliminary observations sufficient for rough prediction.[370]

The RFP also specified that the system had to be sufficiently sensitive to detect fast-moving faint objects and outlined the types of optical systems and photographic technologies that should be considered. Whipple knew well that the Smithsonian's present array of 12 inch Super-Schmidt meteor cameras could barely catch a magnitude 6.0 object (the limit of eyesight visibility) that was trailing at 1 degree per second. If it was moving faster, the Super-Schmidt would be unable to catch it, given the sensitivity of available photographic emulsions. There were hopes of acquiring faster emulsions, or utilizing newly promised electronic image converters and image-tube techniques, but there were too many unknowns and great expense associated with electronic amplification, so for the purposes of their "tentative" program, Whipple and Spitzer limited their discussion to photographic methods, pushing for a larger aperture Schmidt.[371]

In their response to their own RFP, considering various combinations of cameras and mountings, Whipple and Spitzer opted for a larger Schmidt with a flat field that could use roll film that would allow for "extremely quick film changes."[372] This hybrid design would be coupled to a motorized mount with "variable-variable rate drives" similar to, but more controllable than, Tombaugh's rapid scanning system.[373] To get this all done, Whipple had to expand his working group, adding two more astronomers, two optical experts, one mechanical engineer, one photographic specialist, one photoelectric specialist, and one image-tube specialist.[374]

The panel quickly endorsed both the RFP and SAO's "tentative proposal," asking for a formal proposal from the Smithsonian to engage in the preliminary study. Whipple also prepared a budget for the "tentative" program, and tellingly noted that the position of the director of the project would be borne by the host institution. Hardly subtle.

The Smithsonian's "Proposal"

The proposal Whipple presented to Carmichael outlined, in essence, the infrastructure he envisioned for SAO. It was built around a contracting office to manage technologies and services they knew they would want to incorporate into the eventual design. These ranged from instruments for absolute time standards to acquiring expertise to develop a fuller theory for the solutions of the equations for the orbital elements, their variations, and perturbations due to air density and geophysical parameters. All this needed calculating and computing facilities; design and construction facilities for the specular surface of the satellite and its geometry; photographic facilities; optimal observing sites worldwide; and, most critically for Whipple, an effective administrative structure.

Carmichael gave Whipple's draft to Keddy and Graf who immediately set to editing it to make it more forceful and direct, with fewer qualifiers, and, most significantly, shifting it from being a proposal of the "Astrophysical Observatory of the Smithsonian Institution" to a proposal of "the Smithsonian Institution through its Astrophysical Observatory."[375] Over the next several days, more editing placed the Smithsonian firmly as the contractor and tightened up the budget. Carmichael submitted the proposal days later to Alan T. Waterman, the NSF director, while Whipple followed up with Hugh Odishaw to "iron out the details." There was no time to waste. The NSF board would be meeting on 5 December, so this proposal needed to be in Odishaw's hands as soon as possible.[376]

Keddy and Graf worried what the ultimate impact would be of such a program on the Smithsonian. The preliminary requirements study was not large, but the ultimate contracts, which would be in the millions of dollars, were far larger than anything the Smithsonian had yet handled as a single research program funded by external government resources. And further, counseled by Whipple, Carmichael knew that to be competitive for the ultimate job, the Smithsonian had to be ready to coordinate closely with the interests of the Army Map Service, the Hydrographic Office, the Coast and Geodetic Survey, and the Naval Research Laboratory, all the possible competitors.

Despite his staff's concerns, Carmichael agreed to the scope of the proposal, deftly rationalizing it as a continuation of an institutional legacy. As he later described it for the Regents, the program

> follows two traditions that we cherish in the Smithsonian Institution. We are participating as pioneers in the progress of science, comparable to Dr. Langley's original research in the flight of heavier-than-air craft. We are also promoting international interest in the Satellite Tracking Program which is in keeping with the worldwide pattern of contributions to knowledge by the Smithsonian Institution.[377]

Carmichael pointed to the same virtues that Odishaw had highlighted for the U.S. National Committee's Executive deliberations on 30 November 1955, as they finalized their decision for who would be tasked with the optical tracking, and both were taken verbatim from Whipple's proposal.[378]

Sharpened by Keddy's editing, the Smithsonian proposal made clear that it was not presuming to be the selected institution, and to make their proposal sweeter, "an appreciable part of the cost of salaries in the framework of the proposal will be borne by the Smithsonian Institution and will not be charged." The Smithsonian itself fully intended to engage in scientific research through the satellite program, "including the long-range phases" such as upper atmosphere studies and geodesy, and finally, of course, the Smithsonian was an ideal institutional home for international programming, in words repeated faithfully by Odishaw and Carmichael. There was also the fact that Whipple's Harvard infrastructure included the services of a small staff that managed an international telegram service, which "provides for rapid exchange of information about new astronomical discoveries and events." Another advantage Whipple highlighted, presaging his remarks to Harvard alumni in the wake of Sputnik a few years hence, was that a "huge reservoir of scientific and technological talent in the Boston area will be available for convenient and quick consultation on the varied problems."[379] With a ringing endorsement from Carmichael, Odishaw carried it to the executive committee, emphasizing that Whipple's prominence was critical to the operation and illustrating that fact by reciting Whipple's many honors and awards.[380] Odishaw's support won the day. The circumstances and arguments presented throughout this process clearly demonstrate how Whipple became the choice—by moving to the center, by positioning himself, his technology, and his institution as the most likely combination for success. He was playing against many factors, many unknowns, and indeed many technical doubts and concerns.

The Lingering Question of Visibility

Whipple well knew, along with everyone involved with the overall effort, that the object they lofted into space had to be bright enough to be seen visually, as well as optically and electronically. Maximum public impact was the goal, so visibility was critical for political as well as physical reasons. The NRL/Vanguard faction was committed to radio as the best way to obtain physical information, if the little transmitter survived the launch. But optical and especially visual reconnaissance would provide the most widespread political impact.

Vanguard team leaders developing the launcher and payload had always been skeptical about optical tracking, but their own technical advisors in the NRL's Optics Division had checked Whipple's visibility estimates and supported them in theory. Even so, Richard Tousey, the lead physicist in the division, who was experienced in practical aspects of visibility during the war years, felt that the chances of catching anything with present cameras "were only one in a million."[381] There had to be a way to predict visibility accurately enough to increase the chances that the faint fast moving object might be recorded in the restricted fields of even wide-field telescopic cameras. Aside from the weather, there was also only a narrow time slot when the object would be illuminated by sunlight and when the sky would remain dark enough to see it. Bright stars are not visible in daylight and they are thousands of times brighter than what was expected of the first small satellites. And later at night, satellites in low Earth orbit would be dark, shrouded in Earth's shadow. The Sun had to be at least 5 degrees below the horizon to make the sky dark enough to reveal stars, and at more than 5 degrees, Earth's shadow extends to more than satellite height. One possible solution would be to equip the payload with some form of optical beacon, a flashing light as Martz and others were suggesting.

Whipple did not favor flashing beacons, but if they were going to be added, he wanted to take advantage of them. Early on, he asked Robert Davis, a graduate student, to look into such matters, and fostered continuing studies of the use of flashing light systems for geodetic research.[382] His task for Davis illustrates the breadth of Whipple's readiness. In September 1954, Davis, a former Harvard undergraduate, announced he would return to Harvard from military service for graduate work. Whipple immediately recruited him to stop in at the Varo company in Garland, Texas on his way back to Cambridge to meet its founder, Austin Stanton. He wanted Davis to learn more about the Navy's plans to launch a satellite and what their needs for optical satellite tracking were.[383] Whipple was already consulting for Stanton, and suspected that Stanton, like Davis an electrical engineer, would provide deeper insights into possible detector technologies he was then developing for spaceborne miniaturized electronics systems as well as low light level "night vision" devices and various electronics systems for guided missiles and general military application.[384] Whipple wanted Davis to do a bit more digging into Stanton's involvement with miniature vacuum tube electronics and imaging systems for his tracking cameras and also to see whether image tubes were anywhere near viable. All options were on the table for Whipple.

In 1955, however, photography still seemed the most promising technology, but even then Whipple did not close the door on some form of electronic imaging. The one design element that Whipple definitely had decided on was Tombaugh's active three-axis tracking system. Whipple's own proposal did not cite one design over another, only that no "off-the-shelf" optical or timekeeping equipment was at hand. He was ready to devote almost half his budget to contracts for detailed engineering designs and capital equipment.[385]

NSF took only nine days to inform Carmichael that the Smithsonian's proposal for $49,910 was granted and would be paid in full within another two weeks.[386] At 15 minutes before midnight on Friday, 30 December 1955, Whipple gleefully telegrammed Hynek: MONEY AVAILABLE. HURRAY. USE YOUR JUDGEMENT ON PUBLICITY.[387]

Working within the Smithsonian: "Gaining Fiscal Control"

As Whipple recruited staff, he also came to grips with what working with Smithsonian upper management was like. On 30 December Keddy instructed Whipple to show up in Washington on 10 January for congressional budget hearings. There would also be a Smithsonian regents dinner later that month at which Whipple was expected to present his "New Horizons" vision. But he also had a lecture to prepare for the American Academy in Cambridge, at least one PhD defense, an IGY meeting in Washington, lectures in Cleveland and New York before the Institute for Aeronautical Sciences, and finally the Satellite Symposium of the Upper Atmosphere Rocket Research Panel in Ann Arbor at the end of the month. Keddy, reminding Whipple that he had to have "flexibility" in his travel schedule, which meant be available for presentations in Washington, also insisted on prior approval for all travel plans.[388] Keddy also laid out the rules of the road for making long-distance calls. Whipple and Rinehart were using the phone freely. Keddy assured Whipple that for emergencies, the phone was fine without prior permission from the Office of the Secretary. But in normal communications, the directive was that airmail was preferred for precision and accountability.[389]

Keddy exerted his manifest power by strictly enforcing all government rules. All federal procedures and guidelines for staff hirings, travel, office furniture, supplies, and services had to be followed.[390] From the start, then, Keddy demanded that Rinehart and Whipple "set up fiscal control" of the various contracts they were accumulating. Whipple and Rinehart were leveraging support from the U.S. Air Force with the NSF satellite-tracking planning contract to align the Meteor Project with satellite tracking. This required not only flexibility but also leeway securing staff who recognized and were comfortable with dual funding lines from complementary but not identical projects. Keddy, always concerned about limits to any discretionary action "for the prudent administration of the funds involved," sent them directives and boilerplate guidelines that were more confusing than enabling.[391]

Staff Building

Hynek jumped at the chance to take a two-year leave, feeling that Ohio State should be happy to donate something to the IGY effort, "namely me!" He relished the chance to travel the world setting up tracking stations. It would be "frosting on the cake."[392] Hynek did not express specific research interests, though he once toyed with the visibility of stars in the daytime. Whipple well knew that Hynek had eclectic interests; he had achieved some visibility as editor of an influential review of astrophysics in 1951 and evidently was interested in more visibility.[393]

Whipple expected Hynek to drop everything and show up by 4 January. He would have no secretary at first, but his office was ready: "We have been trying for about a month to add to my current staff and still have no luck at all." There was a "pool" available and everyone would have to "temporize for a while" and manage the place because Whipple would be constantly traveling between Cambridge and Washington. On the bright side, Whipple reported, "It is extremely exciting and the interrelationships with the international groups of various types will be one you will enjoy as well as the technical aspects." Hynek could draw on the vast pool of local talent in Cambridge and also had available at least two established stations: a New Mexico site at Organ Pass, "where we can make all sorts of tests," and the Agassiz Station. Above all, Whipple wanted Hynek at hand, even if on temporary duty, to get the program "on the road." He asked Hynek to convey his apologies to Mrs. Hynek and family, "but perhaps she will forgive me in the long run."[394]

After Rinehart, and, of course his continuing dance with Sterne, securing Hynek was his top priority. But he had at least five other major positions to fill. Hynek was able to secure a three-month leave from Ohio, but was duty bound to return to teach a popular course in the spring. Working with what he had at hand, Whipple outlined duties: Rinehart would be involved in instrumentation, preparations for building stations, and overall administration. Sterne would evaluate the old APO solar constant work and articulate what research meant in the new organization, but he would also be involved in the tracking program, most of all the computing and analysis. Hynek would be involved in everything else. From the Harvard side, "Mac" McCrosky was running at more than full speed, helping to organize the instrumentation and identify possible observing station sites.

Whipple tried to attract Gerald Clemence, who led the Navy's Nautical Almanac Office, to provide overall scientific administration, from site selection to the management of the computing and analysis section. Whipple also approached Ivan King, William Sinton, and William Baum, specifically for electronic detection techniques.[395] He failed with all of them, first because there was too much red tape to shift a high-level government employee, and second because, once again, Harvard appointments were not forthcoming.[396] McCrosky, on the other hand, was deeply devoted to the type of work he had been doing under Whipple's guidance, and with his Harvard PhD soon in hand, remained loyal to Whipple as his "scientist in charge of the SAO Meteor Photography and Recovery Program." Over the years, the two had engaged in many fruitful campaigns for the Harvard Meteor Expedition based at the Las Cruces station and elsewhere. McCrosky had distinguished himself in providing many innovative field techniques for rapid data reduction and analysis.[397]

When Clemence proved too hard to attract to organize the orbit computations and find new and more rapid means of analysis, Whipple turned to Leland Cunningham, his old friend from Berkeley, "to derive working theory of artificial satellite in IGY program," as he described it in a hasty telegram. He needed Cunningham immediately to prepare for a meeting Whipple was organizing in Washington with agencies invested in operations related to optical tracking. Whipple knew, of course, that Cunningham did not have the prominence of others he was trying bring to the meeting, but he had the competence, which was critical to convincing these agencies that the Smithsonian could do the job right. Above all, Whipple knew that Cunningham would fit in positively with his staff, and that they would be comfortable with his advice and counsel.[398]

Establishing SAO as the Central Agency for Optical Tracking

Writing to von Braun in early January, Whipple sought his counsel for the meeting and also concurrence with his bid to compete for the big money. He noted that the SAO network would have at least eight stations shared with NRL Minitrack and Army Microlock stations. This collaboration would create a worldwide geodetic framework of at least 30 foot accuracy and be operational by 1 July 1957, Whipple predicted with a flourish, signing off "with . . . best wishes for the conquest of space."[399] Von Braun quickly and forcefully advised Whipple that he should keep his optical tracking separate from the "missile end of Vanguard." He endorsed the optical tracking as a backup system to radio, but more critically, he advised that it should be sold as a means of achieving higher positional accuracy, which was of course of great interest to "our high altitude re-entry missile test program."[400]

To make a final decision, the TPESP had created a new oversight panel for optical tracking, asking Spitzer, who was not proposing on any team, to chair it. Their first report in January described optical tracking as "primarily for scientific objectives," whereas Minitrack would monitor Vanguard itself. It also directed the SAO to initiate the operational program, adding that it was its intention, "subject to further deliberation, to ask the SAO to continue management of the optical program throughout IGY." The report also endorsed the public element, enlisting "professional and amateur astronomers throughout the world."[401]

In effect, Spitzer's panel assumed that unless Whipple and the SAO faltered in some way, optical tracking was theirs to lose. Whipple was well along in establishing the overall structure of a system elaborated from his and Spitzer's original specifications, and finally by February 1956 felt secure enough to call on other agencies to join in, including the Army Map Service, the Coast and Geodetic Survey, the Aeronautical Chart Service, the Naval Research Laboratory, and the Geological Survey.[402] Whipple advised Keddy that the Smithsonian was "contractually committed" to collaborate with them and that the Smithsonian should host a conference in mid- to late February when a number of acknowledged orbital mechanics experts and geodesists would (coincidentally) be in town, including Leland Cunningham, and that this meeting should include representatives from military photogrammetry units.[403] This would not be a public meeting, Whipple advised, but institutional players in Washington needed to be notified, including NSF, ONR, the National Geographic Society, the Office of Chief of Engineers at Andrews Air Force Base, as well as Hugh Odishaw and his staff at the National Academy.[404]

The meeting was a political formality, but it did put critical specialists in contact with each other, and did jumpstart a stream of studies by Cunningham, culminating in a paper, "The Motion of a Nearby Satellite with Highly Inclined Orbit," that he delivered at an American Astronomical Society meeting in Berkeley. Whipple felt this strategy served as a "shot in the arm," not only for his program but also for Cunningham's energies. By the spring, Cunningham had provided a systematic structure that could be applied quickly to utilize the approximate observations of the amateur network to direct tracking cameras to follow a wide range of satellite motions. However, questions remained: What kind of satellite would they be following? And even more pressing, how would Whipple and the Smithsonian manage the operation?

The Shock of the New

At a meeting in April, Spitzer's Working Group on Tracking and Computation endorsed the Smithsonian's overall plan for a two-pronged optical-tracking program. With a total budget amounting to $3,380,610, Whipple, Hynek, and their rapidly growing staff would form two groups: one to find the satellites and one to track them with the highest precision possible. This was not money in the bank yet; Whipple and Hynek still had to submit formal proposals for the project in stages, and each had to be vetted by the Smithsonian. Just at this critical time, however, Hynek had to return to Ohio to meet his obligation to his university and his family. Saying that he was "very grateful indeed" for Hynek's willingness to drop his work at Ohio and spend the last three months with them, still Whipple made it clear that his departure was "very unfortunate" but unavoidable. Hynek did take the leave, continuing to work remotely with undiminished energies.[405]

McCrosky started to explore possible sites for observing stations in Panama and other countries in South America, following the north-south fence of the Minitrack sites. Just getting the paperwork rolling came as a shock. Early the previous March, even before the money was assured, Keddy complained that the "$3.5 million program is $1 million more than the amount required to run our day to day operations per year, so you can see managing and servicing the proposed program will be a Herculean job." Keddy insisted that Whipple secure the services of a full-time business manager, because just then the Smithsonian itself was in need of staff to open its new Museum of History and Technology by 1960. Keddy advised Carmichael that he doubted Whipple could handle such a program, even without the momentary loss of Hynek. Keddy also did not appreciate why it had to be a "Crash Program," suggesting it build gradually from 1956 to 1959.[406]

As Keddy heckled, John Graf dickered with Whipple over wording of the final proposal. Whipple dryly appreciated that Graf had taken so much care with the many revisions to be sure that it was the Smithsonian and not the SAO that was the actual contracting agency, and even tried to reinstate the original wording, but Keddy would have none of it. Frustrated but resigned to it, Whipple admitted that it might make the "contractual operation easier to do."[407] Looming over everything, though, as Whipple warned Graf, were rumors that Congress might reduce the overall satellite program by several million dollars, which Odishaw hoped could be recovered if contracting institutions waived overhead.[408]

Keddy's gradualist demand was unacceptable. Whipple had demonstrated urgency and had shown that immediate action was needed to build two prototype stations along with the core elements of the communications network.[409] To placate Keddy, Whipple emphasized that the Smithsonian Institution was central to all aspects of satellite tracking. Not only would it be deeply involved with the gathering of information necessary for predicting the orbital trajectories of satellites, but it would also take those predictions and use them to produce, through a photographic-tracking system, highly precise orbital data, and from that would analyze the data to obtain geophysical data "available in no other way." Confident that he had obtained the sanction of all possible competing parties, Whipple hammered home that the "present proposal" collected into a "unified whole" all the aspects of the optical-tracking program including analysis.[410]

Shaping Vanguard and Building an Optical System to Track It

The original NRL proposal called for a conical 21.5 pound payload, some 20 inches at the base. This design would minimize weight and be resistive to heating issues. But by October 1955, largely after the TPESP endorsed Whipple's objections, the NRL changed the configuration to a 30-inch sphere, claiming it would better suit scientific studies of air resistance and provide constant optical visibility. But NRL's John Hagan "was irked" by this modification and by others after it.[411] The extra weight of this design also reduced the payload weight for the scientific instruments, which required James Van Allen to fit his cosmic-ray instruments into simpler and lighter cylindrical structures.

The 30-inch sphere persisted through early 1956 but was ultimately reduced to a 20-inch sphere when engineers from the Martin Company who were building the launch system protested. It would not be reduced further, however, given the possibility that the Minitrack system might fail, and they had to ensure optical visibility.[412]

This decision also reinforced Whipple's call for a larger telescopic camera. The Harvard 12-inch Super-Schmidts could definitely track a 30-inch sphere, but not a 20-inch.[413] Whipple at first considered a 16-inch f/1 system but, advised by James Baker (Figure 17), increased the aperture of the Schmidt correcting plate to 22 inches, with a 31-inch spherical primary, giving a 30-degree usable field.

By spring 1956, at least, Whipple felt that he owned optical tracking for Vanguard. He accomplished this goal not so much by competing with other contenders but by co-opting them, incorporating all the best elements from his own meteor network with those from

Figure 17 James Baker and a model of his Baker-Nunn camera. *Smithsonian Institution Archives, Acc. 16-263, Box 6.*

others such as Tombaugh. Whipple also knew how to parlay his elite civilian academic status, responding vigorously to all requests for public statements and pronouncements supporting spaceflight, and in one instance, at least, arguing that establishing a worldwide tracking system for any American satellite would be greatly facilitated by a connection with the IGY, and hence with civilian academic interests.[414] Whipple, however, had also initially envisioned that most of his foreign optical stations would be co-sited with military Minitrack or Microlock stations, so his claim for their civilian status, most likely, was another example of his facile entrepreneurship.

In the year since becoming director of the APO, Whipple had built up an infrastructure that looked and behaved civilian but had roots well-established within military agencies. He had gained the confidence of military colleagues and government patrons, and most critically, by accepting the central role of the Smithsonian Institution as the fiscal manager of the operation, he gained Carmichael's full support. His rapidly growing staff certainly expressed its own enthusiasm for the job when they submitted some 77 entries in an internal contest for what to name the project. What was initially called SPOT for Smithsonian Precision Optical Tracking, however, would very soon pose very serious technical and managerial problems for the Smithsonian and for Whipple, especially after the Soviets surprised the world.[415]

Chapter 6

Satellite Tracking: Building and Operating the Program

SPOT would soon be renamed the Baker-Nunn tracking network to honor the chief designers of the optics and mounting, because the greatest challenge would be to build the first tracking instruments and send them around the world. Although the network and program were originally conceived as serving the finite goals of the International Geophysical Year (IGY), these goals were abruptly recast by the American reaction to Sputnik.[416] As a result, what had been a finite crash program to demonstrate a capability soon became, for Whipple, a roadmap for establishing a new kind of astronomical institution, one based ultimately on a continuing national commitment to spaceflight. In this chapter, we explore how the satellite-tracking network was established and the crisis caused by having to answer the Sputnik challenge: making it work and parlaying it into a bold initiative to create a permanent institution.

Baker's Challenging Optics

When Whipple approached James Baker about the upgrades needed, Baker was painfully aware that the many innovations he had built into the original Super-Schmidts took some five years to make them operational. If satellites had to be tracked that were launched within the IGY, that gave him far less time. In 1956, Baker was working at Harvard on a half-time appointment and was consulting for optical firms like Perkin-Elmer. He was well equipped to generate a design that, in theory, would fit Whipple's needs, which meant quickly operational.[417]

Baker's designs in the summer of 1956 reflect not only the increased demand for optical sensitivity and speed caused by the shrinking dimensions of Vanguard, but also the fact that after almost two decades of refining and extending the basic optical design developed by Bernhard Schmidt in the 1930s, Baker was definitely not out of new and ever-more creative ideas. By the end of the year, he had a small staff of Perkin-Elmer optical specialists on his team, and access to a high-speed computer that could rapidly churn out ray tracings of complex multi-component optical systems.[418]

Baker also knew that, unlike the meteor cameras, the tracking telescope had to take multiple exposures rapidly. This required a flatter field. He had already experimented with a cylindrically curved focal plane that would produce a long and thin field of view on a filmstrip, and suggested this as a direction to take for the satellite tracker.[419] With a supplemental grant for more computer time approved in June, Baker and his colleagues plowed through combinations of primary reflective mirror dimensions and figures coupled with various combinations of multi-element transmissive correctors, some cemented and some air spaced, and finally decided on a three-element, air-spaced corrector feeding a 30-inch aspheric primary mirror.[420] It was far more complex and physically far larger than had been anticipated, and was going to be far more expensive. There were problems from the start.

The German manufacturer Schott provided the glass for the corrector plates, and Corning provided its Pyrex-brand mirrors. As expected, the corrector plates posed the most demanding challenge for Perkin-Elmer in Connecticut, the general optical contractor. Baker designed the plates to be as thin as possible to reduce ultraviolet absorption and required that four of the six optical surfaces have complex aspheric properties. Starting in summer 1957, in what was yet another example of the support provided by military agencies to the IGY, the U.S. Air Force (USAF) Military Air Transport Service started flying the rough optical components from Schott in Germany to Perkin-Elmer. Hardly uninterested bystanders and facilitators, the Air Force would continue to watch the progress of the Smithsonian's satellite-tracking network, ultimately adopting it in the 1960s.

The first camera was promised to be ready for testing by late September 1957 and be on view at the National Academy soon after for a major conference. Carmichael duly approved an "immediate crash" program, which included a $20,000 incentive to Perkin-Elmer. The cash also required that Perkin-Elmer accept the presence of another Smithsonian contractor, Steve Sydor, an expert in the properties of optical glass and manufacturing techniques, at its Danbury plant to help create a special optical facility to turn the glass blanks into optical elements.[421] Sydor's tasks included "daily reports" on all progress.[422] Sydor quickly assessed the situation and warned Whipple that there would be delays. Whipple, however, had already been called to Washington in August 1957, as we have seen, to address the status of solar monitoring and to air growing frustrations with the satellite-tracking program. Perkin-Elmer was complaining, demanding more funding to expedite the optics, and there were rumors that satellites would be orbiting before the end of 1957. Whether they were Army, Navy, or Soviet satellites made no difference. By the time the satellites were flying, there had to be a tracking system in place. Now, despite his earlier reticence, Keddy went on the attack.[423]

Despite his original ignorance of the need for a crash program, at the August meeting Keddy accused Whipple of being an incompetent manager. The delay in optical fabrication was only one problem. There were also serious problems with manufacturing the mountings and with other elements of the satellite-tracking program. Keddy charged that both Whipple and Hynek have "abdicated their jobs," hiring too many people and contractors and giving them free rein over their progress. Indeed, there had been several dozen hirings by then and it was far from clear what many of them were for. Keddy could not distinguish between Sterne and Hynek, for instance, but pointed his finger directly at Whipple for not keeping either of them in check. "Hynek has proved to us," Keddy asserted, that he was not an operator, adding, "We are only continuing an organization that is the cause of the mess we are in." There would be no hope of recovery, he told Carmichael, "until there is some centralized administrative control there."[424]

Nunn's Mounting Problems

Almost everything about the new camera required more than extrapolating known optical and mechanical capabilities. Tombaugh had built and tested small experimental cameras, but nothing of the size or sophistication and standardization needed to drive Baker's cameras was known in the civilian world.[425] All three axes, for instance, had to be driven independently by highly accurate variable rate motors that could be adjusted quickly through wide ranges of speed. Moreover, the precise time of each exposure had to be recorded, unambiguously, to 1 millisecond (0.001 seconds). The best solution was to find a way to imprint the time information directly on the image, which required a source of time information in the form of a visible readout that could be projected directly onto the photographic image.

Advised by Smithsonian Astrophysical Observatory (SAO) staff and Baker's contacts at the Perkin-Elmer Corporation, Whipple engaged Joseph Nunn of Pasadena, California, as consultant to design and construct just the prototype for the mechanical parts of the camera system mounting, not manufacture the dozen systems needed to establish the network. Nunn ran a small design firm that had worked for Caltech to maintain and improve instrumentation for Palomar Observatory and had also recently designed a radically new mounting for a 36-inch reflector that was built by Boller and Chivens of Pasadena, California.[426]

Keddy had placed Earl C. Eisenhart in Pasadena to keep watch on progress. Eisenhart was impressed with Nunn's technical competence, but harbored deep concerns about Nunn's financial and managerial abilities. After a visit to Boller and Chivens, Eisenhart felt that this firm, though also small, was large enough to handle the manufacturing.[427] By late August 1956, however, even this arrangement was questioned, because there were growing concerns in Carmichael's office over the propriety of engaging Boller and Chivens without an open bid, as well as Nunn. This put Hynek in the hot seat because as contracting officer, Eisenhart and Keddy wanted him to clarify just where the liabilities and obligations lay. Hynek blithely viewed Nunn "as a member of our team" and became exasperated and confused when Nunn claimed that he, with Boller and Chivens, had already contributed "novelties of the art" to the design of the mounting and so warranted sole source status. The matter grew only darker when Hynek tried to distance himself from the business or legal side of the problem, telling Eisenhart that "I am not competent to judge such matters." He preferred to let "legal minds wrestle with such problems." Even so, Hynek defended his actions: Both he and Whipple were stretched to the breaking point and neither had the time or energy to keep track of multiple lines of development. On top of that, he was due to return to Ohio to teach in the summer quarter. Hynek well knew that hopes of securing the prototypes by the end of the year were very dim.[428]

Figure 18 Whipple, with Karl Henize and J. Allen Hynek, against a recording studio backdrop of their worldwide network of Baker-Nunn cameras. *Smithsonian Institution Archives, Acc. 16-263, Box 7.*

By the end of September 1956, Nunn provided construction drawings and full blueprints to finalize contractual arrangements with Boller and Chivens in early October.[429] And by the end of the year, Whipple expected the delivery of the first camera from Boller and Chivens in June 1957 and the optics from Perkin-Elmer a month later. He did not raise the question of how long it would take to integrate the two. Meanwhile, Hynek, along with McCrosky and Karl Henize, a young astronomer who had recently been hired, selected the sites for the first 10 stations around the world and installations were expected to commence soon after delivery (Figure 18).

The Stations

Along with the optics and the mounting, site selection for the network proved complex. There were intensive recruitment efforts to find and train observing staff who could be sent around the world to lead the stations, as well as extensive negotiations establishing the many sites on foreign territory. Geographical coverage was paramount, but there were also political sensitivities arising from the need to balance technical requirements standardizing the sites with accommodating the wide range of cultural and technical standards the host countries possessed.[430] During the September 1956 meeting of the IGY in Barcelona, Whipple interviewed representatives from countries that had petitioned to host a station.

He reported that they are "extremely enthusiastic and they are giving us a great deal of support, very real support."[431] The Smithsonian initially wanted to share control with countries that could meet their criteria for the operation. But when some of the selected countries failed to meet deadlines, Smithsonian management again questioned Whipple's management skills. A station in Nainital, India, was especially worrisome owing to construction problems and jurisdictional confusion.[432]

Given the compressed time frame and the delays, there was no chance for prototype testing or even component testing. The cameras were effectively built en masse and "the first one completed had to work." Boller and Chivens devoted over half of its resources to the project, and still had to subcontract out many of the larger components.[433] Finally, months behind and ever slipping, the first Baker-Nunn was assembled for testing in the Boller and Chivens shops in September 1957 but was hardly ready to be demonstrated. This was weeks after the acrimonious review session in Carmichael's offices, where Whipple was lambasted for failing to have an operational camera system on view before the IGY's conference at the Academy in October.

Sputnik

If Keddy needed any more evidence that the satellite-tracking program was in trouble, he got it soon. The upcoming academy conference in late September and early October that was expecting to examine a Baker-Nunn camera was in fact the very meeting that ended with the news that America was not the first to go into space. At about 6:40 PM on 4 October 1957, Moscow radio announced that a Soviet satellite had been successfully launched. A few minutes later, a *New York Times* reporter reached Lloyd Berkner and other conference attendees at a cocktail party at the Soviet Embassy, and Berkner made the announcement to stunned silence. There was much finger-pointing, including some directed at the Smithsonian.[434]

September and early October 1957 had not been pleasant for Whipple, who faced "grueling" meetings and much criticism. Even before Sputnik, the "lean-faced, high-domed Fred Whipple lost his calm." Trying to deflect the concerns of R. W. Porter, chair of the Satellite Panel on 3 October, Whipple asked why the fuss, since Vanguard itself was hardly ready and was behind schedule.[435]

The events of the evening of 4 October, however, forever changed the landscape; there are indeed multiple perspectives on what happened that night, and what it meant to the world, but the general agreement seems to be, for space scientists like Whipple, that Sputnik "drastically altered the debate" not only over the question of support for an American program, but also over the necessity of pursuing all options, from optical tracking to releasing the hold on Army Ordnance to prepare to orbit something.[436]

Whipple had already left Washington that afternoon and was headed back to Cambridge. By the time he reached his office, it was filled by SAO staff who had rushed back to work as the word spread. Immediately deluged by reporters asking for a statement, possibly his most memorable and revealing remark was to the *Harvard Crimson*: "It is a great achievement on the Russians' part, and I commend them for it. We won the first round with the H-bomb; they won the second with the space satellite."[437]

Whipple had been prepared for an event of this magnitude. He had already placed his staff "on alert" for any launch, by any nation. The Harvard Announcement Card telegram network got the word out rapidly. They had not yet installed a new teletype system, but they had telephone patches and the Moonwatch teams had been practicing for months and some had been practicing that night[438] (Figure 19). The only Baker-Nunn in existence was still being tested at Boller and Chivens. It had been assembled for tests, and key components were found wanting, including the optics. Karl Henize was on site and wanted to put it together and start searching, but preliminary predictions from initial Moonwatch data told him that Sputnik would not be visible from Pasadena for more than a week[439] (Figure 20).

Naturally, there were recriminations, and Whipple received his share. No matter, however, because Sputnik now gave people like Whipple, Porter, and literally everyone associated with the American satellite program an unassailable bully pulpit. At the fourteenth

Figure 19 Another common configuration for Moonwatch observers. Each small telescope was equipped with a tiltable mirror to monitor a different portion of the station's north-south meridian line. *Smithsonian Institution Archives, Acc. 16-263, Box 3.*

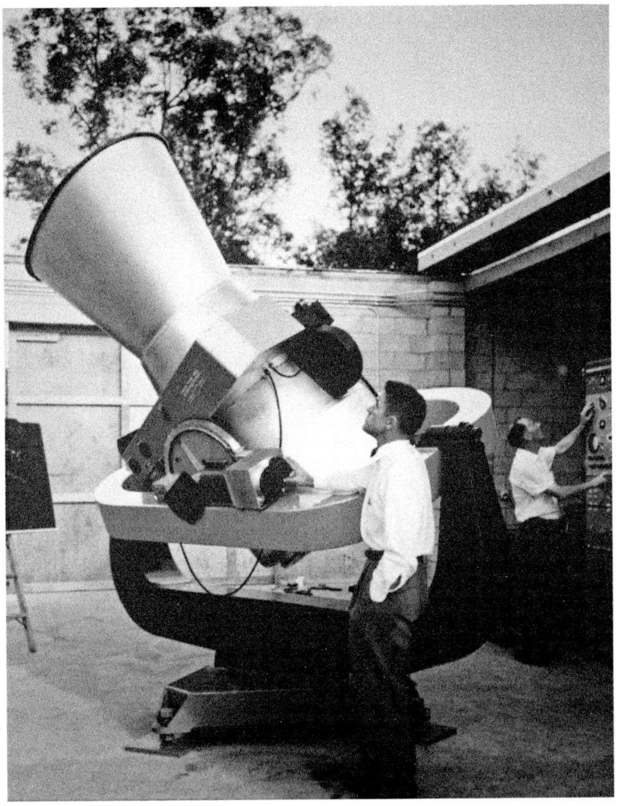

Figure 20 The first operational Baker-Nunn being tested by Karl Henize in Pasadena at the Boller and Chivens plant. A Norman Time Standard Clock is in the background. *Smithsonian Institution Archives, Image # 82-3366.*

meeting of the Technical Panel of Earth Satellites (TPESP) on 6 November 1957, all eyes were on Moonwatch. Hynek addressed the panel, "pointing out that it has developed into a more important operation than was intended or even thought of at the time."[440]

Sputnik accelerated the installation of the Baker-Nunn stations as well as the search for any means to track the satellite. The TPESP called for all available tracking systems at American test ranges to track satellites with whatever they had at hand.[441] For example, Porter pushed Hynek and Whipple to contact the Chief of Ordnance to transfer two small missile-tracking cameras with its essential personnel and appropriate gear to the Smithsonian stations at West Palm Beach and on the island of Curacao.[442] But those closer to the action, such as Karl Henize, knew this would only cause confusion. He advised Hynek strongly that this would "involve a whole new order of policy making; personnel hiring, training, and scheduling; and general management in an organization which up to now has been hard pressed to properly administrate the installation of only one type of instrument, the Baker-Nunn Satellite Tracking Camera." Porter also called for the India and Iran stations to be operational quickly, but both were far from ready.[443]

In the panic, Porter joined Keddy's call that Whipple hire a "competent management consultant" to rectify operational weaknesses at SAO. When Whipple balked, saying no one was immediately available, Porter turned his request into a demand.[444]

In desperation, Whipple tried to switch his Harvard-MIT-Lincoln Lab radio meteor program manager to the larger job, but the man abruptly resigned early in December.[445] When Carmichael learned of this, he added his voice to the call and ordered Whipple to hire a clearly designated executive officer for the observatory. There had to be a second-in-command, especially after the Academy itself called for an IGY-wide management survey. Carmichael urged Whipple to comply, advising that "the selection of the right man for this position will relieve you of the tedious detailed administration which the Director should devolve on the second in command if he is to continue effectively in the scientific field."[446]

In January 1958, Porter and his panel members feared that the SAO Baker-Nunns would not be available "in time for the bulk of even the earlier USA firings," which now included two Jupiter C flights from the Army Ballistic Missile Agency (ABMA) as well as the hoped-for Vanguards. Thus the TPESP, despite Henize's concerns, called for a detailed "action plan" based on Porter's request to have "loaned equipment expedited to key sites," including USAF cinetheodolites sent to India, Iran, and Peru. Small missile telecameras from the U.S. Army facility at White Sands would be sent to Curacao and to Florida, and even some Super-Schmidt meteor cameras, contracted to the SAO by the USAF, would be sent to Argentina and Hawaii. This was an all-out effort to have a worldwide optical system in place by early February 1958.[447]

But who would be in charge of this system? Would the SAO maintain the lead? Setting up the system; equipping each station with adapted makeshift cameras, keeping the data flowing from and between Moonwatch networks; and keeping all lines of communication open between the observers and the SAO's computation center, the radio networks, and the stakeholders in the Army, Navy, and the National Academy of Sciences (NAS) was a nightmare. There were hassles establishing a working relationship within the Smithsonian between the SAO public information unit and its lead office in Washington, as well as keeping the lines of budget authority straight.[448]

Finally, but only after Odishaw and his U.S. National Committee (USNC) for the IGY authorized additional funds, Whipple acquiesced and engaged a management consultant recommended by Porter. Carmichael was relieved, assuring Porter, Odishaw, and the panel that the Smithsonian was actively seeking and was now developing "more and more effective administrative procedures." But these were unprecedented challenges, and a competent consultant was needed to survey the organization, ascertain its strengths and weaknesses, and recommend improvements to ensure greater economy and efficiency in the operation of the SAO, particularly the optical tracking of satellites.[449] Perceptively, Carmichael asked that the consultant's study also include "sufficient time for a personnel study in Washington on the contribution made to the administration of the Astrophysical Observatory by our officers here."[450] The review began on the very day, ironically, that President Eisenhower's office telegrammed congratulations to the SAO for being designated the central nervous system for satellite communications. That endorsement, and the management report delivered by the consultant later in April, may have lifted spirits. But the system still had to work.[451]

Achieving Operational Status

Amid a constant stream of speaking engagements and dealings with Keddy and Odishaw, who Whipple classed together in less-than-complimentary terms, by March 1958 Whipple started feeling that he was making some headway through the National Academy Foreign Affairs offices and other agencies concerned with foreign relations. A big step came when he and Hynek clarified policy on the amount of "self-determination" the satellite-tracking stations would enjoy overseas.[452] Writing to McCrosky, then in Hawaii valiantly rendering a newly arrived Super-Schmidt useful while continuing to oversee the computational efforts, Whipple rejoiced,

> The acceleration in the satellite-tracking stations program is simply terrific. The pressures are almost too much to bear but the results I think will be satisfactory. The third optical system appears to be practically as good as Jim Baker hoped for!!! I will be out in Pasadena to check over this one before it goes on its way.[453]

Beset by all the pressures, Whipple missed having McCrosky in the office next door to commiserate. More than once, Whipple wrote to thank McCrosky for his help with tracking, but added: "Life is rather dull without meteors next door. After all, they really are my heart's blood, and progress in these directions seems somehow more important than some other types of progress."[454] Once the Baker-Nunns were delivered and their quality hailed, Whipple's spirit improved. There were continuing skirmishes with Keddy and Carmichael, but none as difficult as those in June/July 1956 and August 1957, though the effects were still lingering.

Throughout spring 1958, with the overseas stations established and relationships warming with the locals, Karl Henize reported to the NAS and the State Department that "political feelings have subsided and we can now continue with our scientific program."[455] His station managers in India and Iran reported that the latest optical systems were giving "unprecedented performance;" some 80 percent of the light was falling within a 25 micron image.

Next to Whipple, no one could have been more gratified than James Baker. Baker wrote to Richard S. Perkin in May praising "yourself and the Smithsonian staff too" for having faith in his design, in spite of the many well-intentioned doubters.[456] Whipple more than confirmed Baker's triumph, reporting soon that the Air Force is going to "buy a number of them now we have demonstrated that they are the best way of tracking faint earth satellites."[457]

By the summer 1958, one dozen Baker-Nunn stations were operational around the world, staffed by varying mixtures of Americans and locals, and Whipple was becoming a household name (Figure 21). Because there were as yet hardly a half-dozen satellites to track, although all were highly newsworthy, the system, including the tracking, communications, computation, and analysis sections, was able to return critical information even though technical problems remained and bugs were still being tracked down and squashed.[458]

FRED WHIPPLE continued

An astronomer with a desk calculator, Whipple seldom looks into a telescope

ASTRONOMER'S PRIDE is this collection of more than 100 "astronomical" neckties. Here Whipple dons a number sprinkled with rockets and spacemen.

ASTRONOMER AT HOME shares an interest in painting with daughters Lolly and Sandy. The children have the run of Whipple's study. By his values, the family comes before his work.

the world, and 2) science must be explained in laymen's language because "the taxpayer has a right to know." Thus he has been one of the most-quoted men on the Space Age — because he believes in communicating his knowledge.

Whipple likens his own research to detective work. In astronomy, there are few concrete facts and measurements. "The problem is to see what you can deduce with any degree of competence," he says.

His latest deductions appear in a paper entitled *On the Lunar Dust Layer*. This theorizes that, contrary to centuries-long belief, there is no dust layer on the moon — an important concept for moon rocketry. Preparing it, Whipple could hardly wait to drop this bombshell on his colleagues.

He is at this sort of thing all the time. Ideas can pop into his head anywhere — a frequent place is the modest Cambridge cafeteria where he always eats the same sausage-and-potatoes lunch. He will immediately sit down to calculate speeds, forces and orbits. "This is what we call back-of-the-envelope computation," he says. If the figures seem to show promise, the really serious work begins.

This involves collecting all possible data and information on the subject and then spending hundreds of hours putting it into complex mathematical formulas. The tools of his trade are pencil, paper and an ordinary office desk calculator. In his pioneer comet work, he never even looked at a comet through a telescope.

Science for the Girls

Whipple would like to have more time for basic research into what he calls "evolutionary problems of the solar system." But he refuses to give up time with his family — the week-end bicycle trips with daughters Sandy, 8, and Lolly, 6, the summer at a lakeside cottage. (He also has a grown son by his first marriage.)

Both girls have been bitten by the science bug. Sandy is co-editor of a one-page, hand-lettered "newspaper" called *Solar System*. Whipple makes no attempt to push his children into science, but he does stress "scientific method" — the system of making observations, developing theories and testing — in teaching them. He regards science and scientific method as the key to human progress.

"I think that man can and will go on to great things," he says, "not only physically — freedom from pain and hunger — but in the realization of our finest and most beautiful dreams, the fullest utilization of our abilities. The mind has no limits. Most people are using only a small portion of their minds.

"If I could just cut out the waste of time involved in sleep," he adds, "I'd have time to do all the things *I* want to do." A faraway gleam comes to his eye. If anyone can solve this problem, it ought to be Fred Whipple. ■

Parade • June 22, 1958

Figure 21 A page from *Parade*, highlighting Whipple home life in June 1958. Note the padlock on the filing cabinet. *Courtesy* Parade.

Even before the Baker-Nunn network was complete, the Smithsonian's Moonwatch teams were providing a constant flow of data to help direct interim cameras. They produced the earliest knowledge of the behavior of the satellites and hence the first hints of the nature of the high atmosphere and the varying gravitational potential of Earth. By October 1957 there were about 150 Moonwatch groups officially activated around the world and staffed by thousands of volunteers.[459] The SAO had succeeded in establishing its command center for communications, computation, and analysis by the end of the year as well, and, most critically, Whipple and Hynek established the policy, celebrated by Nelson Hayes and his cohorts in the SAO's public affairs office, that "no question would go unanswered if the answer could be found." In the end, certainly also due to the efforts of Hayes's staff, the public came to know the Smithsonian through Whipple's open-door policy to the media. In December 1957, a vice president of the venerable Shawmut Bank in Boston wrote Whipple: "Since [S]putnik, you and Dr. Hynek have been interpreting the astrophysical world to the people through the media of press, radio and television."[460]

In late April through early May 1958, the first solid results were aired at meetings of the American Geophysical Union and in publications soon thereafter. Although overshadowed by James Van Allen's announcement of the existence of a belt of trapped particle radiation around Earth, based upon his team's cosmic-ray counters on Explorer I and Explorer III, Sterne, in collaboration with Gerhard Schilling and Luigi Jacchia, presented revised formulae for determining atmospheric densities from Earth satellites and used them to make some preliminary conclusions about the density profile of the high atmosphere, noting carefully that the reliability of the results could be compromised by ionization effects and the non-spherical shape of Explorer. The densities were also far higher than those determined from rocketsonde ionization measurements and were also higher than the Air Research and Development Command's standard atmospheric tables. Sterne suggested that the ARDC tables required revision.[461]

Jacchia (Figure 22) concentrated on anomalies in the motion of Vanguard I, showing they were similar to those seen for Sputnik I and Explorer I. From this, he concluded that they were not due to the non-spherical shapes of the latter two, but were caused by 27-day density oscillations in Earth's upper atmosphere, possibly linked to the Sun's rotation and to the orientation of the satellite's orbit with respect to the Sun. Most significant, Jacchia and his team, including Naval Research Laboratory (NRL) staff and SAO members George Veis and Stephen Maran, ferreted out secular accelerations and sporadic fluctuations of the satellites' motions caused by terrestrial atmospheric changes induced by solar phenomena like flares.[462]

Along with their atmospheric work, Jacchia, with Veis and Sterne, also developed techniques to derive Earth's gravitational potential. Jacchia's first results were based on data organized and published very rapidly by the Royal Aircraft Establishment at Farnborough, which had been gathering in all observations, optical, visual, and radio, of the Russian as well as the American satellites. They used these observations to show how the orbit of the satellite changed over time, specifically, how the point in its orbit that crossed Earth's equator drifted in space and time. And then they equated these secular changes to derive the

Figure 22 Luigi Jacchia. *Smithsonian Institution Archives, Acc. 16-263, Box 7.*

shape of Earth and how it differed from the ideal case of a rotationally flattened ellipsoid. They were able to separate perturbations by the atmosphere from those caused by Earth's shape, because the former were a-periodic and the latter periodic.[463]

Establishing Institutional Identity

The SAO's press office was already issuing a rapid and continuous flow of "special reports," keeping track of the new satellites. These reports had to be reviewed by Carmichael's office, which became perfunctory for technical works. But Carmichael did insist that popular articles, like one Whipple sent to the *Saturday Evening Post* titled "Eyes in Space," carry only his Smithsonian affiliation, and not Harvard's. Whipple coyly resisted, reminding Carmichael that "only a few months ago . . . you were asking me to avoid mention of the Smithsonian with regard to my public relations."[464]

Carmichael insisted on a clear institutional identity for the work of the Astrophysical Observatory. He was reacting in part to the recently commissioned management study by the firm of Bigelow, Kent, Willard & Company of Boston. Its "Administrative Analysis of the Optical Satellite Tracking Program" evaluated organizational structure and administrative methods at the SAO as well as in Washington.[465]

The so-called Willard Report recognized that Sputnik had accelerated what had been a "carefully calculated timetable" into a "crash" program.[466] It celebrated the "extraordinary enthusiasm" of the staff, which deserved "a great deal of credit for the work accomplished to date." It put a positive spin on the Baker-Nunn history, happy that by February 1958 at least four of them had passed muster. While it lauded the technical side of the enterprise, it singled out what was "a potentially unstable administrative situation" owing to the crash nature of the program that threatened the long-term health of the SAO. Identifying only "external forces beyond the control of the Observatory" as the cause for this instability, the report recommended, first and foremost, that "lines of responsibility, authority, and communication between Washington and Cambridge . . . be re-defined and clearly spelled out in writing."[467] Next, the study recommended that the observatory establish long-term goals and an organizational structure to meet these goals. Finally, operating procedures in finance and communications needed overhaul, and had to be placed in writing, giving the director and his "immediate Executive staff . . . greater, more positive, and more accurate control of the Observatory's operations."[468]

Whipple resisted any suggestion of more shared authority with the central Smithsonian. However, deeper in the details, he found bright spots. The Willard Report identified multiple and redundant levels of review, and some were even conflicting. It just made no sense that the administrative officer at SAO had to approve "all matters passing between Cambridge and Washington, in either direction."[469] It made more sense to change this office to "special assistant, Washington Affairs" to focus "on all matters requiring a special knowledge of Federal or Smithsonian regulations and policies." A newly defined executive officer and assistant, reporting to Whipple, would decide which requests needed attention from Washington, and which could be handled internally in Cambridge. The trivial matters that had clogged operations, such as travel orders and approvals to make long-distance phone calls, would be at the discretion of decision makers and budget staff in Cambridge. Local authority would suffice for hirings of clerical and technical staff and minor contracting, depending on which funding lines were being tapped.[470]

Willard analysts gathered evidence from staff testimony. Gerhard F. Schilling, recently hired away from the National Academy's IGY staff as executive assistant, felt that Whipple was trying to formalize institutional structure.[471] But others testified that there was more structure inferred on the organization charts than actually existed.[472] The Satellite Tracking Program did have an executive officer, but the position was vacant during the review. The Willard Report recommended that the SAO director report directly to the office of the Smithsonian secretary, and that the rest of the organization report to Whipple through an executive officer (EO). The EO would also direct a full and relatively autonomous "Administrative Service Section," empowered to perform triage without burdening the director's office. Another equally significant change was to separate the Satellite Tracking Program from what would now be called the Upper Atmosphere Studies division. But both would report to the same associate director, Hynek, and Hynek would now enjoy the backup of a second executive officer to manage the tracking program itself. There would also be administrative assistants in each of the tracking subdivisions.

Willard analysts also noted the unending flood of public inquiries, from the media, from individuals, and especially from the members of Moonwatch. Of course, they were making the study at the worst possible time, the very month that "1958 Alpha," or Explorer I, was in orbit after launch on 31 January. They calculated that the time staff spent taking phone calls alone amounted to 1.5 labor years. Concerned that this necessary activity, if not managed, would lead to an overworked and irritated staff that might dispense information improperly, the report recommended steps that would prevent damaging "the reputation of the Observatory."[473] Those steps meant: more staff.

But the Willard Report also recommended that Whipple delegate more and limit his number of direct reports, making lines of authority clearer. It found that job descriptions did not accurately reflect roles and responsibilities and how each impacted others above and below. Supervisory ground rules were too loose and ill defined. Staff members were certainly enthusiastic for the job at hand, but Willard analysts sensed that "this enthusiasm is due largely to the nature of the work rather than the surrounding conditions under which the work is performed."[474] This was a very unstable condition and made long-range planning and stability unlikely. In the long run, it would lead to "ineffectual performance and the gradual deterioration of organizational morale."[475]

Most critically, "operating methods and procedures" in finance needed to be "completely re-worked." Accounting and bookkeeping reviews were too infrequent, and there were too many cases in which the numbers differed between Cambridge and Washington. The Willard Report observed darkly that it was "quite apparent that this lack of control has created doubts at the Washington level regarding the total of administrative capabilities in Cambridge." Even with these "doubts," however, the Willard Report did not condemn, as Keddy had urged, "an entire barrel on the basis of a few bad apples." There was certainly a lack of control, but applying the term *poor administrators* to the entire SAO executive hierarchy was "wholly *un*justified."[476]

The Willard Report did not try to rationalize how, why, or where the observatory existed, as a bureau of the Smithsonian or as a department of the Harvard College Observatory. It did recognize that the service and science functions of the institution had to be administered separately, or at least distinctly, and reported independently. Indeed, the National Science Foundation (NSF) regarded support for tracking as a service function and the upper atmosphere work as a research function.[477] It also did not examine relations with Harvard and how that relationship factored in the growth of the institution. Whipple knew he had to walk a delicate line between the institutions he served. The incident over institutional identity with his *Saturday Evening Post* essay was symptomatic. So was his address to Harvard alumni in Boston in late December 1957, noted in the Introduction to this book. There were, indeed, complementarities between Harvard and Smithsonian interests, but there were also discordances, not the least what Harvard would do with an SAO permanently in residence, constantly redefining and extending the boundaries of scientific opportunism stimulated by Sputnik.

The End of the IGY, Transitioning to NASA Support, and the Rise of James Bradley

Sputnik created pressures for some form of "national space establishment," which led ultimately to a bill signed by Eisenhower in July and the creation of NASA in October 1958.[478] As NASA grew by acquiring older institutions in military and government spheres, to obtain the wide range of talents and properties required to execute a national space program, there was a question about what role academic institutions would play, to say nothing of what was still, technically, a quasi-governmental entity like the SAO.

During the last months of 1957 and through spring 1958, stakeholders followed the deliberations over what would become NASA. On various occasions and in various capacities, Whipple was asked to state his case for, first, what this national space establishment should look like (see the next chapter), and then, with NASA established, argue for the SAO's retention of optical satellite-tracking functions. He did the latter successfully, but not the former. Showing political savvy, he finally strongly endorsed the bill that would authorize the creation of NASA. Carmichael relayed Whipple's endorsement to Congress not only in his capacity of secretary, but also as vice chairman of the National Advisory Committee for Aeronautics, the foundation NASA would be built on.[479]

In December 1958, Odishaw informed Carmichael that additional funds Smithsonian had requested for the coming years would be partly provided by the NSF and the balance by the new NASA.[480] NASA thereafter assumed fiscal responsibility from the NSF for the Smithsonian's satellite-tracking programs. Moonwatch remained active, though the initial excitement waned along with the workforce. Those who continued through the 1960s substantially improved their techniques, precision, and productivity.[481] The Baker-Nunn stations, in contrast, increased their staffing and constantly improved instrumentation in the NASA years. Initially, with the end of the IGY, there was "a sense of insecurity among the personnel" at the SAO stations over the future of the program.[482] Whipple darkly appreciated that the emergence of a new federal institution meant a renegotiation of territory and a juggling of power relations.

Publically, Whipple never gave any hints that the program would not continue. Throughout 1957 and well into 1958, as he and Menzel constantly plied Harvard alumni to fund a dedicated administration and laboratory building as a visible laboratory for space research, Whipple highlighted the SAO's satellite tracking, so critical in the ongoing competition between the United States and the USSR.[483] The federal government would support programming and staff, but he hoped Harvard alumni and friends would build the facilities that would house them. Whipple envisioned brick and mortar to house activities well beyond tracking, including the development of a centralized capability for designing, building, and testing Earth satellites, and then operating them once in orbit. Case in point was the space telescope he, Hynek, Davis, and others on the staff were then proposing directly to Army Ordnance, prior to NASA's emergence: what would become Project Celescope at NASA.

There were indeed transitional glitches, and so Carmichael's newly designated assistant secretary for management, James Bradley (Figure 23), soon became critical in helping Whipple through the process. Bradley distinguished himself as an unusually adept and capable administrator. He had been a civil engineer, had valuable experience and insights working in the Bureau of the Budget, and had spent six years as administrative assistant to the assistant secretary of the interior. He proactively kept Whipple informed of all actions, assuring him that the Smithsonian was positively engaging the new NASA administration.[484]

Figure 23 Leonard Carmichael and Assistant Secretary James C. Bradley, ca. 1959. *Smithsonian Institution Archives, Image # 84-13887.*

Bradley quickly won Whipple's confidence, making it clear that he appreciated the fact that even if the tracking waned someday, there was still the question of continuing support for "post-observational projects," namely, reduction and analysis.[485] For the present, however, tracking operations dominated, and keeping them operational and having new staff and new equipment for faster and more efficient on-site reduction and analysis were both paramount.[486]

Bradley also helped to stabilize relations with the press, and, most of all, with Congress. Hynek got into hot water again after being quoted by the *Washington Post* in early March 1959, saying that if the Air Force had been more forthcoming about a recently launched Discoverer satellite it might have been tracked optically by the Baker-Nunn network. Discoverer had been in the news because its radio signals were not immediately detected after launch and when they were, they were erratic, caused most likely by tumbling. Hynek's remarks provided fuel for the reporter to call for a "unified program" in tracking these things, even if they were classified programs, as was Discoverer. "The reasons for suspicion and misgivings are many," the reporter went on, stirring the pot, given that "certain space projects dealing with communications, navigation and weather had been assigned to the military" but now parallel responsibilities were being considered for NASA. This could have led to a "serious and crippling splintering of the American space program," the reporter claimed, and it had to be stopped before the "vested interests" of NASA and the Pentagon became any more entrenched.[487]

Bradley, acting to reduce confusion within the Smithsonian, advised Whipple that it was not wise to criticize another government agency and then equally quickly assured

Carmichael that "Dr. Whipple was quite understanding of your viewpoint and agreed with it readily." He confirmed that Hynek's statements were factual; the Air Force's SPACETRACK had not provided prior information on the launch, the expected orbit, or initial tracking. Still, Bradley also assured Carmichael, Whipple appreciated that Hynek should not have spoken to the press in this manner, and also added that "all concerned [at the SAO] had learned a lesson from this incident." Bradley also reduced tensions by having Hynek assure the House Committee on Science and Astronautics on 12 March that it was all only a simple misunderstanding.[488]

Unlike Keddy or Graf, Bradley made real headway rectifying the administrative problems Whipple faced. He convinced Whipple of their practical necessity, assured him that it would not weaken his autonomy, and then provided Smithsonian slots for SAO administrative personnel. Bradley was Whipple's strongest ally in Washington, helping him, Whipple recalled, to "get the basic support out of the congressional budget instead of out of the grants." Whipple had been working successfully, using what he termed his "brinksmanship principle" wherein "we would finance the entire administration and support activities through the grants and hire the scientists and their support direct on the congressional appropriation." And as grants got more competitive and money scarcer, Bradley knew how to make the Smithsonian more attractive for funding, helping Whipple secure access to more federal slots. With Bradley's help, Whipple found he could get better people if he offered permanent civil service positions rather than finite contracts. So the relationship really started to click by the early 1960s.[489]

Bradley also succeeded in aligning the observatory's management procedures ("personnel, supply, fiscal, accounting, auditing, contract negotiating, overhead") with those of the central Smithsonian. He had managers from Washington train the staff in Cambridge, and then reversed the seminar, inviting Cambridge staff to be in residence in Washington. He promoted this "interchange on the administrative side in order to bring the fledgling organization into an understanding of Federal processes."[490]

It was indeed fortunate that Bradley appeared just as the animosity between Keddy and Whipple was peaking and further when the management consultants delivered their report in April 1959. Feeling frustrated and abused by the intense scrutiny of his management style from Carmichael's staff, Whipple drafted a letter in which he resigned his post. It was never sent, but hints at Whipple's state of mind.[491] There was further relief when Keddy, 69 years old, announced his planned retirement from the Smithsonian in 1960. Bradley replaced him.

Whipple's first full-time business manager was Carlton "Carl" W. Tillinghast, an MIT graduate in nuclear engineering who had worked at Pratt and Whitney, served in the Navy, and spent two years as a research engineer at the Mitre Corporation in project management. He came to the SAO in 1959 to administer the computational division of some 30 staff, and within the year Whipple saw his talent and made him assistant director for management. As Richard McCrosky happily recalled, this was a most rational decision: "Anyone who treated me the way Fred treated me certainly needed a business manager."[492] McCrosky was happy enough being left alone, but knew that some management, especially accountability, was required.

Whipple's hesitation to hire purely management personnel in the early years was, he later claimed, partly a problem of achieving adequate support. But his resistance went deeper. When R. W. Porter had pestered him to find a management consultant for the Baker-Nunn fiasco, Whipple marked the passage of his letter with five dark lines and a question mark, scribbling in the margin, "How does he get that way?"[493] Similarly, calls from Keddy and Carmichael for establishing a management structure did not bring with them funding or slots at first. As Whipple recalled, "The federal funds paid practically nothing for administration. I always got that out of grants and contracts."[494] Increased support for management, first from Odishaw, and then from Bradley, slowly changed this situation, and once NASA was on the scene, project and program management dominated, but they were paid for.

The First Years of the Satellite-Tracking Program: "Big Science"

In a letter to his staff on the first anniversary of Sputnik, Hynek recalled the saga of 4 October in dramatic detail. On that fateful evening, Harvard and Smithsonian staff rushed back to their offices and pitched in selflessly. Charles Whitney, Mac McCrosky, Ted Sterne, Gerhard Schilling, Luigi Jacchia, and others joined Hynek and his staff to execute contingency plans created months prior by Henize and Leon Campbell. Babette Whipple and other wives supplied coffee and sandwiches over the next several days, as John White commandeered offices, telephones, and staff to create a press room capable of handling the reporters who were knocking on the SAO's door. C. M. "Chief" Peterson built a "world-wide nerve center focused in one room on the second floor of Kittredge" across the street, and led a staff that since then had handled, by October 1958, some 5 million words of text reporting on the data from some 535 Baker-Nunn photographs guided by 6,784 reports from Moonwatch teams. "Those were stormy days on the technical sea," Hynek recalled. "Both the ship and crew came through—but not unscathed." Hynek rejoiced that what started out as "cut and try . . . was now perfectly routine" and ended his pitch with a promise: "If it moves we'll track it"[495] (Figure 24).

Within the year, Hynek's staff had grown from 37 people to more than 100. It's the growth of the institution because of the satellite tracking that bears notice here, and how the "crash" program mentality of the first years eventually stabilized into a long-term operational view. Whipple was creating new lines of investigation or reviving older lines of interest constantly. James Bradley described the growth in February 1959, poignantly reminding a NASA Procurement and Supply Division official that they were covering "the reduction and analysis of low velocity meteors, whose velocities approach those of man-made missiles." He also outlined the project to build an astronomical space telescope and Hynek's study of stellar scintillation from piloted and robotic balloons, which involved the "activation of the 'Cat Eye' project for the detection, tracking, and recording, by electronic means, of celestial light sources."[496]

Figure 24 Janice Turner and Chief (Charles M.) Peterson in the satellite-tracking communications center, ca. 1958. *Smithsonian Institution Archives, Acc. 16-263, Box 8.*

By the end of 1958, there were 129 people employed at the SAO, still variously called the Smithsonian Institution Astrophysical Observatory but more and more frequently the Smithsonian Astrophysical Observatory. There were 16 "research professionals," 27 "supporting professionals," 38 "technical aides," and 48 in "administration."[497] By far the majority of staff were either employed directly in satellite tracking or were involved in areas (upper atmosphere research, meteoritics, geodesy) that depended on the data flow.

In October, Whipple revived his campaign to establish a distinct, permanent physical space for his empire. He appealed to Dean Bundy, describing the "symbiotic relationship" that had developed between Harvard and the Smithsonian. That relationship had now matured sufficiently, he claimed, to "visualize the course of the future with some certainty." Invoking Joseph Henry, Whipple asserted that the Smithsonian initiated new lines of research and developed them to the point they were operational, but then turned them over to others rather than "be caught in the trap of continuing work in a scientific area once its basic nature had been well explored." Ignoring the fact that the 60 years of Abbot's APO hardly fit this model, Whipple stated only that "changes in the practice of research [that] have occurred in the last few decades" had extended the scope of the enterprise such that the "structure of the fully completed programs may, literally, encompass the globe, particularly when the research concerns astronomy, geophysics and the space sciences."[498] The message Whipple drove home was that

> research projects on this more massive scale now [seem] the best possible approach to finding the answers to the new problems posed by modern science, because the methodology requires the combined help of so many diverse science and techniques. Thus a large organization is now necessary to provide the modern equivalent of the small laboratory only a decade or two ago.[499]

This was Whipple's blueprint for the SAO, clearly an expression of what historians call "Big Science."[500] Whipple's vision included many complex institutional phenomena such as the scale of the enterprise, the importance of coordinated goals, and the control of the research that is done. In Peter Galison's words, the "'big' in big science connotes expansion on many axes": geographic, economic, multi-disciplinary, and multi-national.[501] Nowhere were these facets more apparent than in the satellite-tracking program, nicely illustrated by an SAO staffer's attempt to illustrate the connectedness of multiple tracking stations with satellites overhead with drinking straws (Figure 25).

Indeed, the new SAO was made possible by a new kind of service function that aligned with national needs and priorities. This new means of life support was very much welcomed by both institutions. In November 1958, Menzel took pains to explain to Bundy just how the Smithsonian has "come to our rescue," providing staffing for a wide range of Harvard functions from research to teaching to administrative infrastructure.[502] The Smithsonian, and satellite tracking, were, in Menzel's mind, at Harvard to stay.

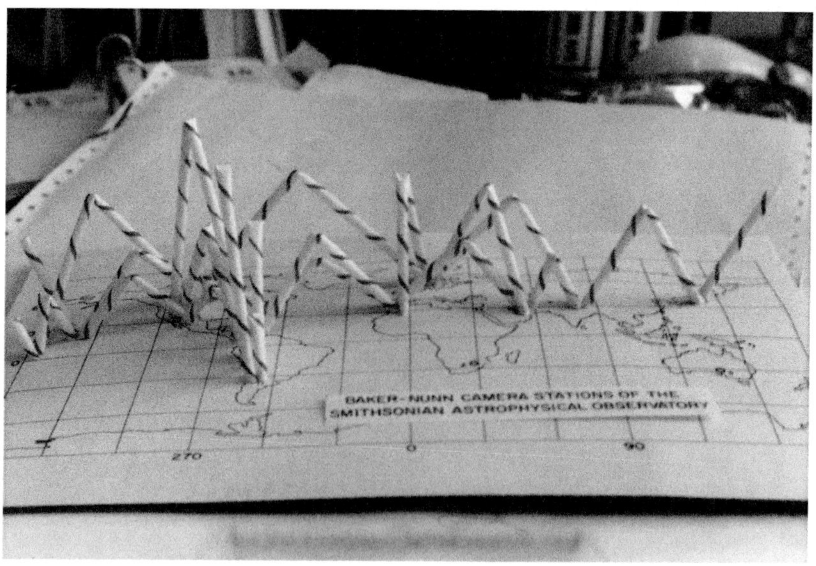

Figure 25 SAO staffers used bent drinking straws to visualize how satellites could be simultaneously "seen" by two or more tracking stations on a Mercator map of the world. *Smithsonian Institution Archives, Acc. 16-263, Box 2.*

For instance, when the first NASA administrator, T. Keith Glennan, faced the House Committee on Science and Astronautics in 1960, he singled out satellite tracking as a bright spot in NASA's efforts toward international cooperation, because "agreements with several nations have been negotiated covering the installation, manning, and use of tracking and data acquisition equipment." He celebrated the fact that "our policy of frankness and our adherence to the tradition and well-understood policy of prompt disclosure of scientific results is building good will throughout the world."[503] The satellite-tracking program continued throughout Whipple's tenure in the 1960s and early 1970s, bringing Whipple notoriety and personal acclaim (Figure 26). Although Hynek had referred to it happily as "perfectly routine" in October 1958, he did so only to convince the Bureau of the Budget and other critics that the Smithsonian was well organized. However, as the techniques stabilized, they were also constantly refined, just as the optical network was constantly repaired and upgraded with the addition of laser-based tracking in the late 1960s, which kept the program effective and competitive. Moreover, once stabilized, the tracking did far more

Figure 26 The family gathers at the White House in June 1963 to celebrate Fred Whipple's "Distinguished Federal Civilian Award." (*L to R*) Earle Whipple, Carmichael, Whipple, Laura Whipple, Babette Whipple, Harry Lawrence Whipple, Sandra Whipple, Celestia Whipple. *Smithsonian Institution Archives, Acc. 16-263, Box 8.*

than provide improved ephemerides. Smithsonian staff would soon employ the data to improve geodetic knowledge, laboring to establish, in true Smithsonian fashion, what came to be known as the "Smithsonian Standard Earth." The orbital data were also employed to improve knowledge of the upper atmosphere, again leading to another large SAO initiative to upgrade and extend the "standard atmosphere."

In the 1960s, these two areas remained very competitive and institutionally profitable, given that they were both critical to ballistic missile operations. But SAO was periodically challenged by other organizations eager to assume larger roles in this lucrative service function. There were also concerns over the classification of this type of knowledge, and by the end of the decade, concerns over relevancy of such functions for what was an astrophysical observatory. Satellite tracking stood alone as a purely Smithsonian activity in what was otherwise a growing effort to combine interests and expertise in a stronger Harvard-Smithsonian relationship. Throughout Whipple's tenure, satellite tracking was central to the continued relationship between Harvard and the Smithsonian, given that its overhead helped to sustain much of the operations of the two observatories. This was not an explicit issue until Menzel retired and was replaced by Leo Goldberg as director in 1964. But before we cover that transition, we turn to the parallel paths Whipple chose for building his institution.

Chapter 7

Organizing for Space Research[504]

On 4 February 1958, just days after Explorer I was launched and declared fully operational, Whipple excitedly called a group of Harvard and Smithsonian astronomers into the classroom in Building A of the Harvard College Observatory for an all-day meeting to explore the Smithsonian Astrophysical Observatory's (SAO) further involvement with satellites.[505] The SAO was already heavily involved, of course; its networks of Moonwatch teams were valiantly catching Explorer and the Sputniks in flight and feeding the information to the SAO's nascent data processing centers. And the SAO's promised optical-tracking network was still embarrassingly months away from full operation. Even so, Whipple wanted a bigger piece of the space pie. The question was, what to do?

What to Do?

Donald Menzel and Cecilia Payne-Gaposchkin joined Whipple and about a dozen SAO senior staffers including Hynek, Sterne, and Schilling, as well as junior staff members such as Robert Davis. Davis was by then completing a thesis in radio astronomy under Bok. Whipple asked all of them to consider how the SAO and Harvard could get involved designing, building, and flying scientific instruments on satellites and how they could organize to track the satellites, receive the data streams, process the data, and publish it as new knowledge. Whipple cast his net wide, hoping to gather in as many ideas as possible and find affinities among his staff. His asked them to form special interest groups and report back in a few weeks.

The sense of the meeting, as Davis recorded, was that "we are looking forward to a real space platform."[506] Whipple's burgeoning SAO staff had expertise in geodesy, as well as in meteoritics, cosmic rays, the outer atmosphere, and meteorology. But the Harvard astronomers gathered by Whipple were also interested in the Sun and stars, as well as galactic structure. They generally agreed to stay out of the competitive fields of meteorology and space medicine. They were certainly strong in geodesy, but Whipple urged them to think further. He drew tables on the blackboard organizing possible problem areas into categories where their combined talents possessed either "a certain degree of proficiency"; some proficiency; or, most important, where "we have an operational advantage as of the present moment."[507]

The group agreed that they would start with a "small space platform" and then work toward a "big space platform." They knew that the Russians had already orbited a 1 ton satellite "and more will undoubtedly follow quickly from both the U.S. and Russia." But they sensed that it would be years before anything large could be done, so they should plan for a step-wise developmental program that included the creation of a university-based infrastructure for the payloads and even the satellites, fed by government contracts. Among many suggestions, Davis recorded that "we will concentrate most on the space telescope, since the people with the most proficiency in this field are astronomers: it is the field that is most apt to be neglected if the astronomers keep quiet."[508] Sometime after the meeting, however, Payne-Gaposchkin cautioned everyone that "a Telescope in Space" was not "effectively related to the program of astronomy" at Harvard, although, she added emphatically, it is a "scientific must."[509]

The group organized into teams to explore some 13 overlapping problem areas, and reconvened on 25 February. The largest team, Payne-Gaposchkin, Whitney, Hynek, and Davis, pondered the space telescope. Others focused on spectroscopy (Payne-Gaposchkin, Hynek); corpuscular and cosmic radiation (Fireman and Krook); solar, stellar, and planetary atmospheres (Menzel); the Moon (Schilling and Rinehart); magnetic fields (Krook, Schilling, and Layzer); space navigation (Sterne); image tubes (Hynek); and meteoritics (Rinehart).[510] Davis once again was the recorder and most of the original group appeared, as well as a few new onlookers, like Tom Gold. They all were given time to speak and two studies were circulated prior to the meeting: "Studies of Lunar Planetology" and "A Satellite-Based Astronomical Telescope."

They debated all aspects of a satellite system—the housing, stabilization, telemetry, and ground stations—and decided that TV techniques were essential for pointing, setting on the object, and focusing. They were well aware that Martin Schwarzschild of Princeton was "building a television system for his future balloon flights, [even though] he has a much simpler recovery problem." Davis added, echoing Hynek, "that we should put all our eggs into the television basket," adding that "present experimentation with photoconductive image tubes looks very hopeful." No one on the staff had any particular history or expertise with TV systems, however, but Davis knew that Whipple, and especially Hynek, had been aware of the decade-long effort by the Air Force's Aeronautical Research Laboratory's Project Cat Eye and the progress they had made creating a system that could amplify light "over twenty billion times" as well as suppress sky brightness.[511] Hynek agreed to manage the project, leaving the detail work for Davis.

Cat Eye was the stimulus that convinced Whipple, Hynek, and Davis to look closely at image-tube technology. The German scientist Radames K. H. Gebel, who worked at Zeiss-Ikon during the war on optical control systems, was brought to the United States under Project Paperclip, and by the early 1950s was leading research projects at the Wright-Patterson Air Force Base under this code name. He was a passionate promoter of image-tube technology who was claiming some astounding successes.[512] In November 1956, Gebel's superiors had contacted Whipple about using the device for satellite tracking. The beauty of the system was that it could suppress sky brightness and so promised the ability to see objects during the day as well as at night.[513] Whipple and Hynek joined Major David G. Simons, chief of the Air Force Space Biology Branch of the Aero Medical Field Laboratory at Holloman AFB, to find ways to apply Cat Eye. Simons was a strong advocate for balloon platforms as staging grounds to train humans for spaceflight. In November 1957, they met in Cambridge to discuss plans to make balloon-borne observations of stars and together they hatched a scheme to employ an f/4, 8-inch, off-axis paraboloid to feed a Cat Eye system to perform, in Simons's words "precision observations from space type vehicles." The tests would be done initially by a manned balloonsonde system flown up to 30,000 feet. The Cat Eye would be pointed to bright stars by the balloonist, and would provide sufficient sensitivity to track scintillation patterns in stars as well as suppress sky background.[514]

Hynek obtained a version of the Cat Eye image-orthicon made by General Electric that used RCA electronics. He attached it to a 10-inch Baker cinetheodolite mounted at one of the Smithsonian's stations at Organ Pass, New Mexico. In his tests, he found that the purported amplification factors were wild overestimates; what had appeared to be a 500 times increase in sensitivity over standard Kodak 103aO emulsions was due to "the phenomenon of memory" affecting the image orthicon. It was not true latent information accumulated over time, and so "real results" were estimated to be about 30 times, and useful amplification only in the range of 10 times. Still, this was promising; Hynek and his co-authors, including William A. Baum from Mount Wilson, were confident they could achieve at least a 100-fold increase in amplification.[515] Their optimism for the new radical technology led Whipple and his staff to adopt an image tube for their space telescope, what became "Celescope."

By spring 1958, SAO staff coalesced around a small space telescope for measuring the ultraviolet brightnesses of stars. It was similar to Hynek's payload for the Simons' balloon. In June Whipple wrote to Hugh Dryden at the National Advisory Committee on Aeronautics (NACA) outlining their plan, and Dryden expressed interest. By 1 July, after contact with William H. Pickering at the Jet Propulsion Laboratory (JPL), Whipple and Davis sent a formal proposal to the ABMA to support a feasibility study for a small orbiting telescope that would secure, in the simplest way possible, ultraviolet brightnesses of bright stars. Their goal was to extend knowledge of stellar spectral energy distributions to better calibrate bolometric data acquired from ground-based telescopes.

Why an Ultraviolet Map? Why Robert Davis?

This was a useful goal, but it was not a part of Whipple's primary research agenda apart from his interest in spaceflight. However, he had long been thinking about what might best be done with first-generation satellites. In 1956 he had asked Davis to construct a map of the sky in the ultraviolet, specifically at 1,249 ångströms, near Lyman Alpha. Assuming stars behaved like ideal "black bodies" according to Planck's energy distribution law, Davis calculated what the brightnesses of selected stars should be, trying to account for interstellar absorption and reddening. He generated a list of 218 bright stars he hoped "should prove valuable in the selection of the most interesting regions for study when extraterrestrial photometry of high resolution, either from rockets or from artificial satellites, becomes possible."[516]

Davis was already familiar with photographic photometry techniques from Harvard classes and also had extensive practical training and experience in electronics from his military work, securing top secret clearances as a first lieutenant. Two tours on the USS *Atka* gave him supervisory experience in communications, electronics, and cryptanalysis.[517] His study of the ultraviolet sky was a typical assignment during his graduate years, along with a stint working for Hynek from September 1957 through June 1958 on the Cat Eye project. This last assignment introduced him to television systems generally and how they might provide improvements over "conventional methods of astronomical measurement."[518] His thesis for Bok in radio astronomy was also a mapping project.[519]

Davis realized that Whipple was preparing for a satellite mission, not for the solution of a specific question about the nature of the stars. While working on the Baker-Nunn project, Davis recalled frequent conversations with Whipple and others about what could be done with a telescope in space. But nothing formal took place until the February 1958 meeting.[520] Given Davis's experience, it is not surprising that Whipple turned to him in early 1958 to manage the details of what was to become Project Celescope. But what sort of world would Davis be competing in?

Organizing for Space Research I: Collegial Competition

After their February deliberations, Whipple prepared a position paper, "Notes on the Development and Operation of an Astronomical Telescope in a Satellite Orbit," which in June he distributed in confidence to trusted professional colleagues. He described the feasibility stages, then development and procurement, and finally operation and data analysis, each with a rough timeline. An 8-inch, off-axis, reflecting telescope would employ a television-type device as an imaging detector with various dispersive elements to create spectra. He hoped to deliver the instrument in configurations for direct imaging and spectroscopy ready for multiple flights starting in less than two years. After breezing through the engineering aspects of the problem, Whipple focused on "coordination and management problems," in which he highlighted the fact that the SAO and Harvard were "well equipped" because of the high concentration of "highly able technical organizations in greater Boston." He said nothing about his planned appeal to the ABMA later that month, ending his "Note" only by saying, "All we need now is the 'green light.'" He sent a copy of his paper to Princeton astronomer Martin Schwarzschild, who passed it on to his observatory director, Lyman Spitzer, with the understated comment, "For your perusal."[521]

Whipple's declaration of interest was prescient. After the National Academy established the Space Science Board in spring 1958, in a 3 July letter the chairman of the board, Lloyd Berkner, appealed to hundreds of scientists to get involved in the space program by proposing problems to solve with instruments aboard spacecraft. Payloads would be in the range of 100 pounds, and payload space could be shared by smaller experiments. He needed answers within a week and payloads for testing by mid-1959. He asked each respondent to list specifics, including purpose, weights, and total costs, noting that time was very tight.[522]

Among the some 200 scientists and institutions Berkner queried, about 100 replied: 30 from astronomers for solar and stellar UV studies and at least 5 were for stellar UV surveys.[523] By September, Berkner reported to Menzel that the "response was voluminous." He had found "laboratories and groups where suitable hardware can be developed to carry on the more sophisticated experiments." So as a result, he felt, "the scientific aspects of the space program are not forced into a single government bureau for lack of competent facilities and men outside the government."[524]

Berkner's efforts confirmed that the SAO was not alone in its interests and capabilities and that there was competition. They had chosen the space telescope option at their first and second meetings in February because they thought if no astronomers spoke up, they'd lose out. Now they knew that others, such as Arthur D. Code at Wisconsin, Lyman Spitzer of Princeton, and Leo Goldberg at Michigan, all very well qualified and some more experienced in instrumentation, were also making bids.

Envisioning the Space Telescope

Before Berkner's appeal, Whipple felt confident that the ABMA would approve the SAO's feasibility proposal, partly because they had already approved a large contract for the SAO's services for optical tracking of ABMA missiles and payloads, and could tack this on as a supplement. He was confident enough to respond to an inquiry from the National Science Foundation (NSF), who was obviously interested as well, that they found "an immediate supporter for a preliminary engineering and scientific study and will be able to talk more firmly on the subject in the fall."[525]

The proposal to ABMA was for $43,090 to support a six-month feasibility study that would lead to launching a small satellite aboard an ABMA missile within two years. The SAO would design and construct the instrument, as well as the satellite structure, the telemetry system, its overall operation and control after launch, and the subsequent analysis of the data. The goal of the project was "astrophysical," the proposal claimed, but it might also "yield new and rich material to advance our knowledge in the fields of nuclear physics and studies of the exosphere. Above all, such observations "in this previously unavailable part of the electromagnetic spectrum will also produce totally unforeseeable discoveries of scientific importance."[526]

Whitney and Davis would lead the effort, with Menzel, Payne-Gaposchkin, Hynek, Gold, and Schilling as advisors. Charles Stark Draper, director of MIT's Instrumentation Laboratory, as well as the physicist T. Tamboulian of Cornell, an expert in ultraviolet techniques, would be external advisors. Support was needed for five additional technical staff members: a part-time mathematician and a computer, and full-time design and electronics engineers. Contracted services involved optical and mechanical engineering, computer programming, and computer machine time.[527]

Whipple had drawn together a team that theoretically could provide all phases of the operation of conceiving, designing, building, and operating a satellite in Earth's orbit—everything save for the launch vehicle. He based his model on the highly successful Explorer program, created under the aegis of the ABMA and the Jet Propulsion Laboratory. In late July, Whipple had secured William Pickering's endorsement, which brought JPL into the circle. The JPL director was "responsive and enthusiastic" but worried that Whipple's preferred electronic detectors would be destroyed by the g-forces and vibrations of launch. They would also need pointing accuracy and stability, as well as worldwide radio contact. On the political side, Pickering strongly advised informing Ernst Stuhlinger at ABMA before they contacted Berkner to get access to the Explorer program infrastructure. Politically, JPL dealt with matters relating to civilian "space," whereas ABMA was responsible for "satellites," although he added that this separation of responsibilities may "change depending upon what happens if and when we hit the Moon."[528]

Both Pickering and Whipple knew there would be significant challenges to the proposal. Whipple also appreciated that Davis was already overworked and needed another electronics engineer to relieve him of the detail work. More personnel were needed for almost every phase of the venture, above all, "someone to make intelligent mechanical decisions." Pickering saw ABMA expertise in the systems design as critical, especially because Whipple's proposal included a "Ground Telescope Control Station."[529] Whipple labeled this last requirement "Political"; it would give the Smithsonian a "first rate Observatory . . . to deal solely with telescope satellites." It would promote rapid data acquisition and analysis and would, most of all, "make plan complicated and impressive." Pickering advised Whipple to make the facility look as astronomical as possible, giving astronomical names "for different sections and parts." Accordingly, the plan was to create a ground station consisting of two antennae spaced 30 to 40 miles apart so that the telemetry stream could be used to "compare refraction of radio frequency with optical refraction" as well as yield precise heights and orbital characteristics. Such a facility would be highly versatile: "Handy for any other project that interests." Pickering also opted for existing systems. No doubt elements of the Navy's Minitrack system could be utilized. And because they both believed that the Air Force's Cat Eye had shown promise, possibly the cost of its procurement could be folded into the proposal.[530]

Later in August the team developed a basic cylindrical design for the payload and satellite, considered as a unit. About 10 inches in diameter, larger than the 6.25-inch cylinder for the Explorer payload, its weight and length were still well within Explorer dimensions, although they hadn't started serious thinking yet about ways to point or even stabilize the telescope. The telescope was kept as simple as possible, quite appropriate for a first glimpse of the ultraviolet sky. Davis's calculations from 1956 served as a constraint to the design of the instrument, indicating the level of sensitivity it had to achieve in order to record a representative sample of the stars in the sky.

By the end of the summer, the basic optical system remained the same. They added a set of three selectable filters (lithium fluoride, quartz, calcium fluoride) through which areas of the sky could be recorded and photometrically compared, overlapping to at least 10 seconds of arc resolution. They wanted to experiment with a suite of 8-inch and 6-inch diameter mirrors made variously of fused quartz, Pyrex, and unspecified ceramic. Each of these would be tested with aluminum, beryllium, and platinum reflective coatings. Clearly they knew they were entering unknown territory and wanted to do comparative tests of all possible combinations of materials that had promise for good stability and reflectivity in the vacuum ultraviolet.[531]

The detector was key to the system and they described three options, including an image orthicon with the specifications of a commercial RCA 6849 tube but modified to admit light in the vacuum ultraviolet, replacing its glass enclosure with quartz or lithium fluoride.[532] The second and third design options were not based on commercial tubes, but specified the needed performance specifications. All had to withstand severe launch stresses. One had to be blind to radiation longward of 3,000 ångströms and the other beyond 2,000 ångströms. Testing would reveal which system was optimal. Tamboulian at Cornell would test and evaluate the ultraviolet capabilities of both the optical systems and the transducers.

On the astrophysical data analysis side, Cecilia Payne-Gaposchkin would certainly be capable of the work. But Whipple wanted to build up his own astrophysical group within SAO's confines. He had already secured Charles Whitney (Figure 27), who transitioned to a joint appointment at the Smithsonian in 1958 after being engaged at various times with Whipple's meteor- and satellite-tracking contracts. Whitney, however, had established a strong research program in stellar atmospheres and was interested in attracting other specialists to build up a group, which Whipple facilitated by forming an astrophysical analysis unit for the space telescope project.[533] Whitney's charge went far beyond organizing expertise to analyze star colors and brightnesses. After NASA was on the scene, he had to develop a data flow interface between NASA and the SAO whereby the signals coming from the space telescope could be converted into astronomical data. Just where that interface lay was critical to how the data were to be rendered. As we will find, it would take several years before that interface would become stable.

Figure 27 Charles Whitney. *Smithsonian Institution Archives, Acc. 16-263, Box 8.*

"NASA vs. Smithsonian"

Through the fall of 1958, Davis, along with Schilling, Whitney, and Henize, assumed that the Smithsonian would be setting up a complete infrastructure for managing scientific satellite operations, including ground-based tracking stations and data acquisition. One can see this clearly in Whipple's development of optical tracking, orbit analysis, and computational expertise, as well as in his reports. Acting methodically, in September 1958 he and Davis met with the consultant group Arthur D. Little and senior SAO staff to plot out management strategies for the project. At the same time—and in response to a request from Congress—Whipple prepared a report on what functions the SAO should retain in light of NASA's evolving role in spaceflight.

The political debate over the creation of a civilian space agency has been addressed in numerous historical works, as has the question of which organizations would be responsible for what—from propulsion to payloads to mission definition—and what would remain military and what would become civilian.[534] In early 1958, as Whipple and his staff deliberated over the nature of their space telescope and engaged in entrepreneurial brinksmanship, building programs with ABMA and NSF funds and housing them with bricks and mortar from private funding, he well knew that the Space Act was circulating through Congress. In late March, the Smithsonian's Office of Legislative Affairs sent him the language drafted by the Bureau of the Budget, proposed by the White House and now bound for Congress. Whipple read it, suggesting a few clarifications. He dutifully advised Secretary Carmichael that the bill "represents extraordinary wisdom and foresight."[535] But at the same time the bill put Whipple into high gear to build his empire before it could be defined and constrained by others.

Whipple was among many scientists called to testify before the House Select Committee on Astronautics and Space Exploration in the spring. The committee invited Whipple to testify because he had direct experience dealing with, and participating in, the International Geophysical Year (IGY) infrastructure and broad experience with military contracting. He had been a co-signer of a December 1957 proposal from the Rocket and Satellite Research Panel for a "National Space Establishment" and he could also describe the contracts and agreements the SAO maintained with military and civilian organizations funding their IGY activities.

On 16 April, quite aware that the basic pattern had been already laid out, Whipple concentrated his remarks on the power relationship between the inevitable space agency and academe. He viewed the new entity as a science agency and argued that it had to be given status comparable to the Defense Department. More subtle than Lloyd Berkner, who had earlier heatedly criticized the Eisenhower administration for "failing to have a 'national policy' on the support of science," Whipple wanted to leverage Sputnik as a way to achieve support at this level, equivalent to the military, but not necessarily separate from it.[536] But in tune with Berkner, Whipple knew that while coordination should reside at the national level, control of this support had to reside in academe:

> I think this national space agency should be the coordinating agency for the activities in space research. That does not mean that it should control this activity, but it

should be in a position of coordinating. I do not see any agency that is in that position at the present time.[537]

Space exploration should not be under the control of the military, Whipple added, nor should scientists engaged in space research be "under the control of the Government." Rather, "space exploration should be under civilian scientific control," as the best scientists resided in academe. Coordination, not control, by an informed civilian government agency, letting contracts to universities and industry, and even to the military, was the best solution.[538]

As the House Committee continued to debate the responsibilities of the new space agency over the summer, Carmichael worried that NASA was probably going to want to swallow up satellite tracking. He asked Whipple to supply him with ammunition to argue why the SAO should retain this responsibility.[539] Whipple's first draft, "NASA vs. SMITHSONIAN," argued that the work had to be done in an academic atmosphere devoid of classification, as at a university. He predicted that if the tracking responsibilities were transferred to NASA, the SAO would lose half of its scientific staff. Conversely, he saw no problem with the SAO acting as a contract service "in an academic atmosphere." Whipple insisted here, as he had in his testimony the previous April, that this was the most effective formula for large-scale scientific activity.[540] His final draft, however, concentrated on the strengths of the SAO's precision optical-tracking network and its popular component, Moonwatch.[541] He reviewed its functions, its successes, and most of all the fact that "the administrative, operating, and scientific personnel have gained a singular know-how and a feel for this project."[542] This was Whipple's core argument: Satellite tracking was the critical agent for infrastructure, and keeping it alive at the Smithsonian would keep the Smithsonian itself alive and active, leading a broad program in the space sciences. Whipple went on to describe that program: a full listing of SAO's ongoing projects.[543] He included brief mention of the SAO's ABMA proposal for a space telescope and presented it as an exploratory engineering study of how to do such things, rather than as a mapping project.[544]

The SAO by now had established an integrated systems approach to doing science in space that included problem choice, instrument development, tracking, data acquisition, and analysis. Whipple recognized its importance to astronomy, but his motives were more to build a capability, establishing the SAO as a central player in space research, much as the ABMA and the National Academy's Space Science Board (SSB) were no doubt doing. There was something of a gold rush—or better yet, a 'land grab'—going on, and although NASA was no doubt going to emerge as the central player, the scope of its mandate, and especially if it was to be a coordinating or controlling agency, was still up for grabs.

NASA Assumes Control

Between July and October 1958, NASA pushed to establish responsibility for, and control over, many of the functions Whipple envisioned as central to the SAO proposal.[545] The process was slow and painful, taking some six years "to achieve a coordinated, mutually agreeable program."[546] And Whipple was in the center of the fray, fighting to maintain a working relationship with NASA through Celescope and satellite tracking. Astronomers had just barely come to appreciate the need for national planning as they slowly adjusted to a dominant federal patron. The NSF had finally gotten astronomers to accept the concept of national observatories, an acceptance still far from universal. Now there was a new player on the block.

The SAO retained its optical-tracking network at first, as the new NASA infrastructure absorbed centers for space activities, such as the Jet Propulsion Laboratory and portions of the Naval Research Laboratory, and one year later, in fall 1959, portions of the ABMA. In view of this situation, the SAO recast its ABMA proposal for a space telescope and presented it to NASA, hoping that SAO would still retain as many functions as possible.

Among many helpful adjustments, Whipple's former SAO executive officer, Gerhard Schilling, was now Chief, Astronomy and Astrophysics Programs within Homer Newell's Office of Space Flight Development at NASA. In January 1959, Schilling and Newell asked the leading scientists who had answered Berkner's call and proposed ultraviolet astronomical instruments to meet with the SSB to deliberate.[547]

The Orbiting Astronomical Observatory

Astronomers had sent Berkner some 30 proposals for all sorts of observations in solar and stellar astronomy. Whipple, Spitzer, and Code all wanted to conduct ultraviolet stellar surveys with space telescopes. Goldberg proposed a suite of solar instruments, Menzel proposed a system of balloon-borne telescopes, and the NRL envisioned a series of ultraviolet scans of the interstellar realm.[548]

They met in February to learn that the sentiments expressed the previous summer by Porter and others on the SSB were still at play: The country would launch a few larger payloads at first because of the lack of launch vehicles and tracking limitations. This fit Abe Silverstein's vision, NASA's director of Space Flight Programs and former NACA leader, to provide a large generic platform for a variety of purposes. As the scientists sent in preliminary updates to their original proposals, however, it was clear that they did not see it that way.[549]

Newell had to get the major proposers to cooperate, knowing that the astronomers were "obviously interested in controlling the project and directing it to meet primarily their own needs."[550] He singled out Whipple: "In the early days of getting going on this activity, Professor Whipple had in a proposal to undertake the management and conduct of the entire job. With this arrangement the proposed satellite would do very well for Professor Whipple's experimental needs"[551] (Figure 28). Newell knew full well who he was reporting to: Keith Glennan, the NASA administrator who viewed scientists as "a bunch of spoiled individuals."[552]

NASA would provide no more than three spacecraft platforms at first to make both solar and stellar observations. These platforms would be managed by a new center based in Greenbelt, Maryland, to be called the Goddard Space Flight Center. Goddard had all

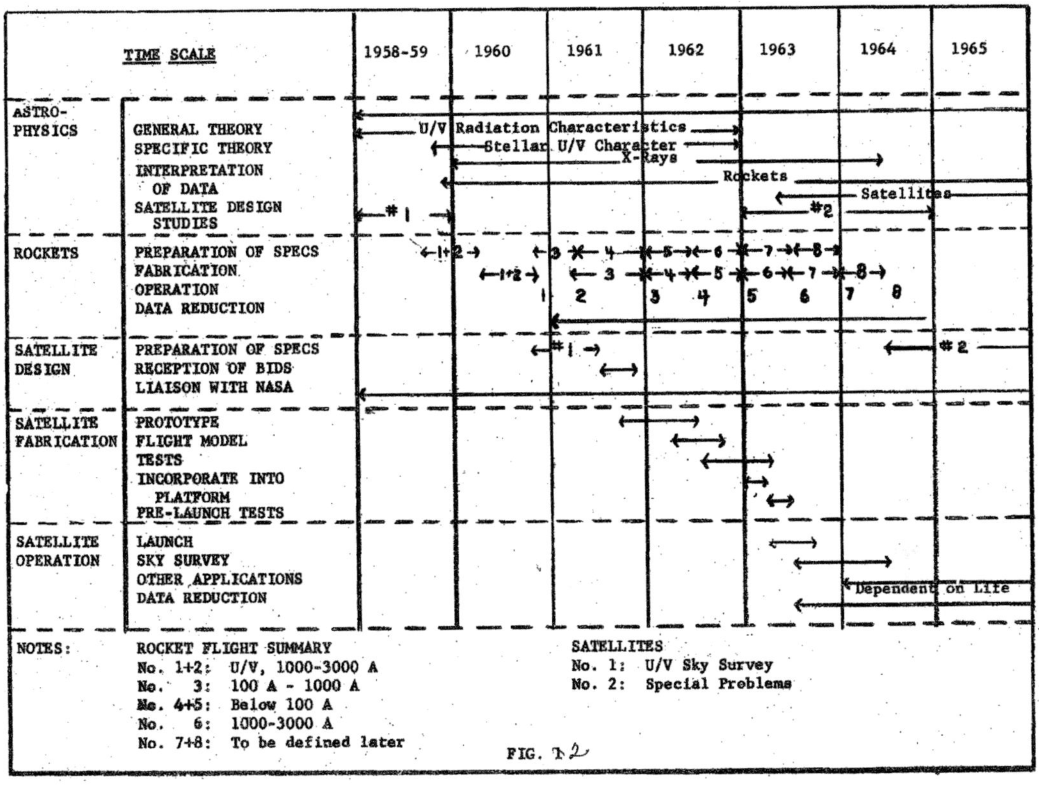

Figure 28 Celescope was Whipple's rationale for building an infrastructure for space science at the SAO, including all aspects of the operation except for the launch vehicles. *Smithsonian Institution Archives, RU 260, Box 8.*

responsibility to manage contractors for a range of missions that involved robotic space science and applications satellites, including the Orbiting Astronomical Observatory series (OAO).[553] The OAO was to be an "all-up" venture; there would be no sequential buildup from simple payloads on sounding rockets and small satellites to the large vehicles NASA envisioned from the outset.[554]

Prior to Newell's meeting, Goddard's James Kupperian, who was to become project scientist and technical officer for OAO, visited the SAO to warn them of NASA's planning. He described a 1,000 pound "space observatory" that would fly by 1961, which would act as a "standard" stable platform for up to six experiments. Kupperian acknowledged that observations of the Sun and Earth required different platform capabilities, but for the moment, NASA offered the single platform. By the time of the general meeting in February at NASA, however, owing to pressure applied by the president's science advisor, George Kistiakowsky, based on appeals from the President's Science Advisory Committee members, Newell had added a fourth three-axis, stabilized orbiting platform, a fully dedicated solar observatory capable of handling a complex of instruments weighing some 500 to 700 pounds.[555] This was consistent thinking at the time, expressed in the Space Act of 1958, getting away from the conservatism of NACA's preferring known technologies to a view of creating "larger, more versatile—and more expensive—tools." In its haste, NASA antagonized some scientists, but gave opportunities to others, like Whipple, who thought big, and were attracted to new technologies and would take risks.[556]

After these meetings, Whipple redoubled his efforts to create a comprehensive development program. He was ready to compromise, but in unison with Spitzer and Goldberg, still resisted the "cookie cutter" model of a universal platform. The astronomers did not share NASA's mission-oriented philosophy—that all sorts of astronomical observations could be performed from a standardized platform. But he also now knew that the original SAO proposal was too narrow in scope. NASA had counter proposed a more sophisticated telescopic array, and that the SAO should leave all the housekeeping to them. Thus, by the spring the SAO came away from the meetings with a battery of four telescopes in a much larger payload. As Davis described it to a possible contractor in March, the package now consisted of "four parallel 8-inch telescopes" that would allow them to look at the sky simultaneously. "We thus have a larger satellite at our disposal than originally planned, and will do in one launching what was originally planned for four."[557] NASA also wanted to create a centralized engineering unit that would collect and process all telemetered data. But reflecting the methodologies astronomers had employed since the discovery and application of the telescope, Davis, echoing Whipple, preferred that "receipt and analysis of data should be controlled by the experimenter."[558] There was movement, but the standoff continued.

In early 1960, after Whipple and Menzel sponsored a naming contest and "Celescope" was selected, the telescope payload had become far more than a single 8-inch reflector.[559] The single SAO telescope was now a battery of four 12-inch telescopes—three photometers and one slitless spectrograph (Figure 29). The three photometers would each be devoted to a single wavelength window as defined by one of the three original filters, and the fourth, the spectrograph, was included partly to fill up the space, just as the Wisconsin team had to do for their suite of payload telescopes.[560] Nevertheless, the three photometric telescopes could now work in tandem, allowing for simultaneous scanning of each region. But this now required a matrixed form of data reduction to manage the calibration drift of the three TV systems.

Through 1960, astronomers continued to press for a rational program, pleading for smaller projects to gain "preliminary experience" in how to build and operate spacecraft instrumentation.[561] But what they got was NASA's largest and most complex scientific spacecraft series.[562] The first OAO was scheduled for launch in early 1963 on an Atlas-Agena rocket, with the Smithsonian Institution sharing space with the University of Wisconsin. The second was reserved for the Goddard Space Flight Center, and the third for Princeton, although each one also carried arrays of smaller instruments.[563] NASA limited the instrument groups to the instrument payloads themselves and appropriate fine guidance systems that would complement NASA's stabilization system. There was pressure on all sides to make each mission payload as large as possible, to think not sequentially, but to think "all up."

Within a year, it was painfully clear that the available electronic detection systems were woefully inadequate. Hynek had been working with Air Force sponsorship since 1958

Figure 29 Celescope did grow over time. The single 8-inch telescope concept was enlarged to four 12-inch reflectors acting in unison. The SAO array would look out one end of the OAO spacecraft (NASA graphic) and the Wisconsin package out the other end. *Smithsonian Institution Archives, Acc. 16-263, Box 1.*

to create the balloon program to develop and test stabilization systems as well as Project Cat Eye. By February 1959, they had developed a "working stabilization system" for small telescopes and were using it to explore both photoelectric and television techniques."[564] But when the Air Force did not continue the program at the end of the year, and NASA acquired responsibility for the platform, stabilization, coarse guidance, telemetry capacity, star field acquisition systems, and ground control functions, the SAO's Cat Eye balloon program died along with its hopes to test systems with sounding rockets.[565]

An Impasse: Acquiring People and Property

Indeed, even as NASA acquired more territory, Davis knew that his challenges were still daunting. In June 1959, Davis had sent Whipple a status report that defined the SAO's limitations. They would focus on stellar astronomy, they would try to purchase rather than develop ultraviolet TV camera tubes, and they would begin hiring staff, especially electronics specialists who would work as contractors at various field sites, including industry and NASA. And they needed more space, for offices and for laboratory space to do the real work of designing, testing, and prototyping.[566]

Obtaining facilities for the latter was another challenge. Davis reported that Steve Sydor, who had been working on the optical instrumentation for the Satellite Tracking Project, felt they needed a dedicated facility for the space telescope. He was ready to donate his own tools, because SAO personnel typically had to travel to other worksites or beg and borrow from current laboratory facilities. Advised by Whitney, Davis agreed that they needed established worksites to monitor contractors for the electronics. They did not worry about the optics because James Baker was available. But astronomers were not conversant with cutting-edge electronics or digital-imaging devices. Whipple was still trying to lure one leading astronomical practitioner of image-tube technology, William A. Baum, plying him with promises.[567] But Baum was just then applying his Carnegie Image Tubes to studying the most distant galaxies with the largest telescope in the world. He could not be convinced to alter course.[568]

The SAO had contracted with Westinghouse in July 1959 to develop an ultraviolet-sensitive television tube based on the vidicon design created by RCA. Called "Uvicons," their development was slow and costly, partly because of the "evident lack of commercial and military applications of these tubes."[569] By June 1960, although some electronics specialists had been secured, there was still no adequate working space, and Davis still felt they did not have enough support, or that no one on the staff was sufficiently adept at image-tube technologies to be able to vet the contractors. Davis reported that this was "the most serious non-financial problem at Celescope" at that time, and something had to be done.[570]

Retaining competent staff and attracting competence in electronics became problem number one by the end of the year. Whipple knew that some of his key people were getting frustrated by the lack of a senior Harvard post and had offers elsewhere. These included Hynek, Sterne, and Rinehart, all of whom were leaving. As he was courting Baum, he learned from one of his staff, who had just returned from an international conference, that although astronomers were highly excited about the promise of space research, few if any were willing to consider changing course to adapt: "Most astronomers steer away from space research unless it bears directly on research in which they are interested."[571]

Whipple knew his institution was woefully inadequate to the job but still made the decision to build a satellite for which they had neither the problem expertise nor the technical infrastructure to execute without expansion. As exciting as the prospects must have seemed in the first months of 1958, through 1959 reality was sinking in. His senior staff was restless, NASA was confining the scope of his endeavors, and the SAO still lacked physical space and, in effect, the visibility and permanency needed to convince new patrons of their promise. In late November 1959, his partner in all this, Donald Menzel, wrote to Leonard Carmichael, laying out the situation, pleading for funds to build a 40,000 square foot addition to the observatory: the conquest of space began at home.[572]

As Menzel petitioned the home front, Whipple took to the national stage, rationalizing the SAO's goals again when the House Committee on Science and Astronautics opened another inquiry in late 1959 into how well government agencies, universities, and private aerospace concerns were being coordinated by NASA and the Department of Defense. Whipple outlined all the ways SAO was pursuing space research—explicitly, the space telescope—noting how it would provide information on "processes in normal and abnormal stellar atmospheres; processes in interstellar space; and abundances of the elements in interstellar space." But he also emphasized that the ultraviolet star catalog itself would be useful for "the practical needs of space navigation."[573] The committee could easily read between the lines here: At the time, and for some time to come, the Air Force would be intensely interested in such information, for the selection and identification of suitable guide stars to calibrate and supplement the inertial guidance systems of intercontinental ballistic missiles.[574] Along with improved geodetic information, and improved knowledge of the resistive characteristics of the high atmosphere, a precise map of the stars as seen in space was of great value to Cold War patrons.

Chapter 8

Permanency and Authority

Leonard Carmichael most likely could not have predicted the transformation the old Astrophysical Observatory (APO) was experiencing at the hands of Fred Whipple and his enabler, Donald Menzel. His hope for the revitalization of solar research at the Smithsonian was loyalty to the legacy of the institution, if not to Abbot, and, moreover, was resonant with his general plan to improve the academic strengths of the institution. He did not know Whipple as Bundy did, the determination and passion of his "dear dream." Nor could he have been aware of the fact that the Smithsonian represented to Whipple the opportunity of endless landscapes on playing fields far larger than Harvard would tolerate at the time. But what Carmichael must have known was that the new arrangement offered Whipple a new dedicated path to congressional appropriations, aided and abetted by government and military contracts.

The hallmark of Cold War science was, indeed, the government contract. Through the late 1950s and into the early 1960s, the Smithsonian Astrophysical Observatory (SAO) secured major contracts from the National Science Foundation (NSF), the Atomic Energy Commission (AEC), the Army Ballistic Missile Agency (ABMA), the Air Force, and then NASA, numbering in the scores.[575] The rate of growth at SAO was such that by June 1961 a weekly *SAO News* started appearing to keep track of new faces among the more than 300 people now employed and to keep everyone informed of the burgeoning scope and constantly changing landscape of the institution. In fact, on the front page of the first issue the lead article proclaimed "Moved Again!"

> Many moves . . . many, many moves . . . so many that maybe some of them have been forgotten . . . after all, most of us are trackers of satellites, not offices.[576]

By then just the offices administering Moonwatch had moved five times. Satellite-tracking facilities were widely scattered; photoreduction moved three times. What was now called Celescope, the space telescope, moved three times, and "Editorial and Publications" moved at least four times. The only stable offices were those of the director and his staff and those within "special studies," in other words: the astronomers, physicists, and chemists engaged in meteoritic analysis, the upper atmosphere, and astrophysical studies. Only in June 1961 did SAO occupy a dedicated physical space, a new building, built expressly as a new "Space Science Building"[577] (Figure 30).

From the beginning of their relationship in 1955, both parties, Harvard and the Smithsonian, acknowledged that office, laboratory, and shop facilities had to be factored in

Figure 30 The new "Space Science Building" opened in June 1961. *Smithsonian Institution Archives, Acc. 16-263, Box 4.*

to the plan. Katherine Haramundanis recalled what it was like after she was hired in 1958. She and her staff were constantly shifted to various rented spaces that were typically not set up for their tasks: Not only was desk space wanting, but the old structure was inadequate for delicate measuring machines. She recalled that someone laid out buckets filled with water to show, from the constant rippling, how unstable the floors were.[578] After she returned from a year's absence in 1961, she found the new quarters much improved, in what was called the "IBM Building" on Broadway in Cambridge. This new environment helped convince her that the SAO was here to stay and was constantly growing.[579]

Permanency

From the beginning, Whipple's and Menzel's appeals, to both Bundy and to Carmichael, highlighted marketable strengths.[580] Whipple described his burgeoning empire as one that could "cross the boundaries" that once separated the disciplines. The SAO would provide an institutional platform for "collaborative efforts of many scientists and technically trained people" from diverse backgrounds. There would be at least a dozen fields in the physical and mathematical sciences that needed to be engaged. These would change over time, depending on funding sources, but they always encompassed many disciplines and required infrastructures beyond what had been normal practice in astronomy. Laying out all their programs, from satellite tracking to meteoritics, to the physical analysis of meteoritic materials, to the space telescope, there was great need for highly specialized equipment and working facilities, such as producing atomic time standards for a wide range of geodetic applications. All this activity required office space, as well as supporting optical, physical and electronics laboratories, a computing center, and specialized machine and optical shops.[581]

Throughout fall 1958, Carmichael and Bundy worked out a plan to rent space in a new building being built by IBM in Cambridge.[582] They needed at least 2,500 square feet to house their IBM 704 computer and office space for its Computation Section of some 20 additional workers. Longer term, the Smithsonian was ready to rent some 40,000 square feet in a new space Harvard would provide.[583] But Carmichael could not commit funds for more than a year in advance, subject to congressional appropriation.

Renting was only a stopgap. Menzel and Whipple kept pushing for new distinctive institutional facilities. Without a permanent facility to point to, they were less competitive for Air Force or NASA funding. By early 1959, they had an agreement and construction plans and an architect's visualization had been completed.[584] The *Crimson* announced in February 1959 that Harvard would indeed build a new building, especially to house yet a larger computing center with an IBM 709. The building was essential, the *Crimson* editorialized, because "ever since the Russian satellite caught the scientists partly unprepared with organizational details and programming plans for the computers, matters have been quite confused and decentralized."[585] Publicity like this generated endowment funds of more than $1 million by the spring, with ground breaking for the "space science" building to start by the end of the year. At last, Whipple rejoiced, Harvard would have a "strong, earthy foundation for space science research." To symbolize its roots, a small speck taken from the first shovel of dirt would be incorporated into the spacecraft "for Good Luck and shall return eventually as a meteoritic spark."[586]

In addition to a wide array of offices, shops, and laboratories, Whipple singled out the "great computing machine" as the key to facilitate the tracking work as well as other applications, as they arose. At the ground breaking, Menzel poignantly pointed to how the "lack of inner space" had hampered their studies "in the field of outer space." Both spoke of how the new building would enhance teaching and make the subject more attractive to students.[587] *SAO News* recorded the dedication and opening of the building on 5 May 1961, calling it a "Great Day for Space!" a nod to Alan Shepard's suborbital flight that day as well as to the presence of the Harvard Visiting Committee.[588]

Through the 1960s, the satellite-tracking empire stabilized but the rest of the institution continued to grow to a size unprecedented in the history of astronomy. Although one would get a sense of stability by reading the annual reports and pronouncements issued by all three institutions—SAO, Smithsonian Institution, and Harvard—this did not translate to the working level in the early 1960s, at least in the minds of most of those on the rapidly growing staff. Recent hires by the Smithsonian were typically on contract, distinct from those few long timers who had lived their lives within the security of the Harvard College Observatory, funded largely by the university and the substantial philanthropic largesse of friendly individuals. As Robert Davis described the situation for his NASA monitor in August 1960:

> Most Celescope professional personnel are working here out of personal interest in the program rather than from any consideration of permanent security (which does not exist in the present set up). Our experience has indicated that the number of people on our personnel budget is just adequate to meet the requirements of the program.[589]

The line workers were mostly contractors, and even though some of them had been around long enough to have experienced how management moved staff from project to project as funding dried up or as new projects demanded immediate expertise, few felt comfortable working in that mode, especially those with academic aspirations. Those who were not part of the core family remained unsure of the permanency of Smithsonian support, knowing that it was subject to annual federal allotments and the vagaries of the political process.

In the first few years, both Whipple and Menzel knew they were working on borrowed time. Whipple's vision for transforming Harvard astronomy, via the Smithsonian, into a "big science" operation incorporated Menzel's concept of a "research reserve."[590] Many who joined Whipple from Harvard, such as Charles Whitney and Eugene Avrett, knew that they had a certain "institutional obligation" to provide a wide range of services, from administration and project management to orbit analysis, all while remaining active in their chosen field of stellar atmospheres, which ultimately informed Whipple's Celescope program.[591] But how long could they afford to do this?

Insight into how staff likely viewed their prospects for the future comes from the queries of a young staffer, E. Michael Gaposchkin (Figure 31) who with his younger sister Katherine (Haramundanis) and brother Peter were raised at the observatory. Mike and Katherine became staff members. Gifted in mathematical programming, Mike was hired as a supervisor for Digital Computer Programs within the Computational Division at SAO. Starting in 1959, Mike's job was to develop the satellite-tracking routines initiated by George Veis from Greece and Yoshihide Kozai from Tokyo. Both had been attracted to the SAO to build up the expertise that would further their scientific goals as well as the service functions that supported the entire enterprise.[592] Specialists such as Kozai and Veis typically stayed for a few years and then moved on to distinguished academic appointments in their native lands. But what were future prospects for the junior workers?

In April 1963, during an otherwise typical administrative meeting, the question of buying into the Smithsonian retirement system came up. Employees working on contracts who were designated full-time Smithsonian could choose a benefits package that was nearly identical to those of federal employees. Mike Gaposchkin then "brought up the question of what effect the possible temporary nature of SAO might have upon enrollment since

Figure 31 Michael Gaposchkin, ca. 1966. *Smithsonian Institution Archives, Acc. 16-263, Box 6.*

SAO's existence is based on a yearly orientation." Carl Tillinghast took his question very seriously; evidently, this was a shared concern of middle managers throughout the program. He assured the staff that there was no longer any basis for discussing the "temporary nature" of the SAO. "Washington," Tillinghast went on, would not approve the long-term functions of the division if they were not intending to support it in perpetuity: "We now have a firmer base than many businesses which depend on one large government contract. We have a small but good base of federal money."[593] Tillinghast wanted to assure staffers like Mike and Katherine, who had been moved between projects as their expertise was required or as institutional priorities changed, that they were secure. This experience became an institutional style in Whipple's still-new SAO.

As the Smithsonian Astrophysical Observatory grew in the 1960s, its institutional style continued. This would create questions, over time, about the nature of the work itself, its magnitude, and its appropriateness as a Smithsonian activity. At first Harvard astronomers and visiting committees started asking questions. Then came voices at the national level who were concerned about the definition and scope of national facilities for science. And it became a question of the allocation of funds for science and the Smithsonian's role in American science.

The Smithsonian's Role in American Science

To appreciate the existence of a Smithsonian facility at an academic institution—in this case, Harvard—and the complex issues raised by this cooperative arrangement, we review here how the Smithsonian rationalized its existence as a research institution. It is critical to bear in mind that, even though it is not strictly a federal agency, whatever it did had to be officially sanctioned and recognized by government. This meant that both the executive branch, within which the Smithsonian lived as an independent agency, and the legislative branch had to agree on just what the Smithsonian's role was in American life. Moreover, they both had to approve what part of that role would be funded by congressional appropriations.

In autumn 1963, as S. Dillon Ripley prepared to succeed Carmichael as the eighth Smithsonian secretary in January 1964, he knew there were numerous problems pending that would require his immediate attention. Two in particular bear mention here. First, he knew that in 1962, the Smithsonian had tried to submit a bill to Congress to "retread" the basic Smithsonian mandate "to put the SI at the center of the Government's oceanographic activity." The bill had been deferred by the regents, but it was introduced "by mistake" by a congressional regent in January 1963. This caused confusion as Ripley moved to Washington, and though it was eventually resubmitted in modified form in the next round, to the 89th Congress, it was again removed because Ripley's advisors felt that the actual successes of the oceanography program more than secured its mandate.[594] This stumble was fortunately undamaging but it became an object lesson for the Smithsonian administration in the following years.

Second, and more disturbing, Ripley knew that there were hearings before the House deliberating over the National Science Foundation's responsibilities. Congressman Albert Thomas had introduced a bill that stated that one "Government agency should not transfer funds for research to another Government agency."[595] Carmichael advised Ripley that the NSF would take this as a "stop order" for funding individual Smithsonian researcher requests. Transfers were possible only when one agency called for services that another agency could provide, in response to specific and explicit requests for proposals. But this did not apply to proposals for autonomous research programs or research promoted by individuals within institutions whose mandate, or congressional authorization, was not specifically indicated. So all proposals engaged by Smithsonian researchers had to fall within bounds that had been sanctioned or authorized.

Ripley was up to the challenge. Celebrated as a leading ornithologist, a professor of biology, and director of Yale's Peabody Museum of Natural History, S. Dillon Ripley II had served in the Office of Strategic Services during World War II and was extremely adept at moving in high-level Washington circles. As a staffer remarked at his death: "While he was a New England patrician and a renaissance man of learning, he had a great popular flair."[596]

Soon after Ripley arrived in Washington, he met with Leland Haworth, the new NSF director, asking that he endorse the Smithsonian's plan to appeal to Congress directly to allow Smithsonian researchers to continue to compete for NSF research funds.[597] He also argued that the Smithsonian's "demi-official" status as a government agency prevented it being "considered as a 'regular' government agency," and needed further clarification. Ripley wanted language in the FY1967 appropriations bill to loosen the legal restrictions that Thomas's bill had set in place. Even though Congress, prompted by the Smithsonian's congressional regents, had helped somewhat to relieve its problems, "Our minute budget gives us no assurances of incremental increases sufficient to measure up to our staff needs for research support."[598] What Ripley wanted to establish, more than anything, was "that by virtue of the Smithsonian's unique public responsibilities to the scientific community based upon our national collections and curatorial responsibilities, research projects sponsored by us automatically fall into the category of national interest programs." With this done, it would be possible to remove the "ban" on support from the NSF so the Smithsonian might once again attract the highest-quality staff to generate valuable research and programming.[599]

SAO and "The Smithsonian Research Bill"

In many ways, Whipple's grand vision for the SAO paralleled Ripley's for the Smithsonian. He defined the SAO in terms of national interests, as in his congressional testimony for building a National Space Establishment on an academic model in April 1958. His comprehensive design for the Smithsonian's management of Celescope fit this vision, right down to justifying the creation of a stellar atmospheres group on the basis that such expertise would be needed to analyze the data. Even this level of activity would hardly be enough for Whipple, who looked for ways to expand into all areas that could somehow be addressed by an astronomical institution. Whipple felt comfortable that the core programs he had fostered—namely, the satellite-tracking and geodetic programs—were relatively safe even under the restrictions of Thomas's bill, as they clearly met national interests and satisfied the mission-based needs of NASA. He was by now comfortable with the political and managerial support shown by the Smithsonian administration. The frustrations he expressed in the late 1950s were now more than assuaged by the effective guidance of James Bradley, who became Ripley's undersecretary. Whipple looked forward to working with both Ripley and Bradley (Figure 32). It was a period of high promise and expansion for science in the United States, but not a time for complacency. They would remain safe only insofar as Whipple could foster sufficient support to make these service functions preeminent and permanent.

As his empire grew in the 1960s—encompassing space astronomy, optical astronomy, radio astronomy, and gamma-ray astronomy, fields quite distinct from his Earth-directed efforts such as satellite tracking and geodesy, and wholly separated from Abbot's solar legacy—so did his concern that it all be sanctioned. Therefore, in the summer of 1964, with a growing awareness of Ripley's ambitions, but sensitive to the fact that authorizing legislation for the Astrophysical Observatory was still narrowly defined and outdated, Whipple asked Bradley to advise on the "Statutory Authority for programs of the Smithsonian Astrophysical Observatory." Bradley sent Whipple's request to Peter Powers, the Smithsonian general counsel, and after a long search, a staff lawyer was only able to find explicit mention of the APO's existence in an 1891 "Sundry Civil Expenses Appropriation

Figure 32 James Bradley's brilliance was recognized by Dillon Ripley with the Secretary's Gold Medal for Exceptional Service, presented 16 November 1966. *Smithsonian Institution Archives, Image # OPA-953-02.*

Act" and then nothing specific until August 1949 in a brief statement: "Appropriations are hereby authorized for the maintenance of the Astrophysical Observatory and the making of solar observations at high altitudes."[600]

The 1949 assessment was specifically directed to the Bureau of the Budget (BoB), which had now declared that the Smithsonian mandate for "increase and diffuse . . . was not authority for the appropriation of Federal funds to provide for the Astrophysical Observatory." The only other entity to be scrutinized in this way in 1949 was the Smithsonian's Bureau of American Ethnology, which was absorbed in 1965 into a combined Department of Anthropology at the National Museum of Natural History. The SAO could not be so easily dealt with.[601]

Reviewing all sources of its funding in late 1964, Tillinghast reported that some 90 percent of the SAO's operating funds came from sources other than appropriated Federal funds. NSF was important, but nowhere near as critical as NASA or the Air Force.[602] No one knew how this ratio would change through the rest of the decade; Tillinghast and Whipple worried how long the support would last before some congressional staffer started asking questions.

Responding partly to Whipple's growing ambitions for both space-based and ground-based national-scale facilities, Ripley and his staff redoubled their efforts to clarify the Smithsonian's role in the federal research establishment. Ripley, Bradley, and others in the Smithsonian central administration, in concert with Whipple, saw these efforts as a means to once and for all establish a strong, broad mandate authorizing the SAO as a national facility for astronomy. By the latter half of the decade, therefore, the SAO became a test case for Ripley and his staff to reconfirm as well as upgrade the Smithsonian's role in federal research.

The Whipple/Bradley/Ripley campaign began in mid-1965 and produced a series of long white papers and proposals by the end of the year. These were then incorporated into a series of draft legislative proposals that, through much of 1966, Sidney Galler, assistant secretary for science, as well as Bradley and Ripley, used in discussions with congressional regents, BoB staff, the U.S. Civil Service Commission, and the White House. The process of constructing these white papers reveals that the central administration strategy was in considerable flux, whereas SAO's was fixed.

Whipple engaged his senior staff heavily in the effort. Chief players were Tillinghast, as well as J. A. Coffey, manager of the Satellite Tracking and Data Acquisition Department, and Charles Lundquist, Whipple's assistant director for science. In the midst of preparing these statements, in late summer 1965, SAO was explicitly cited by the president of the United States for "its exemplary role in the fruitful international interchange of ideas and knowledge."[603] Support like this emboldened Whipple's staff as they prepared statements titled "Goals and Programs" and "The Role of the SAO" and circulated them in the late summer. In mid-December, Bradley sent a draft legislative proposal to the Smithsonian's general counsel, Peter Powers, that laid out the situation, showing that the current SAO program was now far larger than its authorization. It now consisted of "radio astronomy, gamma-ray astronomy, meteorites and cosmic dust, theoretical astrophysics, optical observatory, flight experiments, geophysics, meteors, exobiology, and lunar research." Everyone in the loop in the Smithsonian agreed that the 1949 statement was inadequate, but everyone also had an opinion about what to do to change it, as this was a test case. Galler asked if the SAO had proved its national worth, as had the Smithsonian's oceanography programs. Galler agreed that the draft would be "helpful to the SAO" but worried that it was not sufficiently persuasive to overcome the issue that "Congress and the public appear to view the SAO primarily as a service organization to the NASA, somewhat similar to the SIE's service to the NSF."[604] Would this be sufficient to justify the SAO's continued expansion into areas that clearly were highly competitive with other organizations, private and federal?

Coffey, Tillinghast, and Lundquist fine-tuned the document while Whipple expanded Bradley's listing of SAO research areas to emphasize how "world-wide basic geodesy" depended on their core tracking activities.[605] In late January, after the regents initially ratified his effort to update the Smithsonian's research authorization, Ripley sent Joseph Califano, special assistant to President Johnson, a new proposal for "legislative reorganization" and reauthorization of the research scope for the SAO, adding that the National Museum of Natural History would be similarly re-proposed in the near future. Here, once again, the Smithsonian re-ordered the research area priorities, this time making them sound more mainstream. The SAO "is concerned with theoretical astrophysics, gamma-ray and radio astronomy, meteorites, planetary, terrestrial and lunar research, and exobiology." The geophysical areas Lundquist and Whipple cared so much about were subsumed in less service-oriented and more basic disciplinary fields. Bradley had been warned that SAO staff would grumble, wanting more specificity, but he resisted.[606]

This was, after all, an exercise that had all of the Smithsonian's welfare at heart. Ripley's mandate was to maintain the "flexibility of the Smithsonian's basic charter" to preserve the institution's "freedom for new departures, employing appropriated funds, so long as an objective is within its charter, not in express violation of any statute, and legitimate in terms of the public interest."[607] One of his staffers, Phil Ritterbush, advised that they had to do all that was possible to counter BoB's original questioning of the "appropriateness" of Smithsonian activities. Among many causes, he speculated that when BoB raised its old concerns in late 1963, Ripley was not yet installed and the regents were not very strong. Nor had the Smithsonian "mobilized the kind of conspicuous consensus among scientists like the NIH, NBS, or the NSF have." Since then, they had made some progress rectifying these weaknesses, like creating an elite academic SI Council in January 1966, organizing various high-level academic reviews, and improved networking by Galler and Ripley. But, Ritterbush observed, looking at the various drafts of the legislation, they had still not been very successful saying why the work must be conducted by the Smithsonian and not by another entity. Still, Ritterbush felt the time was right: "Mr. Bradley might want to try a Smithsonian Astrophysical Observatory bill this year, experiment with some of the administrative language, and submit a more general bill next year with which we could tidy up any deficiencies revealed by our experience with the SAO bill."[608]

By March, after confirming that the BoB would be wary of highly specific proposals, Powers recommended that only the most general wording be used. The proposal now covered "research in astrophysics and other natural sciences, in history and the arts." The SAO's mission was described only as "astrophysics and related sciences contributing to knowledge of the universe." Powers admitted that this might appear to be too spare, "but the earlier version sounded to me like authority for an organization half as large as NASA."[609] By the end of the month, Ripley was campaigning with congressional regents and by May he was in direct contact with the BoB for its reaction to their draft proposal, asking if it adequately addressed any concerns for how to create a bill to "clarify and extend the basic authority of the Smithsonian Institution" and how it would play at the White House.[610]

Congressional regents also sought out advice from congressional lawyers. Some of the regents worried that more specificity was in fact desired, but an assistant counsel in the House of Representatives' Office of the Legislative Counsel advised that it was not a legal requirement. In fact, "The hallowed words" to "increase and diffuse" should be legally sufficient for BoB. Specificity could be helpful, the lawyers advised, in order to gain political constituents and regional support. But specificity had to be clearly labeled as "basic research" if only to "define and illustrate the types of basic research intended."[611]

Bills drafted in June had maintained the generality favored by Ripley's staff, but they still dithered over what to include. Bradley grew impatient, reminding Powers, Ritterbush, and Galler that the overall purpose was simple: "to provide fresh support for our programs." He added, likely reflecting Whipple's requests and pressure from Ripley, who was pushing more and more for history and art: "I have no brief for pet language—it's the objective that counts. . . . Let's move on some acceptable bill."[612]

Galler, however, still worried that explicit statements could be interpreted as absolute and restrictive. Soon, the BoB proved him right, responding officially to Ripley in late July and advising him that "changes along the lines proposed whether accepted or rejected by the Congress could serve to limit rather than clarify existing authority." In fact, any specifications could "be construed as exhaustive rather than exemplary. . . . We therefore suggest that consideration be given to deletion of the specific fields of science."[613] The BoB even advised against qualifying the type of research the Smithsonian conducted: whether it be "fundamental" or "basic" was not their concern. If any problem areas were specified, the BoB feared, some of the proposed changes would give "an excessively broad mandate for the Smithsonian." In sum, "we request that you reconsider your proposed legislation."[614]

Smithsonian officials thus went back to the drawing board, as the SAO's newest campaign to build a new huge radio facility continued, in concert with MIT, Harvard, and other northeast universities, as well as with Lincoln Laboratories (see Chapter 14). Through November, however, senior staffers still worried that the stalled "Research Bill" was "our biggest internal problem." Everyone knew it would only be a matter of time before the SAO's expansion would be questioned, especially because of the radio facility, but also because Whipple's new ambitions for a large optical observatory on Mount Hopkins in Arizona were becoming apparent (see Chapters 12 and 13), and his various "service" areas, for example, satellite tracking and geodesy, were becoming more visible.

The white paper Whipple and his staff prepared best illustrates the magnitude of Whipple's ambitions at this point. It identifies the SAO's role in the expressed series of ever-widening scientific and societal spheres he wanted to play in, and provides a template for the programs and facilities Whipple envisioned for the rest of the decade.

The View from Cambridge

It is not clear who actually crafted the framework and content for "The Role of the Smithsonian Astrophysical Observatory in the Federal Structure/Astronomical Community/Cambridge Complex/Great Society." Certainly the Smithsonian central administration and the SAO shared in the exercise. It is safe to assume that Whipple oversaw the process from Cambridge, and that it was executed largely by Coffey and Tillinghast, with input from Lundquist. But equally likely members of the SAO's Public Affairs Office, such as John P. White or James Cornell, SAO's public information specialist, or members of their editorial staff, were involved, as they prepared the annual reviews of the observatory for *Smithsonian Year* as well as for relevant professional societies and knew the art of written persuasion.[615] They were also preparing an extensive white paper for the BoB and NSF that presented the radio telescope initiative.[616] This fact will become clear as we move through the document, examining the rhetoric rationalizing the SAO's existence.

First came *Role of SAO in the Federal Structure*, in which the thrust was to show how the Smithsonian had a talent for anticipating the needs of modern society and meeting them as a federal facility. "Earth-Solar relationships" was the first area, in which the SAO, at its seventy-fifth anniversary in 1965, could look back on its pioneering legacy of monitoring the "effects of solar radiation upon the Earth and its atmosphere." Other agencies initiated "related investigations," but "the Observatory retains leadership in furthering understanding of solar effects on the high atmosphere, using analysis of satellite tracking data." Then came geodesy, in which the Smithsonian claimed to have initiated inquiry and "has been recognized for its innovations and results." Other emerging problem areas fostered by the SAO that claimed relevance to the federal structure included gamma-ray astronomy, the joint optical-radio observations of flare stars, all under the umbrella of "astrophysical problems that are judged to be undeservedly neglected by other investigators." Meteor astronomy fit under this umbrella.[617]

The white paper cast the SAO as a federal agency that "responds to requirements from mission-oriented agencies" such as NASA and the Department of Defense (DOD). Often, the goals of the DOD or even NASA mission agencies were narrower than the SAO's goals, so supplementary funding had to be found. This could be from private sources, or from the NSF. But in distinction to "the passive function of the National Science Foundation to support basic sciences and science education" generally, the SAO is "a center of active research within the federal establishment." In distinction to NASA, the NSF or DOD, then, this section concluded boldly:

> The Observatory is virtually unique in its ability to pursue pure astrophysics directly, rather than obliquely as an offshoot of an applied mission. Hence, the Observatory must be viewed as the "national astrophysical observatory" and should be funded accordingly.[618]

The SAO niche was as a hybrid making it effectively distinct from potential competitors. The designation of "national astrophysical observatory," no matter how bold, would never survive scrutiny. Yet this is exactly what Whipple, aided and abetted by Ripley, would eventually push for in their campaign for both optical and radio facilities.

Second was *Role of SAO in the Astronomical Community*, in which the white paper celebrated the research output of SAO staff, including publishing more than 100 articles per year starting in the late 1950s in some 30 peer-reviewed international journals, entertaining scores of professional visitors providing access to data and facilities and expertise, and training well over 100 graduate students by then, from Harvard and other universities. The SAO also had been operating two international bureaus, the International Bureau for Astronomical Telegrams (Figure 33) and the Central Bureau for Satellite Geodesy. And early on it created its own publications facility that by then had produced hundreds of "Special Reports," not only in the space sciences but also in its "Smithsonian Contributions to Astrophysics," which Whipple started in 1956.[619]

Role of SAO in the Cambridge Complex situated the SAO within an "evolutionary cluster," borrowing the term from Margaret Mead, "carrying on the business of promoting man's progress." Harvard and MIT dominated, but were enhanced by scores of smaller institutional bodies, of which the SAO was a "significant link in the exchange and interaction

Figure 33 The Central Bureau for Astronomical Telegrams brought worldwide public and professional attention to the SAO. Bureau members in the late 1960s included (*L to R*) Richard Southworth, Barbara Welther, Brian Marsden (seated), and Owen Gingerich. *Courtesy AIP Emilio Segrè Visual Archives,* Physics Today *Collection.*

of ideas in this community." The action of moving SAO from Washington to Cambridge "fulfilled the historic dreams of John Adams, who hoped to use the entire Smithson bequest to found a national observatory."[620]

And finally came *The Observatory and the Great Society*, invoking President Johnson's words at an invocation that year purportedly honoring the "Smithsonian's bicentennial" [*sic*].[621] There, Johnson charged the Smithsonian: "To split the atom, to launch the rocket, to explore the innermost mysteries and the outermost reaches of the universe—these are your God-given chores."[622] To this, the white paper celebrated how "the Observatory's study of meteorites teaches something of the vast stretches of time that are the history of the universe." Then came encomiums for radioisotope analysis, satellite tracking and geodesy, and finally the newly defined field of exobiology, exploring the possibilities of extraterrestrial life: "No one can neglect this basic question, for it is at the very heart of man's philosophy of living."[623]

Clearly an emotional appeal, the white paper's penultimate paragraph declared that "the Observatory carries on a lively and informative dialogue with the Great Society." With this as preamble, a second document, "Goals and Programs 1967–1976," got down to specifics, and ended with an estimated budget line for appropriations for the second decade of the observatory.

Looking to the Second Decade of the SAO

The white paper lumped meteors, satellite tracking, and upper atmosphere research together in a broad mandate that Whipple called "The Earth as a Planet," giving form to what were largely becoming, by the late 1960s at least, routine operations of data acquisition. Accordingly, attention turned to making use of the data to gain geodetic knowledge. It would soon became a core activity, driven by the continuing flow of funding from federal resources in the first decade.

Over and over, the white paper described growing ambitions in optical, radio, and gamma-ray astronomy, initiated in Celescope and elaborated in the 1960s when the SAO took steps to establish broader capabilities in optical and especially radio astronomy to study quasars, flare stars, and interstellar hydrogen and hydroxyls, areas where radio astronomy was especially promising. The SAO had already enhanced Harvard's radio capabilities at the Agassiz Station, and would use it as a stepping stone to a far more ambitious facility, directly linked to Harvard and MIT as a "regional, radio astronomy facility in the next five to ten years."[624]

Gamma-ray astronomy in 1965 was "still in an embryonic state." The SAO wanted to explore what was to be expected in that region of the spectrum, extrapolating from known sources, and engaging theoretical expertise within its ranks, locally, and among those few groups interested in the problem. After experimentation with a local solar furnace facility, the SAO would build dedicated atmospheric Čerenkov detectors on a mountaintop in southern Arizona, utilizing photoelectric sensors fed by very large mirrors of "modest optical quality."[625] The SAO at the time envisioned establishing this optical facility at an "observing site somewhere in Western North America" and equipping it with "versatile modern astronomical telescopes." A 24- to 36-inch conventional reflector might be the first instrument, followed by larger ones, ranging up to "moderate size" in a decade. Not mentioning Kitt Peak or any national facility directly, the white paper acknowledged that the staff did receive some observing time at other optical facilities, "but the demand for observing time now exceeds that available."[626]

Research areas in geophysics, meteors, and exobiology would continue, and funding would be requested for maintaining these smaller programs. The Baker-Nunn trackers would, however, require refinement and augmentation with laser-ranging equipment working in parallel from several sites. One new area, mentioned at the end of the white paper, was "Lunar Research," which in 1965 was confined mainly to improving the lunar ephemeris in support of trajectory studies. But Whipple did not leave out other possibilities, indicating that the SAO's "specific role remains to be defined."[627] And indeed, earlier on in the discussion of "Meteorites and Cosmic Dust," areas closest to Whipple's specialty and most closely tied to his professional reputation, one finds explicit mention of building up a staff of specialists who would operate "laboratories that are equipped to investigate the petrography, mineralogy, metallurgy and isotopic composition of meteorites and cosmic dust," adding prophetically that "the same techniques can be applied to lunar samples obtained perhaps during the late 1960's and early 1970's."[628]

Much of what was projected in the white paper became real—some very successful, others stillborn. Even though this series of documents had been crafted to establish authority for SAO programs, facilities, and research areas, they also reveal that none of them was envisioned as being phased out after some appropriate goals had been reached. The only research area abandoned to date was that of the legacy inherited from Washington, the monitoring of solar radiation.

The point of the "Role" white papers and the accompanying "Goals and Programs," in their various forms in the latter half of 1965, then, was to demonstrate the centrality and the leadership of SAO research and service activities to national life. As a detailed plan of action, however, it went far beyond anything the central Smithsonian, BoB, or Congress would likely want or tolerate to establish authority for appropriations. Its purpose, therefore, must have been more to offer a vision of a scope of work within areas where the SAO believed direct appropriations would be necessary. Even so, what it lacked, quite clearly, were boundaries.

The chapters that follow will explore how the established ventures continued through the 1960s, like Celescope, and then the new ventures Whipple initiated during his second decade as observatory director. They will, in turn, express the breadth of his empire and illustrate growing tensions not only between the SAO, Harvard, and the Smithsonian, but between the Smithsonian and government bodies concerned with the federal support for science. Indeed, the white paper did not address the SAO's role in Harvard's own future. At the time the paper was written, the Harvard College Observatory was also undergoing profound change. When Menzel (Figure 34) retired in 1964, Goldberg replaced him as director and within a few years relations between Goldberg and Whipple deteriorated to the point at which a series of deep reviews, both internal and external, led to a profound restructuring of the relationship into what is today called the Center for Astrophysics.

Figure 34 Donald Menzel, lecturing ca. 1962. *Smithsonian Institution Archives, Acc. 16-263, Box 7.*

Chapter 9

Breaking Cognitive Barriers from Meteors to Lunar Samples

Whipple's personal research interests in meteors, their cosmic origins, and their interactions with Earth's atmosphere, all found vastly expanded expression in projects and programs he initiated as he leveraged the Smithsonian as his institutional base. His pre-Smithsonian Astrophysical Observatory (SAO) Harvard meteor programs led directly to satellite tracking, which led to geodesy and to the Smithsonian Standard Earth, whereas his meteor research led first to sampling and analysis of meteoritic samples and eventually to the analysis of lunar samples. The Smithsonian's white paper in 1965 brought all these lines together in "The Earth as a Planet," which was dependent on project funding, mainly though NASA and the Air Force.[629]

Whipple's leveraging, creating an institutional capability far broader than anything he could accomplish on his own as a Harvard faculty member, not only expanded his scope of techniques bearing on his particular interests, but also opened up new ways to encourage a wide variety of hybrid talents and interests, non-traditional at least in the astronomical profession. These became the areas where the Smithsonian saw its greatest growth in the 1960s. They were also areas that remained least integrated into the academic life of the Harvard College Observatory.

Interdisciplinarity at the Harvard College Observatory

Harlow Shapley advocated interdisciplinarity in the first half of the twentieth century: "The big opportunities are obviously not in pure astronomy or physics or chemistry or geology, but in mixtures." For Shapley, meteors and meteorites harbored clues to galactic evolution. Before Whipple found otherwise, Shapley, influenced by Öpik, was convinced that meteors had hyperbolic orbits and so were interstellar. In the early 1930s he parlayed this belief into a multi-disciplinary "Institute of Cosmogony" using various schemes including his Harvard Summer School in Astronomy, which he designed to draw in physicists, geologists, geophysicists, and meteorologists to expose his faculty and graduate students to interdisciplinarity.[630]

Whipple adopted Shapley's interdisciplinarity, but not his entrepreneurial strategy. Shapley was comfortable and effective with corporate and personal philanthropy, but not with government or military options. By the early 1950s, however, Whipple and Menzel, through Air Force and Navy projects, were pulling in more funding for astronomy than any other astronomer, and Shapley's observatory was a beneficiary.[631] Whipple, of course, was clear in his convictions ever since he had reported in 1947 to the Joint Chiefs of Staff that "all basic upper atmospheric research is fundamentally and inherently of value to the general problems of national defense."[632]

The military certainly promoted interdisciplinarity, but all three astronomers also appreciated, and were wary of, the unsteady nature of military funding, especially dangers of depending on a single patron.[633] When the Navy dropped support for Whipple's photographic meteor program in the early 1950s, it was due to the fact that it did not provide them with information on temperature profiles or other meteorological conditions as they had hoped.[634] What it did provide was evidence that seasonal effects in upper atmosphere densities were not due to the atmosphere, but to the densities and tensile strengths of the various meteorite swarms that encountered Earth—some were meteoritic, some cometary, like the Geminids. And after losing Navy support, Whipple, with Menzel's aid, convinced Air Force program officers to take it on to gauge upper air wind patterns from the observations of meteor trails to better inform re-entry profiles, and so his growing interdisciplinary empire continued.

Physical Sampling: From Meteorites to the Moon

In 1959, Whipple published two major reviews of meteor studies and the nature of the lunar surface that nicely illustrate how his penchant for promoting interdisciplinarity resulted in institutional growth at the SAO, supported by multiple patrons. The first, in the classic discipline-based *Handbuch Der Physik*, presented what was important about studying meteors, and what could come from the effort. The second, in *Vistas in Astronautics,* was titled "On the Lunar Dust Layer" and linked meteoric studies to eventual in situ studies on the Moon.

Whipple teamed up with Gerald Hawkins, who had been leading the SAO's radio meteor effort, to make the point that the problem area has "emerged from its isolation as a purely academic subject" and was now prospering in a far larger and more complex arena.[635] Among many applications, they identified how meteors can affect the lower ionosphere and perturb radio communications; that meteoritic dust may have an influence stimulating heavy rainfall on a worldwide scale; and that meteoritic materials might well be diffused throughout interplanetary space, causing phenomena such as the zodiacal light. Meteoritic dust, they added, could be of concern to space travelers, both robotic and human, and knowledge from meteoritics would aid ablation studies and terminal ballistics studies that would help to better understand the formation of craters and Earth's crustal history. In consequence, according to Whipple, echoing disdain for the touchy rock-hunting community of meteoriticists of the Nininger-Leonard era, meteors "had passed from the hands of the amateurs" to a wide range of professional workers, including optical and radio astronomers, orbit specialists, ballisticians, electronics experts, and mineralogists.[636] In other words, it was now legitimate practice.

For Whipple, the most fundamental cosmic question about meteors had been well answered by 1959, following his own efforts in the 1930s that showed there were no observed hyperbolic meteors. There were just two major classes that came from two distinct parts of the solar system: cometary and asteroidal. So a broad range of disciplinary attention was now needed to probe the physical processes generated by encounters with Earth's atmosphere, a range that did not, in fact, include his personal disciplinary strengths in celestial mechanics. A full picture of the interaction required studying meteors of a range of sizes, from the sub-micron scale to the meter scale. It also required chemical and physical analyses of meteorite fragments found on Earth via new methods that would isolate primordial properties from the effects of weathering and erosion.[637]

Whipple and Hawkins knew that if physical samples were the primary source of data, there was a strong selection effect. Stony meteorites weathered far more rapidly than irons and soon became virtually impossible to identify. This conclusion came from sampling statistics: Observed falls were all stony, whereas unobserved falls were irons.[638] Unsaid here, but articulated in a proposal to the Office of Naval Research (ONR) less than a year later for what would soon be called the "Prairie Network," Whipple argued for a new more powerful meteorite recovery program, this time to be optimized to acquire "freshly fallen meteorites for laboratory analysis."[639]

This was not a new idea. Whipple originally tried to sell it as the "Texas network" in the 1940s, "hoping that I could get a rich Texan who would like to back a network that would cover the entire state of Texas (which is about a reasonable size), and photograph all the fireballs to find out where they came from."[640] Then his goal was to compute orbits and establish origins. But by 1959, it was also to improve chemical and physical knowledge useful to understanding the forces experienced by the intruder. The ONR had, of course, responded strongly in the 1940s which led to Harvard's Super-Schmidt program. But now Whipple hoped for a network that was far larger. When the ONR balked in 1960, he applied to NASA and secured funding—not all he hoped for, but enough to create a wide network of less sophisticated cameras. Unlike the Super-Schmidt systems, which were designed to capture as many trails as possible, bright and faint, for orbital statistics, the Prairie Network was optimized for bright fireballs, ones that were most likely to produce recoverable debris after impact, and trails long enough to assess deceleration rates with height.[641]

Whipple knew he was not alone in proposing a collection strategy, especially for fresh samples.[642] In September 1959, James Bradley had to step in to broker a competing proposal from Edward Henderson at the National Museum of Natural History (NMNH). The result was a series of regional and national meetings sponsored by the SAO, orchestrated by Whipple, that brought in different groups, especially the meteoriticist John Russell of the University of Southern California, then president of the Meteoritical Society and a moderating force in the "small, feud-riven" organization. Whipple also asked Ed Fireman to organize another special session, bringing in leading workers familiar with cosmogenic isotopes. He even managed to link the theme to the traditional Astrophysical Observatory (APO) preoccupation, namely, the question of the constancy of the solar flux.[643]

Fireman reported from his special session that attendees agreed that the Smithsonian was the ideal coordinating body for a worldwide collecting plan. To date, collecting had been "very haphazard," hampered by many factors: lack of manpower, legal ownership when the object landed on private land, and even the idiosyncrasies of the "personality of the successful collector."[644] They also called for a coherent large-scale network of linked cameras to track falls and determine their points of impact for rapid collection. Only with such a mechanism in place could they hope to gather fresh samples within days or weeks that would yield reliable mineralogical, metallurgical, and isotopic analysis.

Indeed, the Smithsonian was the ideal coordinating unit. The SAO had an enormous international footprint, not only from its Baker-Nunn stations but, significantly, from its many dozens of teams of Moonwatchers who were already primed to report falls through their existing communications networks. Henderson, too, had his own network of active meteoriticists in need of samples. Accordingly, the first order plan was to direct everyone to report falls to Fred Whipple, or to Henderson.[645]

Fireman left to the imagination how Whipple and Henderson would cooperate. That was worked out in a second meeting in Washington, again brokered by Bradley, between the SAO and NMNH scientists, as they prepared a proposal for outside funding. Henderson, they decided, would draft the proposal, which Whipple then would critique. Henderson's proposal was very modest: where field searches would be organized only after casual reports of falls. Whipple was far more ambitious, calling for an active plan that would increase the chance that more falls would be detected and retrieved. Henderson had no problem with that, but left the active element to the SAO. The stumbling block was where the retrieved specimens would be housed and shared. Henderson naturally pushed for his museum, where specimens would be properly housed and inventoried, and distributed equitably through some form of peer process. Whipple vehemently rejected this plan, insisting that the SAO have "at least an equal stake" in the collection and analysis process.[646] Their differences were never reconciled and caused much confusion at times in the program.

Whipple's second review paper, "On the Lunar Dust Layer," appeared in the newly established spaceflight-oriented series *Vistas in Astronautics. Vistas* started as a conference publishing venue for various sponsors, including the Air Force Office of Scientific Research and aerospace corporations, and was one of many new initiatives designed to create identity, inform policy, and foster communication. Whipple had contributed papers to the first volume on "The Meteoritic Risk to Space Vehicles" and "Orbit Accuracy and Ranges from Ground-Based Optical Tracking." In the second volume, he employed his knowledge of meteor impacts to ask whether it was safe to land humans on the surface of the Moon.

It was a simple but highly provocative question, even for 1959, "whether the physical surface can provide structural support for the landing of space vehicles, the erection of shelters, and physical exploration by members of an expedition."[647] His goal was to test Tom Gold's 1955 assertion, based on thermal studies from the late 1940s, that the solid-looking maria were actually deep, unconsolidated layers of soft dust that would not support landing craft, and would foul any mechanical instrument that attempted to soft-land using conventional rocket thrusters.[648] Whipple disagreed, drawing on extensive observational studies measuring the Moon's albedo in infrared wavelengths and short radio waves, and estimates of meteoritic bombardment of the lunar surface as well as the chemical reaction to this bombardment as interpreted from explosive cratering experiments called "sputtering." There was dust to be sure, but it was neither deep nor soft because these processes would "cement together dust grains on the lunar surface." Whipple also presciently argued that large impacts would propel a good fraction of the ejecta to high velocities, causing a portion to spread across the lunar surface, and even some to escape lunar gravity completely.[649] Concluding his statement, Whipple mused, "The author waits with the greatest eagerness for firsthand information concerning the correctness of these conclusions, when the first landings on the Moon will provide the answers."[650]

Indeed, these were exciting times for Whipple. Multiple lines of support, led by NASA and Department of Defense (DOD) resources, and buttressed by the direct Smithsonian pipeline to Congress, made it possible for him to initiate many collaborations to explore just about everything on the forefront of meteoritics. On an Air Force contract, he enlisted the talents of an Argentine student at the Harvard College Observatory, Carlos Manuel Varsavsky, who had been working at the High Altitude Observatory and then with Menzel on the solar corona and opacities in stellar atmospheres. Whipple turned Varsavsky to the origin of tektites, which had been debated throughout the 1950s, and he became one of the first to argue for their lunar origin.[651] In a related line, Whipple engaged Super-Schmidt team members led by McCrosky and Hawkins to show that none or few if any of the doubly photographed meteor trails captured with his Super-Schmidts could be of lunar origin.[652]

Whipple's institution building expanded his passion for meteor studies from traditional celestial mechanics methods to aerodynamics, hyperballistics, and mineralogy. This journey took him from the vast open plains of America's western prairies to the lunar surface.

The Prairie Network

Richard McCrosky organized and led the effort to operate an array of small cameras to track and triangulate meteor trails to their impact points. One of Whipple's most adaptable and versatile colleagues, "Mac" McCrosky had been active in Whipple's meteor studies for more than a decade, exploring meteor fragmentation using optical and radio techniques (Figures 35 and 36). In a 1965 report on the Network, he described how its goals differed from the Super-Schmidt program: Now, rather than acting like a "bulimic meteor astronomer" maximizing sensitivity to catch the faintest trails, the Prairie Network staff would gather "the most data on the most bright meteors" limited only by finances.[653] It would be the brightest meteors that had the greatest chance of reaching the ground, so their trajectories had to be precisely determined.

Funding from NASA provided for staff and operations and an expansion of Fireman's laboratory for mineralogical analysis, but left little for instrumentation, which in 1961 the Air Force relieved by giving them enough T-11 aerial mapping cameras to supply at first 13 small stations designed in an array of three nested hexagons with stations separated by some 225 kilometers. More stations were soon added, providing three stations at the center

Figure 35 Richard McCrosky examines bright meteor trails in a time exposure from a Prairie Network station. *Smithsonian Institution Archives, Acc. 16-263, Box 4.*

Figure 36 The Prairie Network time exposure shows star trails, a faint meteor, and a bright exploding fireball, a bolide. A rotating sector in the camera produced multiple exposures to record speed and brightness changes in the meteor trails. *Smithsonian Institution Archives, Acc. 16-263, Box 4.*

of each hexagon and increasing the perimeter station distances to 250 kilometers to provide an optimal combination of double and triple stations.[654] Each station operated automatically, driven by sensor-driven circuits that could open or close the cameras depending on cloud cover, sky brightness, precipitation, and winds. There were two programmed exposure times, a short twilight mode, and a "nighttime" mode, with a standard exposure time of two hours.[655]

The first station opened at a dark site in Havana, Illinois, in March 1963, and the sixteenth station in May 1964. The sites had been surveyed by the U.S. Geological Survey and local inhabitants were employed to check in on each station every few days. The exposed film canisters (390-foot rolls of 9-inch wide photographic film) were sent to a central station in Lincoln, Nebraska, headed by Gunther Schwartz.

Managing the field work for McCrosky and Schwartz was Charles A. (Chuck) Tougas, a resourceful ex-Marine who had learned to fly an airplane so he could service the network, which ranged over seven states from Illinois to South Dakota to Oklahoma. James Cornell recalled that Whipple regarded Tougas as his "strong man during the frontier era of the observatory," and to Cornell he was the "field guy who'd go out there and get things done under adverse conditions, lay up the network, find the surplus material, get the roads [ready]." He rented land from farmers and did all the local hiring. "Chuck was here almost from the very beginning, and he was a robust, hard-working guy, but a bit of Sergeant Bilko."[656] After his work on the Prairie Network, Tougas moved to Arizona to build the Mount Hopkins facility.

The chief question in Carmichael's mind was, where would the recovered meteorites end up? Initially, he favored the SAO as the central coordinator, given its success with Moonwatch, but the SAO was, after all, not a museum. He knew that both Whipple and Henderson were campaigning to set up schemes for analysis. Henderson understood that Whipple's chief goal was to retrieve and analyze the physical and mineralogical character of the sample as quickly as possible, especially the relatively delicate outer surfaces that would reveal the most information about its interaction with Earth's atmosphere.[657] But Carmichael also knew that Henderson's philosophy, beyond analysis, was to share samples far and wide, as he had been doing for years, garnering a powerful community of allies among his users. Add to this, Carmichael was very sensitive about where the traditional core of Smithsonian research in the geosciences lay. Therefore, Carmichael suggested that Whipple keep in mind Henderson's interests, asking that "a portion of each meteorite collected" be given over to NMNH's Department of Geology "as a contribution to its work and permanent collection."[658] Whipple diplomatically agreed that he would provide Henderson with samples, "as much . . . as can possibly be spared from immediate study."[659]

Carmichael wanted to support Henderson's facility to make it competitive with Whipple's within a new, larger venue called the Department of Mineral Sciences, adding new staff and equipment, also funded by NASA but with significant direct support from the home institution. Whipple, of course, was not happy with the obvious favor shown by the administration, and so redoubled his efforts, recruiting friends at NASA to reassert SAO's centrality. This led in 1966 to a tripartite committee of Natural History, NASA, and SAO staff, joined by representatives of their legal departments, to decide on specific responsibilities and modes of operation. At stake was who would be designated as the national clearinghouse.[660]

During the first 12 years of the project, from 1963 to 1975, some 334 fireballs were recorded and evaluated. But very little was retrieved.[661] The lone specimen, from what is known as the "Lost City Meteorite," fell in Oklahoma in 1970 and was retrieved by Schwartz less than a week later. It was the first of four from that fall, and the collection became something of a test case between SAO and NMNH over who would actually analyze the fragments. Plotkin has reviewed the process and concluded that "practical and scientific considerations easily took place over jurisdictional prerogatives, resulting in a truly cooperative undertaking."[662]

By then, however, members of Fireman's group and others within Whipple's empire were preoccupied with analyzing the Smithsonian's share of lunar dust, extrapolating from what they had learned from terrestrial samples that came not from the Prairie Network, but from a major fall in Mexico.

Allende and Chondrules

Two key staff in Fireman's laboratory were Ursula Marvin, who began work in 1961, and John Wood, who arrived in 1965. Marvin, Wood recalled, was a critical player, having established an interdisciplinary and interinstitutional "Meteorite Discussion Group" to reach a wider domain beyond SAO and Harvard.[663] In both cases, Whipple provided the means.

Ursula Bailey Marvin trained in both history and geology at Tufts University, and then took graduate training in geology at Harvard University-Radcliffe, earning an MS in geology in 1946. After six years of exploration geology in Brazil and Angola, she returned to Cambridge in 1958 and secured a job at the Harvard Mineralogical and Geological Museum (now part of the Harvard Museum of Natural History) working for its curator, the noted mineralogist Clifford Frondel. Frondel wanted to keep her employed, and so soon asked Whipple if he could support her on Fireman's grants, and soon she was retained as a permanent civil servant.[664] She remained at the museum to prepare meteorite samples while pursuing graduate studies leading to the PhD in 1969. In the early 1960s, she collaborated with a variety of researchers in the Cambridge area, from Harvard's Department of Geological Sciences to the U.S. Army's Cold Regions Research and Engineering Laboratory in Hanover, New Hampshire, exploring the physical and chemical properties of dust and rock samples believed to be of extraterrestrial origin using a wide range of laboratory facilities in the area, typically electron probe machines at the U.S. Geological Survey (USGS) and Fireman's isotopic laboratory.[665] In all this, she, Wood, and others at the SAO were preparing for the arrival of lunar samples. In early 1969 they got a chance to test their readiness, studying recently fallen fragments from Allende.[666]

A very bright fireball, seen as far north as Arizona, exploded over the village of Pueblito de Allende in Chihuahua, Mexico, in the early morning of 8 February 1969. Within days, scientists from NASA's Lunar Receiving Laboratory in Houston, Texas, rushed to the site, gathering many specimens, citing the effort as "an excellent rehearsal" for the lunar work.[667] It was a Type III carbonaceous chondrite, considered very primitive and revealing of conditions present in the original solar nebula. The chondrite family includes stony condensates with comparatively low levels of iron, nickel, and silicon but rich in calcium and aluminum, often containing calcium-aluminum inclusions called chondrules. These are light colored and whitish, roughly spherical bodies typically smaller than a centimeter in diameter. They are thought to be the earliest condensates in the contracting solar nebula.

Alerted by the SAO's newly organized Center for Short-Lived Phenomena, senior curators Brian Mason and Roy Clarke from the National Museum rushed to Northern Mexico to retrieve samples. Meanwhile, Robert Citron, who headed the center, used his connections to secure a sample from locals on site and had it shipped to the SAO. The center was a "new service" of the SAO, created by Whipple at Ripley's request, that combined SAO's satellite-tracking infrastructure with several Smithsonian research bureaus that specialized in rapid collecting.[668] This was yet another instance in which Whipple exploited the Smithsonian's ability to alert diverse communities to all sorts of unpredictable events, from meteorite impacts to volcanoes to floods and earthquakes.[669]

When Citron's Center for Short-Lived Phenomena got the news, it contacted the Air Force, which dispatched a B-57 over the Gulf of Mexico to collect airborne dust suspected to be lingering there after being ablated off the original body. Citron also notified people who had recovered specimens and kept everyone informed of progress. Within weeks, the U.S. National Museum (now the National Museum of Natural History) had collected samples and was in the process of distributing them widely. Fireman received an 85 gram sample from Citron within a week of the fall and larger samples soon after. Allende turned out to be the largest carbonaceous chondrite to be quickly retrieved.[670]

As with Marvin, Wood's career and experience, and his seminal role in analyzing meteorite and lunar samples, best highlights Whipple's interdisciplinary institution building (Figure 37). Trained in geology at Virginia Tech and then MIT, receiving his PhD in 1958, Wood also took a minor concentration in astronomy, coming into contact with Whipple. At Harvard he learned meteorite thin section techniques from Frondel, and decided eventually to pursue dissertation research in meteorite petrography, supported by Whipple with SAO funds. Inspecting the meteorite thin sections, Wood recalled, "I loved the chondrules, each a unique igneous rock, and tried to picture how they could have formed in the early solar system."[671] Whipple placed him under Rinehart to study ablation, hyperballistics, and radioisotopic techniques. He also learned methods of applying x-ray fluorescence techniques and metallography from Marvin.[672] Wood was exposed to many diverse analytical techniques and became adept at bridging both disciplinary and cognitive boundaries.

Variations seen in chondrules, Wood suspected, might reveal the nature of the solid-state processes that led to the formation of these bodies. He pursued these interests at

Figure 37 John Wood and Ursula Marvin in their lunar sample analysis laboratory. *Smithsonian Institution Archives, SIA Acc. 16-263, Box 5.*

Cambridge University and then the Fermi Institute at the University of Chicago, where he studied under the radiochemist and meteoriticist Edward Anders. There he became familiar with techniques in metallurgy and the use of a new technology, the electron microprobe. Anders in particular encouraged Wood to focus these techniques on carbonaceous chondrites for his analysis of their metamorphic nature, which he described through a physical system of classification. By the end of his postdoctoral years, Wood was becoming known and respected in a widening circle beyond the bounds of meteoritics. Whipple, of course, had kept in close touch with him through his postdoctoral years and frequently reminded him that he was welcome back at the SAO at any time. By the time he returned to the SAO, he was a member of the Space Science Board's committee on planetary surfaces and interiors.[673]

Campaigning for Lunar Sample Returns

As a member of this committee, Wood helped to draft and deliver a petition to NASA in January 1962, urging it to restore the planned robotic *Prospector* mission. These were heavily instrumented landers and rovers including one to scoop up some soil and return it to Earth for direct analysis. When it was cancelled, Wood was among those objecting strongly, stating that returning lunar samples was "of greater *scientific* importance than the achievement of sending a man to the Moon and back."[674]

The cancellation relegated any possibility of sample return to human lunar lander missions.[675] So over the next several years NASA issued ever-more-detailed, open-ended requests for proposals for lunar surface science experiments, including sample return analysis. A fourth announcement in May 1966, asking for definite proposals from anyone ready to perform "sample analysis by the spring of 1968" attracted hundreds of proposals, including Wood and Fireman.[676] Wood proposed that the SAO would develop a laboratory facility to handle Apollo materials while he and his colleagues continued with their analysis of chondrites and their metamorphic properties. He recalls that there was enough support at the SAO for regular projects, but to gear up for *Apollo*, he had to start seeking funding to build a larger team and a dedicated facility. Apparently, he had little problem obtaining support and definitely no problem recruiting people because, as he recalled, there was an atmosphere of open interdisciplinary interaction. In the atmosphere generated by Marvin and Whipple, Wood recalled that "collaboration seemed like the logical and gentlemanly thing to do."[677] One of the first people he asked to join his NASA-sponsored group was, in fact, Marvin, whom he first encountered through her meteorite discussion group.

Whipple's networks were informal, yet they were so complex that Wood did not realize for some time that Marvin was already an SAO staffer. With Marvin, Wood's group obtained samples from the Allende meteorite fall. This work, and especially the fact that the meteorite community was warming to Wood's metamorphic classification, gave him a certain status that made him competitive for Apollo samples. And though he succeeded in being designated as a "lunar sample Principal Investigator," he proceeded cautiously, acting as if he, Marvin, and their newly hired postdocs "understood clearly that our position in the lunar program was at the bottom of the totem pole." He rationalized this as due to his age, "lowly stature in the field, and maybe the fact that I wasn't on a university faculty." He sensed he was now in a field much larger than the meteorite community, namely, the "terrestrial geosciences, a very different population from cosmochemists." He also felt like an outsider, claiming that all he could hope for were leavings.[678]

In fact, though there was a general perception that status distinctions existed between geochemistry and field geology in the study of the Moon and planetary surfaces, opportunities for analysis were spread far and wide.[679] Among those obtaining samples from the first successful Apollo mission, there were elite American, Canadian, European, and Japanese universities. But there were also dozens of groups selected from the U.S. Geological Survey, Oak Ridge National Laboratory, and NASA's own centers. There were four separate programs approved for the Smithsonian alone. In addition to Wood, Fireman was waiting for samples to conduct isotopic studies. There were also projects at the NMNH and at Harvard under Frondel. Wood's program, with Marvin identified as Co-PI, was titled "Mineralogic and Petrologic Studies by Optical Microscopy, X-Ray Diffraction and Electron Microprobe Measurements."[680]

Through the years that NASA issued various requests for proposals, 1964 through 1967, Wood's reputation had most certainly become very solid. His recent papers had already garnered some 70 citations and he was firmly established in his specialty with two review papers in respected volumes.[681] This was not only an indication of the quality of his work, but also the size of the community now interested in the problem areas being addressed.[682] And of course, Whipple widely reported that Fireman, Marvin, and Wood now had "laboratories for the isotopic, mineralogical, and petrological study of lunar materials to be returned by the Apollo mission."[683] The SAO was ready for the main event.

Returns from Apollo

NASA had devised a complex process for proposing and then evaluating scientific experiments for NASA missions, starting in 1960 with the creation of the Space Sciences Steering Committee and its numerous subcommittees, and in spring 1964 with the creation of selection criteria that provided metrics for choices. There was to be sure much "stress and strain" and it took some time to achieve equanimity or balance between scientists, their institutions, and NASA.[684] It was within this overall process that the new Manned Spacecraft Center in Houston formed committees and subcommittees to evaluate proposals. In March 1967, Homer Newell informed Wood and Marvin that they were among the selected groups and invited them to Houston in September where all parties would get a chance to examine the new Lunar Receiving Laboratory. There, they would make sure that "all problems of handling and distributing samples" were properly aired and discussed.[685]

About a year after the Houston meeting, as they built up their laboratory, Wood and Marvin stated what they wanted: 1 cubic centimeter "or less" of "lunar soil" to study "on a grain by grain basis." They were still gathering in equipment, including an x-ray radiation source capable of producing an extremely fine beam some 0.1 millimeter in width, for "micro-techniques new to our laboratory."[686] Both Marvin and Wood expected to devote at least half their time over the next year through FY 1971 on this project, with outside support through current funding. Marvin would have primary responsibility for setting up and operating the x-ray probe, whereas Wood would focus on mounting and analyzing the samples on the electron probe, with the aid of postdoctoral fellows.

Despite Wood's claim that they were left with the detritus: not the rocks, but the soil and the dust, there is good evidence that he well knew that it might yield the most important clues to the nature of the lunar highlands, and indeed to the overall nature and history of the lunar crust and the existence of a magmatic ocean, if the fragments retained the structural complexity of their origins. After all, Whipple had made this point, envisioning impact cratering sending debris everywhere. Their Allende work had been geared to studying the same regime, tiny mineral grains, and they had been using these samples in their simulations of how they would analyze the lunar materials. In addition, among her many interests at the time, which included assessing continental drift rates from satellite geodesy, Marvin was a key player in Fireman's efforts to collect and analyze dust from the Greenland icecap, and she was very familiar with dealing with dust.[687] So, when they got their share of lunar regolith after the return of Apollo 11, they were ready.[688]

Apollo 11 lunar samples arrived at the Lunar Receiving Laboratory (LRL) in late July 1969, just as a Conference on Lunar Science was being held there, which brought many if not most of the hopeful samplers to a sweltering Houston. Wood recalled seeing the rocks through thick glass, but it would not be until September that he could revisit Houston to receive about a tablespoon full of sieved material, dust and soil that had passed through a 1 or 2 millimeter sieve fraction. And as it did with other groups, NASA's LRL also lent Wood various thin sections for inspection and analysis.[689]

Wood's team quickly found happy surprises. Indeed, they did not find only mineral fragments, but also tiny fragments of rocks, "sufficiently fine-grained for almost all of them to be polycrystalline, polymineralic, and fairly representative of the rock it had been broken from."[690] In effect, they had a "huge collection" of coherent rock types, and so they started looking for population types. His team soon detected impact-melted glasses in the rocks, declaring "the truth," as Wood described it: "That's an anorthosite composition!" To a meteoriticist, this was a big deal, and as a cosmochemist, he speculated that these specks, distinguished by their color that was lighter than the majority, "might be pieces of the white highlands of the Moon."[691]

They found anorthosites, intrusive igneous rocks characteristic of magmatic origin, in abundance at the Apollo 11 landing site. This fact indicated at once that this granular material came from all over the Moon, and was a result of continuous meteoritic impact and erosion by cosmic rays, just the inferences drawn by Whipple in 1959. Wood knew that other groups at Chicago and Goddard and California would be able to make these conclusions, but these groups would be looking first at the basalts of Mare Tranquillitatis: "The terrestrial geologists among lunar investigators thought it their first duty to understand that basalt, and they worked in that direction."[692]

Wood felt he was in the right place at the right time, and that he and his group had the right training, equipment, and experience. They were meteoriticists, not terrestrial geologists. USGS workers, in addition, were subject to hierarchical review and to conservatism, and would not have been free to make such speculative leaps. In tune with followers of meteor trails like Whipple, and meteorite hunters like Marvin, Wood lived in an institution that knew that the "samples had to have a context and a history."[693]

So instead of local origins for the white pieces from some outcrops and the dark basalts from the site, Wood speculated that they came from an original magma ocean. He had plenty of clues, from *Surveyor* lunar lander analyses, to the detection of asymmetric contractions of mass inside the Moon, discrete "mascons." If indeed the Moon possessed a magmatic interior, the lunar highlands had to be a product of isostatic compensation, just like on Earth.

Wood's group continued their analysis of Apollo 11 samples. As later Apollo missions brought back a wider sampling, they were able to confirm and strengthen their view of the highlands as thermally differentiated clumps of less dense material floating on a magmatic ocean that had been created by repeated asteroid-sized impacts early in the history of the Moon. Wood attributes his success bringing all these clues together to the mixing of disciplines at SAO fostered by Whipple. In their various reports, his group acknowledged both Whipple and Lundquist for "stimulating discussions of lunar theory," and well as both SAO and Harvard for facilities to conduct microprobe analysis, handling and mounting the microscope slides, and extensive computer studies of heat flow.[694]

Wood, Marvin, and their teams of associates continued through the 1970s along lines established during the Whipple years. They continued to gather correlative evidence to develop a refined picture of the many processes engaged in the formation of the early solar system. This included a better understanding of the nature of the parent bodies fostering meteorite falls, their origins and ages, and what they could say about the nature of the original interstellar nebula that gave rise to the Sun and planets, very much a mainstream topic in modern astrophysics.[695]

Chapter 10

"Project Celescope": SAO's Orbiting Astronomical Observatory

In his August 1958 *Saturday Evening Post* essay "Eyes in Space," energized by SAO's satellite-tracking success, Whipple proclaimed that from space, "Great telescope mirrors can focus ultraviolet or infrared radiations on television pickups, to send back pictures of the universe that the human eye will never see directly."[696] New technologies were beckoning and Whipple introduced the "Cat Eye." As a new technology, it symbolized both the promise and the risks of spaceflight. Embracing both, Whipple marshaled forces to build a space telescope for humanity.[697]

Unlike the Smithsonian Astrophysical Observatory's (SAO) meteoritics and lunar-sampling efforts, Whipple's role in Celescope was as facilitator. It was, for him, a way to build up staff and resources and to build an institution. It was not a personal scientific passion. Celescope also had a very different history than the analysis of lunar samples by Wood and Marvin. First, it was just as well that NASA assumed the bulk of the responsibility for the mission development beyond the instrument packages. The scientific groups that had been selected to provide instruments knew that they had their hands full just delivering their payloads. Because it was an institution builder for Whipple, the technical and managerial choices the SAO made, and the subsequent problems these choices raised, created considerable friction between Goddard and the SAO in the mid-1960s. Following the course of the friction will provide insight into the risks inherent in technology choice, as well as the challenges Whipple faced as he built his empire. It will also shed light on NASA/university relations.

At the outset, both the Smithsonian and Wisconsin groups lacked in-house resources to build the instruments themselves.[698] However, Wisconsin had already established a Space Astronomy Laboratory in 1959, and its leaders were highly adept at the photoelectric technologies they would apply. They also contracted locally with small-instrument specialists to keep control over everything and opted for telescope designs not so different from point-source photoelectric techniques that were familiar tools of the observational astronomer.[699]

Whipple and his staff, however, made different choices. They certainly knew that they suffered from a profound lack of in-house facilities to create and test the components, but they also chose a particular technology that no astronomer in that day had mastered. Every facet of the project required breaking new ground, obtaining staff with expertise in areas not before encountered, building new laboratories, and creating a communications and liaison structure able to keep in touch with contractors as well as NASA. And, in stark contrast to the Wisconsin group, and the local expertise surrounding Wood and Marvin, they lacked local personnel with practical knowledge of photoelectric technologies, and chose to contract with large corporations at great distances from campus.

There are many facets to the Celescope story characteristic of the challenges facing emerging technological fields: the challenge players face to construct specifications for an instrument adequate to the task, and then to deliver that instrument with those given specifications to perform the task on time. This was true for almost every facet of the project, and it was shared by SAO staff, NASA staff, and contractors alike. This applied to the satellite payload design, the television detectors, the television telemetry antenna systems, and even the choice of the optical systems.[700]

Celescope presented myriad dynamical, mechanical, and optical challenges, most if not all of them common to first-generation space projects, such as persistent electric arcing and the shorting-out of the high-power supplies that plagued the early Orbiting Solar Observatory (OSO) missions and weather satellites.[701] What was unique to, and important about, the Celescope program was the challenge of building suitable low-light-level image detectors that provided reliable photometric precision over the life of the mission. Whipple's decision to employ this new technology and the results of that decision are, once again, key to his institutional goals.

Technical Challenges: The Detectors

Celescope needed a detector to measure the ultraviolet brightnesses of the stars. Such a system had to meet three prime requirements: extreme low-light-level sensitivity; full calibration over time of photometric, spectral, and spatial information; and sensitivity to the far ultraviolet with relative insensitivity to visible radiation. Choices were photography, which required either the physical return of the film, or a processing system on board the craft; fixed or variable iris photometry with conventional point source, photoelectric photometers; some form of two-dimensional electronic scanning like television; or electrostatic imaging. The latter two required a combination of on-board electronic storage, a high-speed data stream, and appropriate electronic algorithms for data processing.

In the early 1960s, even the most optimistic provider of electronic imaging devices, J. D. McGee of University College London, urged caution. In 1958, he felt that "there was still a long way to go," and as late as 1962 he still felt that "television signal-generating devices are still of limited use to the working astronomer."[702] He envisioned that they would eventually come into their own for satellite observations, but they were not ready yet. As experienced instrument-oriented astronomers like Wisconsin's Art Code knew well, adapting vidicons to precision photometric UV detection at the time was "beyond the state of the art."[703] McGee actively promoted the new technology, but often expressed private frustration that the "problem of making tubes still remains a chancy business" or, quite bluntly, "no satisfactory tube has yet to be made."[704]

There were, to be sure, television systems in use in satellites in 1960 for high-light levels. These were applied variously by NASA and by both American and Soviet military reconnaissance programs, and were successful on Luna 3 in October 1959 and by the weather satellite series TIROS. But none other than the military had the peculiar demands that astronomical application required, mainly extreme low-light-level sensitivity, spatial and photometric stability, and tight physical calibration. And even the military, as far as is known, did not require detectors for the ultraviolet. To make a commercial tube sensitive to the UV required special transmitting optics. Quartz would be good only to 2,000 ångströms, and beyond that something like a lithium fluoride window was needed. Also, the composition of the cathode had to be optimized for the UV range, and preferably made insensitive to visible light.

Given the state of the technology, Davis had his hands full just searching out commercial sources that were willing to develop detectors. This was only one of his many responsibilities, especially as Hynek, who had championed the electronic image detector option and Cat Eye, soon left the SAO for the Dearborn Observatory of Northwestern University. Starting in spring 1958, Davis found manufacturers willing to discuss the technical issues and provide estimates of costs and timelines and delivery schedules, but none was assuring that success would be at hand without considerable experimentation over a long period of time. One, however, offered an interesting alternative.

R. P. Haviland, at General Electric's Missile and Ordnance Department in Philadelphia, was very anxious to be of service, but did not think much of electronic detection and telemetry. Haviland was part of a large team at G.E. that was developing "a 50-pound film package returned from an orbiting satellite" that would be ready "within 9 months." G.E.'s "device" could keep orientation to within 1 minute of arc of the true zenith, and give the coordinates to 1 second of arc and cover the whole sky from its polar orbit "by moving the film to keep up with the motion of the image."[705] Davis was intrigued, but both he and Whipple remained wedded to electronic data retrieval. Davis even admitted that "photography would give higher quality information" but still opted for the telemetry scheme, which he felt had a better chance of success.[706] SAO astronomers also rejected the option of on-board photographic recording, processing, and optical scanning, akin to what was to become part of Lunar Orbiter in the mid-1960s.[707]

Davis also contacted vendors ranging from RCA's new Astro-Electronics Product Division, various divisions of Westinghouse (Air Arms Division and its Research Laboratories), Perkin-Elmer, the University of Rochester and Kodak, the CBS Research Laboratories, and Bausch and Lomb. Some, for example, Fairchild Camera and Instrument Corporation, like G.E., suggested physical retrieval of photographic film or "the use of film and storage with subsequent readout." Even when some, more familiar with the electron-optics field, like Rod Scott at Perkin-Elmer, openly questioned Whipple's claim that off-the-shelf detectors and electronics could be found, Whipple and Davis remained firmly focused on image tubes.[708] In late 1959, they contracted with Westinghouse Research Laboratories to produce an electrostatic detector modified to work in the far ultraviolet. Their decision is another indication of Whipple's ultimate goal—to build up institutional capabilities unavailable elsewhere. After all, NASA was encouraging new technologies, and Whipple wanted to be as competitive as possible. Whipple also had the support of his closest staff. In June 1960, McCrosky eloquently advised: "Observatories of the stature and size of SAO and HCO [Harvard College Observatory] should be actively participating in this next generation of observing techniques."[709]

Westinghouse adapted its "EBICON" tube (electron bombardment induction conductivity) by installing an ultraviolet transmitting window, but after six months of effort had little to show, which worried Davis.[710] The first sample EBICON they were shown in August was only a dummy, and had serious leakage problems in the seams between the ultraviolet transmission windows and the glass walls of the tubes. The engineers knew that whatever they came up with had to withstand 10g with vibration levels of 2,000 cycles, the conditions of launch and orbital insertion. By late August, the SAO admitted that the detectors "are presently the major obstacle to the progress of the program."[711]

Westinghouse soon changed the tube's name to "Uvicon" to better market a new cathode formulation and to suggest a new and fresh course of action. But reliability and sensitivity continued to elude them. Even though the SAO did not need long-lived tubes for its tests with sounding rockets lofting a small photometric telescope, which they hoped for at first, Westinghouse still could not provide suitable examples.

Meanwhile, Whipple managed to complicate matters by proposing to fly x-ray telescopes on the sounding rockets to perform quick surveys, collaborating with P. C. Fisher, an experienced nuclear physicist and x-ray investigator at Lockheed Missiles and Space Co.[712] But this additional goal increased costs, and NASA flatly objected during budget negotiations in Washington on 20 September. SAO had asked for some $1.32 million for FY 1961 but was approved for $700,000. The meeting grew contentious when Nancy Roman, program scientist from NASA HQ for the Orbiting Astronomical Observatory (OAO), questioned the size of the SAO staff, and Goddard's James Kupperian flatly stated that a project of this magnitude was better within NASA's walls at Goddard, believing that in-house projects were less expensive than out-of-house contracted projects.[713] Not only was the program far too expensive, but the reduced budget would also be locked up in problems with NASA's overall 1961 budget.

Throughout the rest of 1960, nothing seemed to go right. Westinghouse was still unable to produce a stable detector that could resist acceleration and vibrational forces. Two new SAO engineers who visited Westinghouse in October to inspect 10 Uvicons found only one that was "ready to work in the laboratory." It could detect ultraviolet light but there were defects, including random arcing, lack of sensitivity, and loss of resolution. They tried to be optimistic, reporting to Davis that although there were serious issues, "no insurmountable problems are being faced."[714] On NASA's side, some sounding rocket flights from Wallops were in the works, but they too were delayed. NASA had momentarily agreed to fly an SAO payload mainly to test its own pointing controls on their Aerobee sounding rockets.[715]

The story remained the same through early 1961. When Westinghouse failed to meet promised delivery times, or reported a failure in a certain development process, SAO staff began looking for alternative sources of acceptable image-tube detectors. They went back to the Cat Eye at one point, still being tested at SAO's Organ Pass station for satellite tracking.[716] They also were interested in an orthicon being developed by General Electric for the Air Force that could reach 16th-magnitude stars at the focus of a small telescope.[717] By March, however, Westinghouse reported progress in hardening the tubes, and despite NASA's call for some kind of quantitative reliability factor, which it did not define precisely, and Westinghouse's initial willingness to speculate on it, SAO engineers knew that it was not a meaningful effort: "This is a new tube; it has not been built before, other than in extremely limited quantities, and statistical data is not available in any way, shape or form regarding this tube." They needed something, but what? "It would be equally foolhardy, of course, to simply blindly say to Westinghouse, 'give us some tubes to put in here.' "[718] The SAO was on a steep learning curve, and it was not a smooth one.

All sides were griping. Spring brought budget reviews and more NASA scrutiny as it prepared its FY 1962 budget. The Goddard OAO project manager followed up Nancy Roman's September 1960 threat that the whole program should be brought inside. Davis remained obstinate, but even his own staff complained, writing an open letter warning that there were problems everywhere, and that what the project needed was a general contractor the SAO could trust.[719]

Westinghouse also demanded a longer lead time and more money. They claimed they now had to produce more detectors than they had planned for, since the OAO required four telescopes, not one. But when James Kupperian reviewed Westinghouse's request, he realized that the development costs had more to do with commercial production than to providing an acceptable detector for Celescope. Members of Davis's team had known this, but the SAO did not act. This caused all sorts of trouble, leading to a meeting in June in Washington between NASA, SAO, Westinghouse, and the central Smithsonian. NASA added lawyers because Westinghouse was now making claims about patent rights. At the meeting, as Davis reported to Whipple, Westinghouse admitted it was padding the budget, but it was a matter of a complex "transfer of information from a research to a manufacturing organization." Initially the scale-up from one to four telescopes had been a mutual consequence of NASA's payload requirements and the SAO's ambitions, and Westinghouse had eagerly agreed. Now, however, realizing the magnitude of the challenge, Westinghouse balked. Only after a good bit of haggling did Westinghouse reluctantly agree that despite the number of tubes needed by Celescope alone, they had to treat it as a research activity and not a manufacturing enterprise.[720]

With the original development contract due to lapse in October 1961, and with every technical report negative, NASA once again threatened to pull the project inside. In desperation, the SAO decided to place the manufacturing contract on open bid rather than sole source.[721] Westinghouse expressed "surprise and distress" but still bid, and in the end offered a significantly better deal. Goddard project managers still felt it was too expensive. At \$2.3 million, "Celescope is the largest and most expensive aspect of the Orbiting Astronomical Observatory experiments," Davis reported. "We get the impression, although NASA continually denies it, that the scientific purpose is no longer the controlling factor in this experiment."[722]

Westinghouse won the contract again, but this time it was backed up by General Electrodynamics Corporation, suppliers of various forms of vidicons for commercial use.[723] Westinghouse performance improved slowly, very slowly. And there were other problems.

Predicting the Brightnesses of the Stars in the Ultraviolet

Throughout these years, the theoretical analysis group under Whitney and Krook was using Davis's rough brightness estimates to create stellar atmosphere models that could predict the ultraviolet behavior of the stars observed by Celescope. "An important aspect of Celescope," an early report stated, "is its ability to give observational confirmation to these results; the Celescope Program is being continually re-evaluated with these models in mind."[724]

The big question was, would the Uvicons be sensitive enough to detect the stars? At first, Davis's estimates gave a safety factor of about 20 times. But in May 1961, Davis learned from the Wisconsin astronomers that ultraviolet fluxes for hot stars (blueward of 1,600 ångströms) gathered from sounding rocket flights were "at least 30 times fainter than expected." Davis remained optimistic: "I don't feel that these new numbers appreciably affect our expectation of success with Celescope."[725] But, in fact, no one knew for sure. Still, Davis remained committed (Figure 38).

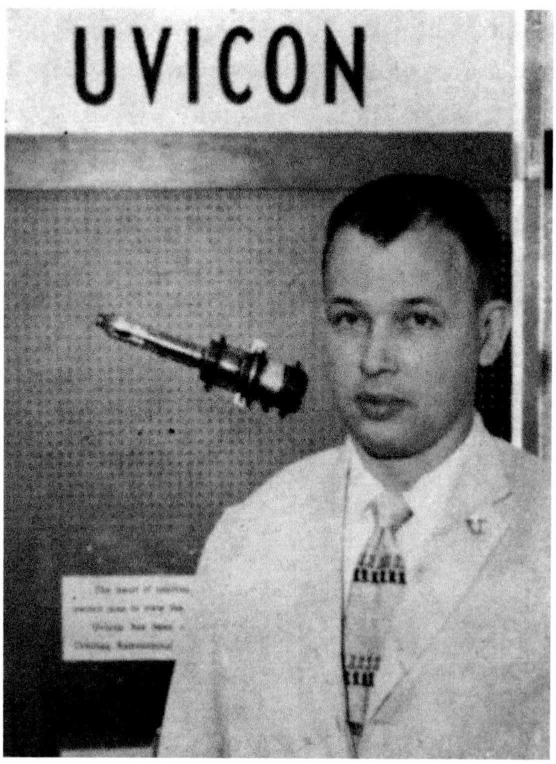

Figure 38 Robert Davis in front of his display of Uvicon detectors at the General Assembly of the International Astronomical Union, Berkeley, August 1961. *IAU News Bulletin* no. 5, 21 August 1961, 9. *Courtesy IAU.*

Senior SAO staffers did not share Davis's initial optimism. They called for an increase in Uvicon sensitivity at least by a factor of seven to meet Celescope's goal of building a representative star catalog. But there were bigger problems at hand. In March 1962, the first payload was ready but was waiting for an acceptable detector. In June, there was a reassessment of technical and financial problems because they still had not been able to get all parts together and in working order at the same time, which made it impossible to gain a specific slot on an Aerobee at Wallops Island. And then NASA cancelled the Smithsonian's Aerobee flights.

The failure to test the payload on a balloon, or an Aerobee, was symptomatic of the problems Celescope faced, right up to launch. The loss of test flights was a significant blow to gaining experience in integrating components into a flyable package, but just as important was the fact that first glimpses of the ultraviolet universe now belonged to others, at Goddard and Wisconsin, using Aerobee sounding rockets. The Wisconsin group preparing the telescope array that would complement SAO's Celescope had been testing its components on balloon flights since 1961, and had secured successful tests of their photometers with Aerobees in 1962 and flights aboard NASA's X-15 rocket plane.[726]

The First Stable Uvicon: Too Little, Too Late

In June 1962, Westinghouse reported that it had a tube with good resolution and uniformity with long lifetime. It still did not meet NASA's requirement for a one-year operational lifetime, but there was hope. Westinghouse also promised to deliver some tubes soon to SAO's contractor for integration, the Systems Division of Electro-Mechanical Research, based in Sarasota, Florida. EMR was responsible for incorporating the optics, electronics, and mechanicals into the actual flight payload, along with general systems design and integration of all subsystems.[727]

Relieved but wary, Whipple called for a full assessment of the program that summer, coming away hopeful that success would come in 1963. But budgets were getting tighter, there was less room to maneuver, and Westinghouse could not produce a sufficient number of tubes that survived testing. The tubes remained the project's nightmare throughout 1962 and into 1963. By August 1963, all the components were now on hand and ready for integration, save for the tubes.[728] "Reliability reviews" called by NASA that had sent Booz-Allen examiners to Westinghouse made no substantial difference. In response, NASA Goddard started pressing the SAO team again. A Goddard contracting officer criticized the SAO, claiming it was not managing contractors adequately and was inept at technical management.[729] He was not confident in NASA oversight either. Charles Lundquist, then newly hired as Whipple's assistant director for science, agreed, but saw an even darker problem.

Lundquist, hired away from NASA's Marshall Space Flight Center in late 1962 by Whipple, brought deep insight into NASA project management. He advised Tillinghast that much of their problem with specifications was due not only to Westinghouse's or EMR's limitations, but also because the original specification directive issued by NASA "is a remarkably poorly written document." The problems they were having now with EMR stemmed from that fact and should have really been corrected by NASA. It was now up to the SAO to demonstrate to both NASA and EMR the proper course to take. Lundquist knew it would be an "onerous task" and he was "by no means optimistic that it will be concluded soon."[730]

Lundquist's prediction was correct, and his guidance led to more effective oversight. By the end of the year, Westinghouse finally succeeded in producing a sample tube "as yet unique," which provided sufficiently high target potential and sensitivity. Throughout 1963 and well into 1964, the major focus remained on the tubes, but EMR's integration of the many steps—testing them, integrating them into their flight casings, and the casings integrated into the telescopes—was still elusive. Progress reports no longer mentioned sounding rocket tests.

Coming to Terms in Late 1964

In August 1964, just as the SAO team received Westinghouse's samples for flight acceptance testing, Nancy Roman, Newell's deputy, made it painfully clear that she was worried. Speaking before the European Space Research Organization, she publically criticized the program's inability to make good:

> Five years ago it did not appear too difficult to build a television camera that would work in the ultraviolet. There were good television cameras for the visible and photographic regions and it would be necessary only to change the photocathode. Experience has proved this assumption to be naive.[731]

Roman was expressing NASA's general frustration with OAO because it was behind schedule and over budget. The first OAO was originally scheduled to launch in late 1963, which now slipped to sometime in 1964 or 1965, with attendant cost overruns and unforeseen technical challenges.[732] This only sharpened, as Goddard director Harry Goett warned Whipple in July 1964, "the gravity of the situation with the SAO experiment." He wanted the SAO's full assurance that it was rectifying all problems and that the flight package would be ready for space.[733]

What little optimism Davis and his team had at this point was wiped out in October when three separate accidents damaged both the prototype package (at Goddard and at EMR) and one of the mirrors (at Bausch and Lomb).[734] This brought another site visit to the SAO, led by Harry Goett. Goett threatened once again to move Celescope to Goddard and also to slip it from the first OAO flight. Whipple ruefully agreed that a slip would well be necessary, but it would be painful and would raise the specter of flying a payload with ageing components. By the time of its flight, many payload components would be more than two years old. And the alternative, to provide a substantially improved package and deliver it later, he felt, was prohibitive in time and cost. So Whipple pleaded that NASA fly Celescope at its earliest opportunity, promising to make an "all-out push to complete the flight Celescope experiment."[735]

Five days later, Goett reported that the Wisconsin experiment was now at Goddard and being integrated. For the SAO to remain on the first flight, the SAO had to deliver the Uvicon assembly to Goddard no later than February 1964, and the entire experiment by late September 1964, less than a year hence.[736] But Goett was soon to be overruled by the NASA Space Science Steering Committee, which summarily removed Celescope from the first OAO spacecraft in late December 1964, "to avoid an additional six to nine month delay."[737]

A Goddard-supplied gamma-ray payload joined x-ray and gamma-ray telescopes from MIT and an x-ray package from Lockheed to fill Celescope's berth on the first flight. These were payloads tested on sounding rocket flights. One of them was, in fact, a payload similar to what Whipple had originally wanted flown on an Aerobee as a prototype Celescope package. These groups were by then well-experienced and in fact, along with the Wisconsin group, were the most prepared to provide payloads that would not further delay the already seriously delayed first flight of the OAO.[738] The SAO was losing out.

Revised Plans

Charles Lundquist gently and firmly took the lead in 1965, strengthening project management overall at the SAO and improving communications with NASA. One of his first tasks was to attend a status meeting at Goddard in January to set a new timeline. If the SAO promised a flight unit by 1 April 1966, NASA could launch in the third quarter of 1967. What worried Lundquist was that by then, ultraviolet packages would have been flown by both Wisconsin and another Goddard package, a moderate resolution ultraviolet spectrobolometer.[739] Thus Lundquist knew that many of the goals for Celescope would be superseded by the time of its flight. At the meeting, Lundquist rhetorically asked himself, "Are we willing to wait until 1967 for results? The answer is a definite 'no.' "[740]

This message went up the ranks in the Smithsonian to the Secretary's desk along with a plan to launch the Celescope package immediately on a series of balloons, followed possibly by sounding rocket flights. Being first was all important.[741] These flights would not only test the detectors but would also search for "interesting objects" for pointed experiments on the first OAOs. Access to the near ultraviolet spectral range at typical balloon float altitudes might yield as many as 800 hot blue O and B type stars. Most of all, they pleaded, "it is essential to the scientific vigor of the program to have actual, useful data available for scientific studies before 1967."[742]

NASA refused to play ball, judging the proposal to be a new project or an unsolicited proposal, not a change order. NASA also rejected the SAO's proposal for a documentary film on Celescope for the historical record and for educational purposes.[743] These SAO actions, whether they were tactics or stress-induced appeals for attention, failed to impress Goddard managers, who remained focused on the payload problems of the ultimate satellite. There was no clear sailing through the spring and summer of 1965. Review and planning meetings at Goddard continued to raise serious management issues with the SAO's prime contractor, EMR, and deepening communications breakdowns between the SAO and NASA.

Goddard managers took a firm line, Davis reported, that henceforth "GSFC will decide among themselves the manner in which to document our agreements concerning these past costs, and then inform the SAO what further action is required of them."[744]

In June 1965, when the SAO submitted a new cost estimate to complete the entire project, which included costs for terminating EMR, Goddard called for another showdown.[745] The very idea of shifting major contractors at this late date angered Goett. Some two hours of interrogation ensued at Goddard, with Davis, Lundquist, and Whipple facing Goett, John W. Townsend, head of the space science division at Goddard, and their staff. Goett came down hard, but Townsend admitted that many of Celescope's problems were shared by other projects, like the high-voltage arcing problems that were plaguing OSO flights. Townsend pointed out that there were few industries with "experience in orbit. Around here this seems to be a sine qua non. Experience in orbit is necessary." This gave Goett his opening to threaten once again to bring Celescope "in-house" at Goddard so they could perform an "in-house review by someone aware of what gets others into trouble." [746] Despite Townsend's observation, Whipple asked for help finding contractors who could do the job and also thanked Goett for "stirring up [EMR] top management" in a way that SAO had not been able to do. Goett, however, shot back that the SAO needed greater commitment: "You have to be manned more heavily than I think you are for managing them." He then vented: "A year ago we decided that SAO should be left on their own, sink or swim."[747] And clearly they were sinking. So once again he tightened the screws, calling for a new "review committee" of both Goddard and SAO staff with Goddard now effectively in control.[748] Goddard now controlled all actions, including who called visits with whom: "we have found that uncoordinated visits are detrimental to effective project management."[749] In December, Newell confirmed that all project management was now centered at Goddard, "including all contractual and spacecraft interface matters."[750]

The OAO-1 Failure and Reassessment

Goddard had its own problems with manpower shortages, especially because by spring 1965 NASA Headquarters was focusing on the manned spaceflight program. Goett hotly resisted, but lost when Newell pronounced that OAO was now only Goddard's third priority, behind developing and maintaining the "Manned space flight network" and developing the TOS (TIROS Operational Satellite) system.[751] Headquarters now had not only downgraded space science programs, but had also replaced Goett with John Clark in July.[752] Clark was preoccupied with manned spaceflight matters, and left OAO to program managers like John Townsend, now Goddard deputy director, and James Kupperian.[753]

OAO-1 was launched on 8 April 1966, carrying payloads from Wisconsin and Goddard that would have performed ultraviolet, x-ray, and gamma-ray observations. After seven minutes of orbital flight, however, severe electrical arcing and battery failure ended the operation before any of the instruments saw first light. Now Goddard came under attack from Headquarters. Management overhauls and a slipped schedule facilitated not the science, but the "primary objective of demonstrating flight operation in support of the experiments for a minimum period of 30 days and a secondary objective to obtain scientific data."[754] No longer expecting a one-year lifetime, the launch of OAO-A2 was set for the second quarter of 1968 to give Goddard enough time to go through "extensive reevaluation."[755]

Congress also took notice. In March 1967, Newell testified before the House Subcommittee on Space Science and Applications, trying to give OAO a positive spin. But Rep. Joseph E. Karth (D-Minn.), chair of the subcommittee, did not share Newell's optimism.[756] Karth attacked the program at many levels, from the design philosophy to the performance of the prime contractor, Grumman. He even told an industry reporter after the hearings that "if we had to do it all over again . . . we would propose using greater numbers of single spacecraft in the Explorer class and sounding rockets to do the work of observatories."[757] Karth pushed to establish stronger congressional oversight of the space program while harboring sympathy for robotic Earth-orbital missions and low-Earth orbit human missions.[758] James Webb and Newell responded in various ways, tightening lines of project management while trying to appease the affected scientific disciplines.[759]

The OAO disaster led to many months of failure analysis, which included recommendations for changes in some of the Smithsonian components. And once again, the component of greatest concern was the Uvicon. Some 11 samples had been selected for flight, and they had already been subjected to new rounds of testing through summer 1966. More than half failed because of arcing, outgassing, and vibration, just the issues that plagued the first OAO payload. One SAO staffer stated the obvious: the Uvicon was "the focal point of concern . . . because it is a major new sensor development which will not have the depth or breadth of ground-based observatory experience as has, say, the photomultiplier tubes."[760] With Celescope's launch slipped to 1968, at least they had more time, and maybe, just maybe, wouldn't be scooped. Westinghouse was constantly refining the design, finding ways to improve photon efficiency. But implementation would cost more, with overruns mounting into the millions.

The SEC Vidicon

By the mid-1960s, Westinghouse researchers at its Pittsburgh Research and Development Center were excited with their new Transmission Secondary Electron Multiplier, an elaboration of a new design first created in conjunction with the Electronics Laboratory at the Wright-Patterson Air Force Base, Gebel's Cat Eye. In testing since the early 1960s, it was still noisy and was subject to electronic "avalanche" effects.[761] But Westinghouse found ways to reduce the noise by looking more closely at how electrons migrated through various formulations of thin films and targets. In what would be called the "SEC" (Secondary Electron Conduction), the backside of a lithium fluoride entrance window was coated with a UV light-sensitive layer that released electrons when UV energy struck the front. The emitted electrons were accelerated and magnetically focused onto a thin-film target that amplified each of the impacts by releasing secondary electrons that, depending on the potential difference, amplified the stream by 200 to 300 times. These secondary electrons then conducted through the permeable layer to a metal screen that stored the accumulated charges and duplicated the original image on the entrance window. The rest of the system was essentially a TV scanning electron beam producing an electrical signal that varied with its position on the film and could be read out in sequence.[762] By the mid-1960s samples were being tested by astronomers at Princeton and Allegheny Observatory, and the results were positive.[763]

But would the SEC Vidicons be ready by the time Celescope had to be delivered? This question could not be answered by SAO staff, who focused on getting the rest of the payload ready at Goddard and then at the Cape. One bright spot came when EMR secured Department of Defense (DOD) support for some 80 percent of the additional costs of final testing and monitoring near launch. The SAO still had to worry about building communications and data processing in Cambridge to handle the science operations once the telescope was in orbit. What had been eventual requirements became immediate needs. Every hand was needed, and more.

Another Path to Space?

NASA's zeal to place humans in space prompted a four-month study in the mid-1960s by the OAO prime contractor, Grumman's "Advanced Systems/Space Science Group," to propose flying the "OAO in Association with a Manned Space Station." The "Smithsonian Sky Mapping Experiment" was considered an ideal candidate for a one-year mission.[764] The Grumman study concluded that OAO had to be a free flyer for undisturbed observing runs, but it would be periodically serviced from a nearby "hangar" made from parts of the "Apollo Extension System" lofted by a single Saturn I booster. OAO would be deployed on a tether after orbit insertion, tested fully, and then released, servicing it every two days to retrieve and resupply film canisters.[765]

One person who really liked this idea was Robert Davis. It would be hard to imagine a more difficult, frustrating, and complex job than being project scientist for Celescope—especially so if the holder of the job had barely completed a PhD thesis and then was thrust into the effort, not only to manage the science, but also initially to define it and see to it that the technology served the science. Regarded by colleagues as a very hard worker, conscientious, and serious, Davis paid attention to everything in the project, deeply concerned about the lot. As Charles Whitney observed, "He felt tremendous responsibility for this project."[766]

A sense of the pressures he faced can be gleaned from his response to an October 1964 NASA recruitment call for astronauts. He stated in his application that while the scientific objectives of the Celescope mission had remained unchanged since its inception, "my responsibilities in regard to the hardware and the administrative aspects of the Project have developed along lines not at all foreseen at that time."[767] Sensing a shift in the nation's space program from robotic craft to human-tended platforms, he did not want to be left behind and so applied to be a part of a "more flexible program of astronomical observations from orbiting laboratories, with on-board participation of the astrophysicist."[768]

Davis was over the age limit in the first round, but reapplied in September 1966 when the National Academy took over recruitment and relaxed the rules. He passed the first screening in February 1967 and told the National Academy that he would like to "extend my present studies of ultraviolet stellar radiation" because "despite many recent improvements in electronic measuring techniques" there were still many problems best addressed by photographic methods. Further, as an astronaut, he felt he could transcend "many of the present operational limitations of the OAO" by examining firsthand how "to accompany and control that satellite in orbit." Davis also speculated about investigating the use of astronomical telescopes on the Moon.[769] He became one of 69 finalists but was not selected.[770]

Davis later rationalized his motives: "Remember, 1967 was that period when Celescope was having all the problems, and this was a good way to escape from it." He thought this effort would succeed, whereas Celescope would most likely fail.[771] Celescope had absorbed over eight years of his life by then; all of his publications to that point had concentrated on some aspect of Celescope's operation and the problems it would address. None addressed new contributions to astronomical knowledge. And now, what was still facing him on Earth, by the latter half of 1967, had taken on a life of its own—and not an easy one.[772]

Flight and Post-Flight: Performance Evaluation and Review

In the months leading to launch, the SAO had still to transfer the payload officially from EMR to Goddard, and had to have staff in residence there. The SAO and Goddard also had to establish a technical presence at the Florida launch site. Not surprisingly, once these groups were formed they had difficulty working together. Roles and missions were not clear; lines of authority were not distinct. Telegram wars commenced. Continuous fault blaming and a general lack of clear lines of communication even between work shifts from one team to another remained unresolved management problems right up to launch. And, because the payload had to be sealed and delivered long before launch, detector development had to be frozen, even though, as Davis later reported, they knew there were problems with charge buildup on their still evolving SEC targets.[773]

NASA finally launched OAO 2 on 7 December 1968 with an Atlas-Centaur, inserting the Wisconsin and SAO payloads into a nearly circular orbit some 760 km high. With all systems "go" within a week, SAO staffers threw themselves a party as *SAO News* cheered "Celescope Works! OAO Is A-OK" (Figure 39). By late June 1969, Whipple reported that "the reliability and performance of the Celescope experiment in orbit have followed almost exactly the prelaunch predictions." By then, Celescope had secured 2,500 star fields in 900 positions, a good fraction of the total (some 1,250) needed to complete the first selected area survey.[774] One of the three cameras failed after 77 days, however, and by then the remaining cameras were "exhibiting loss of sensitivity, owing partly to the effects of space radiation on our optics and partly to the effects of prolonged operation on the performance of the camera tubes." Preliminary data indicated, oddly, that stars in areas where there was more the sao NEWS

Smithsonian Astrophysical Observatory *Cambridge, Massachusetts* *Vol. IX* *No.1* *January, 1969*

CELESCOPE WORKS! OAO IS A-OK

At 2:49 am, Saturday, December 14, The Smithsonian Astrophysical Observatory opened a new window on the universe.

At that moment, the Number 3 camera of SAO's Celescope system aboard the NASA's second Orbiting Astronomical Observatory photographed in ultraviolet light three 6th magnitude stars in the head of Draco, the Dragon.

Later the evening of that same day, a group of very happy people gathered at Dr. Fred L. Whipple's home in Belmont to celebrate the success of a Smithsonian project that had begun more than a decade efore.

Co-hosts of the party were the Dr. Charles Lundquists and the Dr. Robert Davises, and special guests included Dr. Leonard Carmichael, Secretary of the Smithsonian Institution when SAO first suggested the concept of an orbiting observatory; plus, the several dozen scientists, optical experts, data processors, administrators, secretaries, and engineers who had been associated with Celescope since its birth.

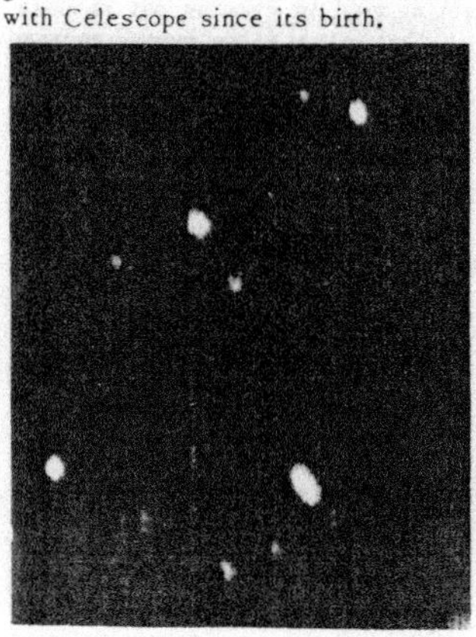

Celescope's UV view of stars in Draco.

The success party was particularly joyful. Indeed, almost at the same time as the last guests were leaving Elizabeth Road in Belmont, Bob Davis was already back at Goddard Space Flight Center supervising the turn-on of Celescope's remaining cameras.

EARLY RETURNS

During its first days of operation, Celescope reported seeing stars that appeared four times brighter in ultraviolet than previously seen in visible light. And an unexpected bonus of Celescope is its ability to observe and make systematic studies of Lyman Alpha lines of hydrogen emission near the Earth.

During the next two days, each camera was turned on in succession until finally, on Monday, December 16, all four cameras were operating simultaneously. On Tuesday, the Celescope sky survey technique was tested and then control of the spacecraft was given back to the University of Wisconsin.

For the next six months -- and possibly a year -- Smithsonian and Wisconsin will control the 4400-lb. spacecraft on alternate weeks. Both experiments are designed to observe ultraviolet light from stars, with the SAO making a general map of the sky with four cameras and Wisconsin stu ying the spectra of individual stars with seven cameras.

Celescope was born at the Harvard and Smithsonian Observatories just after the launch of Sputnik I and many months before the establishment of the National Aeronautics and Space Administration.

The original concept of Whipple, Davis, and others called for a single, 6-inch telescope. Eventually, their dream of a tiny stargazing satellite evolved into a full-scale program for several orbiting observatories as the OAO program became one of the major efforts undertaken by NASA.

The project was also one of the most complex and controversial ever to be undertaken by NASA. To look at stars in ultraviolet light meant that new and totally different types of observing systems had to be developed. And because astronomers -- at first, anyhow -- could not ride aboard the satellite, observations had to be transmitted back to earth for analysis.

To solve these problems, Westinghouse developed a special uvicon television tube sensitive only to ultraviolet light. And EMR, Inc., designed a system – in effect, a small space television studio -- to receive the pictures and transmit them back to earth.

Understandably, the project often bogged down with the pure problems of creating entirely new instruments and observation techniques. Scheduled to fly aboard the first OAO in 1965, the Smithsonian -- luckily -- was unprepared for flight.

(Continued On Page Four)

Artist's conception of OAO in flight.

Figure 39 *SAO News* celebrates Celescope, safely in orbit and producing data. *Smithsonian Institution Archives, Acc. 13-120, Box 1.*

obscuration, in the galactic plane, were brighter than expected and hence more of them appeared than was predicted. This was certainly an unexpected result and, in fact, caused considerable confusion in coming months and years.[775]

Just as he was tasting some success, Whipple learned that NASA was going to cut back his tracking programs. Bradley had been assured by the Tracking and Data Acquisition office that NASA still "considered this work to be valuable" and so maintained its budget in the face of "severe budget pressures." Now, however, the FY 1970 budget projections looked really grim, so NASA wanted to phase out SAO's optical-tracking network, though Newell hoped to ease the pain as much as possible by letting the SAO retain some staff salaries for other projects.[776] But this did not please Whipple, who was just then trying to improve the network using new laser-tracking technologies. The SAO was momentarily successful in this, but it put a pinch on other programs, including Celescope.

The media dramatized the first successful OAO. The *Washington Post* claimed on 19 April that it had "uncovered so many unsettling facts about the stars that astronomers themselves are bewildered by what they all mean."[777] Indeed, there was confirmation that the ultraviolet continua of the hottest stars were fainter than predicted, but this had been shown repeatedly by sounding rocket flights. A nearby galaxy had a nucleus that resembled a quasar spectrum, leading to speculation that quasars are more common than expected. The Pleiades became the "brightest cluster of stars in the sky," which was not surprising because it is the nearest cluster of very hot stars. Speculation on the cause of the increased brightness included the discovery of an extended cloud of material surrounding the Pleiades. Wisconsin Experiment Package astronomers Art Code and Bob Bless were quoted multiple times, as was Goddard's James Kupperian. No one from the SAO was quoted in print, although the breathless article was adorned by Celescope's image of the Pleiades star field, without attribution.[778]

Later notices were more balanced. The media, prompted by NASA, made much of Celescope's observations of a liquid oxygen (LOX) dump from Apollo 13's booster rocket on 11 April 1970. NASA had requested this service, giving only a few days' notice.[779] Otherwise, the routine became pretty stable throughout Celescope's lifetime, which lasted through 26 April 1970.[780] More than 8,000 frames were exposed, covering some 3,800 fields. Data reduction began slowly in early 1969, and by the end of the year enough material had accumulated to begin checking data consistency. Quite soon, the team did not like what it was finding. And this came just at the time when there were reductions in force and restructuring because budgets had been expended and NASA threatened more cuts.[781]

Davis's chief responsibility after launch was, as he recalled, "getting clean data, somebody else can figure out what it means."[782] But getting clean data was not a straightforward, linear process and required the expertise of astrophysicists who could perform boundary work between observation and theory in the realm of stellar atmospheres. These astronomers would also become the ultimate beneficiaries of the intended product, the *Celescope Catalog*.[783]

The model atmospheres group Whipple had judiciously formed, his Menzelian "research reserve" led by Charles Whitney, was quickly activated. Eugene Avrett recalled that when they compared Celescope data with their predicted fluxes, the "accuracy was a little bit disappointing." The greatest problem was calibration: The tubes behaved erratically and changed constantly during flight. Even the on-board calibration lamps and the preplanned technique of exposing overlapping fields to intercompare brightnesses were not sufficient remedies to keep everything consistent.[784]

Haramundanis managed the initial reductions, organizing the data from the four cameras, roughly matching them up and confirming coordinates and star identifications with entries in the new *Smithsonian Star Catalog*, a project she had recently completed.[785] She recalled numerous problems: The telescopic fields were not perfectly aligned, "which complicated the star identification part significantly."[786] There was always the fear that a color could be way off because the wrong star was being compared. Through the summer of 1970, the first data releases rendered a +/−0.4 magnitude confusion circle, which was unacceptably large. With more data they beat it down, but had a long way to go to their goal of a +/−0.1 magnitude confusion circle, measuring brightness to 10 percent.[787]

By the end of the year, Lundquist knew that the core team needed help. He and Whipple tapped senior staff to form a Data Processing Implementation Task Group.[788] The astronomers in the stellar atmospheres group teamed up with mathematically adept specialists in data processing from throughout the Observatory. Avrett remembered weekly meetings held to avoid, what they all felt, would be a "near disaster . . . because we were faced with the enormous embarrassment to the Smithsonian: 'Here we spent all this money and we have all this data, and we can't do anything with it. Sorry.' "[789]

Just about the entire Gaposchkin family joined in, save for Sergei. Haramundanis's husband, Edward, a mathematician, her brother Mike Gaposchkin, and her mother all got involved through the rest of 1970. Once, Haramundanis recalled, when the group was contemplating some form of "back analysis," which was to model the poorly determined parameters as random variables and then treat them statistically, "My mother was horrified, and she walked out of the room. I have never seen her do that before."[790] Payne-Gaposchkin, for all her brilliance, was still a classically trained astronomer comfortable with a minimum of data massaging. What Celescope called for, however, was a staff willing to deal with messy data, and who had the talent and training for the task at hand.[791]

Whipple, uncharacteristically, attended every review meeting of the Data Processing Implementation Task Group. Lundquist worried over the details and kept lines of communication open with NASA. Lundquist was especially useful because his PhD thesis at Kansas University was in theoretical radiative transfer. He "could talk with the people that were doing stellar atmospheres."[792] The weekly meetings, Avrett recalled, were more like daily activities for Lundquist: "It was a daily problem getting up, and what's the plan for tomorrow" was the drill. He brought not only experience in crisis management but also a certain calm confidence learned under von Braun at Huntsville, having had the task of directing a real-time modeling effort during the 1958 Project Argus nuclear explosions in space to ascertain their effect on Earth's magnetic field.[793] This was far from Lundquist's first encounter with Celescope, of course, frequently stepping in to deal with contractors. By mid-1969 and through to 1970, he knew once again that he had to step in directly because, as the calibration problems began to overwhelm the process, he could see that the stress on Davis "was terrible."[794]

The task group attacked the problem from two directions. The SAO atmosphere theorists generated models of stellar energy distribution. The observationalists used empirical methods, plotting colors and brightnesses in various combinations and then using these correlations to assess the consistency of the data.[795] Davis had selected some 50 young bright stars in Orion as a test field, and reported that the Orion stars had ultraviolet magnitudes at least 1 magnitude brighter than predicted by the latest theoretical models. Dave Latham confirmed this, presenting color-color plots from computer models that seemed to agree with Davis's for stars hotter than A0, but were systematically off by 1 magnitude for stars cooler than A0. The "source of the difference must be traced down," Latham urged, especially because there were large variations between Davis's magnitudes and those that had been measured directly during a rocketsonde flight by Naval Research Laboratory astronomer George Carruthers. Further analysis did not remove the anomalous brightnesses, so Rudolph Schild, as Lundquist reported in March 1971, "packed up his gear, and headed down the trail to Arizona" to make ground-based observations of these Orion stars.[796] His efforts led to an important new understanding of the nature of the hot stars embedded in the Orion nebular complex. But they also demonstrated that more than one method of data reduction had to be tried.

Lundquist agreed: "I'd been around the game long enough to value redundancy."[797] He brought in S. Ross & Company as general consultants because they had expertise in making first-order approximations from rapidly changing multi-dimensional data sets, which corresponded with the changing spatial stability and photometric sensitivity of the detector targets. And Lundquist also tapped Mike Gaposchkin to apply a method of differential corrections to the data using least squares for an independent analysis "so that we could check one against the other."

Avrett led a group looking at the interface of model atmospheres and the telescope. They interrogated the entire instrument and software analysis process from detector to transmitter to ground station to computer. Meeting constantly to sort out irregularities, they slowly and frustratingly gained some control, "figuring out how to map the sensitivity as a function of space and time."[798] Lundquist asked Avrett to lend W. A. Deutschman to be deputy project scientist, strengthening the computational side. Deutschman had experience at the High Altitude Observatory where he had performed laboratory studies in the extreme vacuum ultraviolet and would assist Davis in all technical aspects of the program.[799] From Avrett's perspective, once Deutschman got the stellar magnitudes into one consistent global database through repeated computer runs and iterative corrections, accounting for and smoothing the discontinuities as much as possible, they went as far as proper practice would allow. That happened in the spring, when the team declared, "Well, okay. Now it works, and this is it"[800] (Figure 40).

NASA was also eager to move on. Just before the November 1970 launch of the next OAO, a Goddard telescope to observe ultraviolet stellar energy distributions, Nancy Roman stunned everyone by sharply asking why the SAO team bothered at all, saying, "Let's cancel Celescope because this is so much better."[801] But OAO-B died when the payload shroud of

Figure 40 Dubbed the "Celescope Crew" by *SAO News*, Katherine Haramundanis, W. A. Deutschman, and Y. Nozawa pose with a celestial globe marked with the primary Celescope fields. *Smithsonian Institution Archives, Acc. 16-263, Box 1.*

the Atlas-Centaur launcher failed to open and to separate.[802] So OAO-2 data remained the only game in town until the launch of the European Space Agency's polar-orbiting TD-1A in March 1972, and then OAO-3, "Copernicus," in late August.[803]

It was hoped that something could be learned from the experiences of Celescope because, at the least, it survived launch and did operate as a space mission. As Davis noted at an AAS symposium in August 1971: "The development of these Westinghouse Uvicon tubes from a starting point where the SEC principle was a new laboratory discovery, to final successful flight operation, is a technological triumph of Celescope."[804] Lundquist, too, hailed the efforts of the some 50 key people who were aided by 100 more over the 13-year life span of the project who successfully isolated the "target charge buildup" problem, and took the steps to compensate for it. It was the most complex problem in the calculus of corrections and its isolation became one of the most significant products in the failure analysis conducted for Celescope.[805]

The Science

NASA viewed the OAOs and OSOs as national observatories. Before launch, NASA's Office of Space Science and Applications instituted a "guest observer" program whereby any astronomer could apply through NASA to study specific objects at times when the instrument scientists could not conduct their own observations. Davis alerted all SAO staff to the opportunity, and collected some 16 responses. Celescope was, of course, a mapping program so did not allow for great flexibility. But the array of telescopes would be available at times when none of its "selected areas" was accessible.[806] SAO proposals included the study of subdwarf atmospheres, silicon-rich stars, A-peculiar stars, studies of the influence of line blanketing, and studies of star-forming regions. SAO astronomers also proposed studies of pulsating variable stars and stellar rotation. There were also proposals for targets of opportunity: comets, novae, or supernovae. Many of these general topics were shared by the Wisconsin team and by guest observers who came from 11 American and Canadian institutions, and three from Europe.[807]

The August 1971 meeting where Davis spoke was sponsored by NASA and the American Astronomical Society (AAS) as a joint symposium on results from OAO-2. Arthur D. Code, who led the Wisconsin Experiment Package, organized the two-day symposium around the specific celestial realms that had been impacted thus far. These included solar system objects, interstellar matter, stellar observations, and galactic and extragalactic systems. Code asked his staff to prepare technical descriptions of the Wisconsin telescopes and its operating system, and invited Davis and his team to do the same for Celescope.[808]

Forty-two scientific papers covered the topics identified by SAO personnel. But only five of them applied Celescope data, and of those, only one was by an astronomer external to SAO.[809] SAO astronomers Rudolph Schild and Frederic Chaffee recounted their involvement in Data Processing Implementation Task Group discussions that dealt with the classic problem of recognizing when a new class of phenomenon, or new class of object, actually existed. In this case, there were no theoretical predictions, and the observational evidence itself was produced through methods and means not completely trusted.[810] Still and all, by the time of the meeting they were able to report that late B-type Orion stars had small Balmer discontinuities—in other words, they were brighter in the ultraviolet than field stars of otherwise the same spectral type. They were by then also able to discriminate between this phenomenon and the possible influence of space reddening, which preferentially dimmed stars in the ultraviolet. Both effects were acting, Schild and Chaffee concluded, and they were both quite real, indicating a new class of "Orion" stars. Lundquist hailed the finding at SAO's briefing at NASA Headquarters the previous March: "The anomalies in the Orion stars are typical of other areas of research which can be expected to be revealed to the astronomical community as more *Celescope Catalog* data becomes generally available."[811] But by the time the catalog was issued in 1973, Haramundanis and Payne-Gaposchkin "discussed the results and concluded that the errors were so large in the data that there was very limited information that you could obtain from these observations."[812]

Still, as with the detection of the Orion class, the Celescope data found other applications. One of the persistent interests of SAO staff was mapping the characteristics of the interstellar medium, trying to determine whether the regions around supergiant stars were different from regions not under the influence of these vastly bloated gas spheres. The Celescope observations indicated at first that a peculiar "extinction bump" was indeed caused by the stars, but this was shown not to be the case by the Wisconsin results.[813] This led SAO astronomers to examine the distribution of material in the interstellar medium, comparing new ground-based photometry from Kitt Peak and Cerro Tololo with the Celescope data. From this, they concluded that the interstellar medium was homogeneous in the direction of a spiral arm of the Milky Way and inhomogeneous perpendicular to the arms. These broad but useful conclusions gathered 174 citations in following years.[814]

By the end of the 1970s and through the 1980s, Celescope data were cited here and there, mainly because the data remained the largest body of ultraviolet data available. But there were soon much better data from newer ultraviolet space telescopes, like the smaller International Ultraviolet Explorer, which indeed employed SEC Vidicons that, although vastly improved over the first-generation Uvicon, owed much to the lessons learned from Celescope. By 2003, in a compilation of some 800 publications on the ultraviolet colors of early type stars, Celescope data were acknowledged as the last of the "classic" catalogs starting in the 1950s.[815]

A Summation and Assessment

Celescope started as a project within the Upper Atmosphere Division and quickly became a division in its own right while Whipple competed for control over functions soon absorbed by NASA. That happened quite quickly, so by 1963, Whipple had redefined Celescope in another new division, Stellar Observations, that included programs related to the mission but also discipline-based and technology-based problems. By then, some 12 professional staff led dozens of specialists and technicians in problem areas identified as "UV sensitive TV cameras," "UV pointing sources," "UV identification catalog," and "SAO Star Catalog." In addition, at least one program in the Stellar Theory Division covering studies in stellar atmospheres closely aligned with Celescope. Finally, as Whipple's organizational structure continually changed, in 1961 Celescope ended up in a division explicitly devoted to "Space Science." It was to coordinate the SAO's many studies of the behavior of satellite orbits and other problem areas addressed by spaceflight, such as geodesy.

But by the time of Whipple's annual report to the AAS at the end of 1963, the technologically or functionally organized groupings disappeared. Personnel continued to accumulate, and work continued in those areas for some years on a project basis but the organization was defined differently.[816] Most of the programs survived, but were also redefined, not in terms of a centralized capability for national purpose, but in terms of traditional problem areas more familiar to the astronomer. This shift was likely the result of an emerging institutional maturity for SAO, but it was most definitely an institutional retreat from Whipple's early ambitions to establish a national facility for space research.

OAO-2 was indeed a watershed for the traditional optical astronomical community.[817] It was an early peek at the wavelength band closest to the hearts and minds of most practicing astronomers, but it also demonstrated the investments required in staff and energy, and the risks, of doing astronomy in space. The contrast between the Wisconsin Experiment Package and Celescope could not have been greater: The former, highly successful, built on expertise and technologies already within the grasp of astronomers, was led by astronomers who had already committed to the specialty. Celescope, however, was successful mainly in NASA's official view: The machine was launched, inserted into a proper orbit, and actually worked as designed for over a year, more than the announced lifetime. Of less concern to NASA was the science. In June 1971, assuming that astronomers would not object, NASA threatened to remove remaining funds for data processing and turn off the satellite. Astronomers did object, alerting Congress. NASA released remaining funds.[818]

The risks taken by the SAO, choosing an experimental electro-optical detector technology in the hope that it could be quickly altered into a controllable and reliable astronomical device, returned only partial success. Yes, the Uvicons worked, but were not sufficiently stable to provide useful data consistently over the lifetime of the mission. Yet the experience gained by industry led to a stabilized SEC-Vidicon design that did work and was continually applied to astronomical satellites over the next decade, like the International Ultraviolet Explorer (IUE), until the technology was replaced by solid-state devices.[819] The risk was also justified insofar as it was in line with recommendations of a 1965 Woods Hole conference "Space Research—Directions for the Future," which concluded that television tubes were then not ready for astronomical use, but that "a reliable image tube with significantly greater sensitivity than photographic film and with electronic readout would be extremely useful in optical space astronomy."[820]

After the Celescope program was completely ended in 1973, Davis remained at the SAO for the rest of his career, joining other astronomers in various projects to observe globular clusters, x-ray sources, and OH masers, using the new space observatories such as IUE. Whipple had charged him with a responsibility that no one in 1958 could practically fathom. That he persevered at all showed a remarkable tenacity not only on his part, but a strong social and moral consciousness on the part of those around him. The fact that SAO staff were mutually supportive throughout this time can certainly be seen in the story of Celescope. Recalling the incredible pressures of that time, Robert and Ruth Davis gave observatory secretary Helen Beattie top marks for managing the tension in a way that rallied everyone to the cause[821] (Figure 41).

Speaking before a national audience of astronomers in late 1971, Whipple reflected on the costs of Celescope, admitting that the emphasis and funding placed on spaceflight during the past decade suggested that "vested interests have indeed affected our national decisions to some extent." He openly admitted to having "contributed to this imbalance" by promoting these interests himself. But he didn't offer excuses, nor did he back away from his view that science had come "first with methodology and technology second." He tried to rationalize the process, taking into account the costs, "both in dollars and in scientific manpower," to make the observations as valuable as possible. Only this would make the effort worthwhile, and would "compensate also for the fact that time itself is a factor for the individuals who must put many years of their scientific and technical lifetime into such a project."[822]

Figure 41 Helen Beattie joined Davis, Carmichael, and Whipple to pose for photographs at a celebration for Celescope. *Smithsonian Institution Archives, Acc. 16-263, Box 1.*

Indeed, in the almost 14 years since Whipple brought his staff together to contemplate ways to build his empire and realize his "dear dream" in space, Whipple had returned to Earth, telling his audience that "one of the gross imbalances in the space astronomy program has, of course, been the lack of ground-based equipment, particularly large radio and large optical telescopes." But these were, as we will see in the following chapters, just the new arenas he was directing his attention to in Arizona and New England.

In retrospect, even the chief player at NASA, Homer Newell, looked back at the OAO era with some regret. "Exposed directly to the outside pressures to match or surpass the Soviet achievements in space," Newell observed near the opening of his memoir *Beyond the Atmosphere*, "NASA moved more rapidly with the development of observatory-class satellites and the larger deep-space probes than the scientists would have required."[823] In words ironically in step with Whipple's own in his 1971 Philadelphia address and repeated in his 1977 oral history, reflecting even Roman's remarks at the European Space Agency (ESA) summer meeting in 1964, Newell finally admitted near the end of his book that "OAO was a bit out of step. . . . It had delayed satellite optical (visible plus ultraviolet) astronomy for about a decade, whereas a series of cheaper, simpler satellites could have kept research moving while work on a larger instrument proceeded."[824]

Whipple, of course, initially saw the effort as means to larger institutional ends—acquiring an infrastructure for space research—rather than pure knowledge production. Even though NASA assumed much of that infrastructure, the vastly expanded scope of the project it demanded left Whipple with significant room to grow. One of those areas of growth became what arguably has been Celescope's most lasting institutional legacy.

Legacy of the "Research Reserve"

The theoretical group built up by Whitney remained a core asset to the SAO in future years. It provided expertise during the planning for and processing of Celescope data, because the leaders in the group appreciated that they had been hired for a variety of tasks addressable by their talents and training. But what this produced was in fact something more that would outlive Celescope itself. Through 1961, for instance, Whitney, Krook, and at least a half dozen staff and Harvard students contributed dozens of papers to the literature on solar and stellar atmospheres, variable stars, Earth's atmospheric density profile, and geodetic information based on the first satellite orbit calculations. This productivity continued, creating the infrastructure for attracting and training new generations and thus meeting another goal originally set by Carmichael when he agreed to move the APO to Harvard.

Owen Gingerich's involvement provides insight into how the infrastructure was built and applied. Returning to Harvard as lecturer in late 1959 after teaching in Beirut and at Wellesley, Gingerich had a PhD thesis to finish in non-gray stellar atmospheres he had started years earlier working under Cecilia Payne-Gaposchkin and would continue under Whitney. "And when I came back the world had changed. Smithsonian had moved to Harvard along with a big computer."[825] The expertise Gingerich acquired, applying ever-larger computers to modeling stellar interiors and atmospheres, made him a "local expert" useful to other projects. He was asked to "plan the computing aspects of the data flow" from what would become Celescope, which included fine pointing the instrument, an operation that had to be piggybacked on the primary rough pointing of the spacecraft from Goddard. Establishing the interface proved to be elusive, largely because, as Lundquist well knew, "the [NASA] engineers were so vague at that time as to how exactly this interface was going to work. They could talk about it for a whole week without giving us any specific instructions of what we actually had to do."[826] This only increased Gingerich's responsibilities exploring computer techniques for a wide variety of applications. He recalled that it was very clear even to a junior staff member why the atmospheres group had been formed, and that over the years waiting for the flight they did what they could to improve knowledge of ultraviolet stellar continua through improving opacity models.[827]

Large, high-capacity computing facilities were the key to the stellar atmospheres effort.[828] The computers existed because of satellite tracking and made the SAO an attractive place to conduct highly intensive computer applications, like studies of Earth's upper atmosphere through satellite orbit analysis, or the creation of a Standard Earth model.

SAO was the first observatory in 1958 to acknowledge the use of its own IBM 704. In the early 1960s, IBM 709s and the first 7090s were limited to only a few major centers such as MIT.[829]

The stellar atmospheres group provided a career pipeline for Harvard graduates. As Eugene Avrett completed his PhD thesis under Max Krook on temperature distributions in non-gray stellar atmospheres, he was thinking about a job. He knew what was happening at the SAO, and through Krook was aware that there were broad-ranging opportunities for theoretical studies. Whitney soon hired Avrett to perform non-gray stellar atmospheres modeling.[830] Like Avrett, Wolfgang Kalkofen came to SAO via Harvard physics and was mentored by Krook. He and Avrett in turn mentored Harvard students, many of whom took positions in the theory group. By the late 1960s, Avrett graduate student Robert Kurucz led a project called ATLAS that provided a model atmosphere computer program for hot stars in both convective and radiative equilibrium that became very highly cited and much applied over the years.[831] And in the next year, Gingerich, Kalkofen, and Robert Noyes built "The Harvard-Smithsonian Reference Atmosphere."[832] It became a standard reference, using extreme ultraviolet observations of the Lyman continuum of the Sun from OSO IV to create a powerful new model of the solar photosphere. This collaboration was also enhanced by the expertise of at least a half dozen other specialists working on both the Harvard and Smithsonian sides of the house.

Noyes came to Harvard as a postdoc scholar in 1963 from Caltech. He was attracted to the work Whitney was doing and during his postdoctoral year became excited with Goldberg's solar satellite project. Indeed, this crossover behavior was encouraged; members of the stellar theory group enjoyed considerable flexibility in constructing ad hoc teams to attack various problems, both in preparation for the analysis of Celescope data and generally for studying problems related to the solar atmosphere and stellar atmospheres. There was considerable institutional overlap in support and services. The Smithsonian, for instance, supported members of Goldberg's solar satellite project,[833] and, of course, Harvard expertise informed the analysis of Celescope data.

In addition to their publications and computer programs, the group also attained visibility organizing a series of widely attended conferences on stellar atmospheres. Starting in 1964, they attracted specialists from all over the world to deliberate over standards and practices, seeking an "archetypal" model that could serve the community. The conferences grew over time; Gingerich led a third conference in April 1968 that attracted more than 100 workers from 19 countries. Overall, led by Whitney and Krook, Gingerich felt that "it was the stellar atmosphere work that was really giving the Smithsonian Observatory its scientific reputation."[834]

Chapter 11

Building the Infrastructure for Satellite Geodesy and Aeronomy: Work of the Research and Analysis Division

The funds flowing to the Smithsonian Astrophysical Observatory (SAO) via NASA and the Air Force, in the millions of dollars annually, made possible a large SAO computing bureau and workforce in computationally intensive fields such as geodesy and aeronomy.[835] The SAO also received large grants for Celescope during these years, amounting to millions more. Celescope made it possible to hire technical specialists to build the instruments, and astrophysicists to manage the data. By the late 1960s, large computer centers became core facilities in astronomy, as well as in the rest of the physical sciences, and in effect defined competitiveness almost as much as telescopic aperture. Their presence and continued expanding capacities were regarded as key assets to the Harvard campus as well as to government sponsors.

Through the 1960s, Whipple's annual reports describing SAO activities evolved from building a capability through service functions to addressing intellectual problems. The large and expensive formal infrastructure that contained many staff, however, remained and grew into divisions called, variously, Communications, Data, Photoreduction, Computation, and Research and Analysis.[836]

Members of the Research and Analysis Division were on tap to assist in a wide range of problems across SAO. Whipple identified them not by that division, but by the problems they were addressing. In 1963, for instance, George Veis was listed along with staff in the "Planetary Sciences" area under "Determination of Tracking Station Coordinates." But he also appeared in the "Stellar Observations" area, assisted by Haramundanis, producing the SAO *Star Catalog*. Luigi Jacchia headed the meteor reduction section within "Meteoritic Science," but also the "Atmospheric Drag on Satellites" group within the "Planetary Sciences" area. Others in the division, including Don A. Lautman (Meteoritic Science/Zodiacal Light) and Imre G. Izsák (Planetary Sciences/Earth's Gravitational Field), appeared in several other problem areas.

Whipple rationalized his infrastructure in terms he felt made it marketable for intellectual ventures. After all, at various critical periods, especially at the close of the International Geophysical Year (IGY), he was faced with the challenge of converting what had been a finite service activity into a permanent fixture on the landscape, competitive for funding from sources that were potential contenders. By late 1958, the SAO was no longer responsible for tracking every satellite that was launched. NASA had parceled out assignments to a number of networks, including SAO, to track only specific satellites, based upon the capabilities of each network. So as he plied NASA for continued and even expanded funding in 1960 to maintain and upgrade equipment, he also argued for "the need to strengthen our scientific staff . . . to realize the full scientific potential of this research program."[837] His senior staff, more or less, worked within this model, looking for ways to apply the SAO's growing infrastructure to the most interesting or pressing intellectual and practical problems, arguing that the SAO was the place for both.

Without the cash flow brought in by the network, where would support for the science come from? In a 32-page proposal prepared by Whipple's staff for renewed annual funding to the tune of over $4 million for tracking in 1964, they emphasized both scientific objectives and service functions. As part of the latter, they argued that the infrastructure SAO had created for the IGY had to be available for the upcoming International Year of the Quiet Sun. In like manner, the infrastructure they had built up for geodesy had to be preserved to support a joint program between NASA and the Department of Defense (DOD) to build a special flashing beacon satellite called Project ANNA, as well as its proposed extensions.[838] Maintenance and upgrade, therefore, had to have specific goals that were marketable. A good example was a suggestion by the SAO geologist Ursula Marvin in 1966.

Uses for Satellite-Tracking Data

Marvin had many interests in parallel with meteoritics, such as Earth and lunar science. Anticipating her collaboration with Wood that found that the lunar highlands were the product of isostatic balance, she had earlier been interested in using satellite data to monitor continental drift. In a review of the problem in 1966 she argued that "the ultimate solution to the continental-drift problem may come from satellite geodesy," even though, she admitted, "the weight of the evidence available today favors continental drift."[839]

Still, drift had not been unambiguously detected by the mid-1960s, so Marvin's argument gave the tracking network data another purpose. By then, the SAO Baker-Nunn network (now called "SAO astrophysical observing stations" to make them seem more relevant to the work of an astrophysical observatory) could pinpoint each station to an absolute accuracy of 10 meters, and there was the promise of an order-of-magnitude increase in accuracy if and when laser ranging facilities were added to each station, assuming that future satellites would contain reflective properties, as Whipple had hoped for OAO-II. This capability, Marvin predicted, would make then-present drift rates detectable by laser measurements of satellite motion from at least three stations situated on two separate land masses sometime in the next 5 to 30 years, depending on the actual rates of present drifting.[840]

Marvin's proposal was part of a larger effort at the SAO in the summer of 1966 "to review and explore the research opportunities offered by satellite tracking." Organized and led by Lundquist on Whipple's behalf, participants prepared arguments describing what could be done, and how the Smithsonian was the place to do it. After a summer of weekly meetings, each led by a different staff member, more than a dozen staff responded with detailed essays and proposals. Marvin's was among these, and others suggested everything from mapping the geopotential and assessing its variations over time to better understand the rigidity of Earth, to studying relativistic effects on the many motions of Earth's spin axis, to a further refinement of the structure of Earth's upper atmosphere and its oceans.[841] Introducing the summary volume of essays on these subjects, Lundquist invoked Whipple's original 1958 objective:

> To tie together the observing stations and the center of the geoid to a precision of the order of 10m, . . . to add appreciably to knowledge of the density distribution of the earth, particularly in the crustal volumes, . . . to provide precise information as to the density of the atmosphere . . . and periodic cyclic effects that may occur in the earth's high atmosphere.[842]

Lundquist then reviewed the Smithsonian's efforts in each of these areas, remarking with pride that "the objectives established some 10 years ago by Whipple have been achieved and most have been surpassed."[843]

After a decade of effort, SAO's satellite-tracking programs had made it one of the most visible astronomical organizations on the planet, indelibly linking astronomy with spaceflight. The SAO consciously promoted this visibility with adept and creative public affairs staff aiding the public information efforts of the International Geophysical Year through assisting the National Academy of Sciences and later to some extent NASA. In the first six years of its existence it not only capitalized on but also helped to create the popularity of Moonwatch using its outreach resources in newspapers and popular science magazines.[844]

As the SAO's infrastructure and expertise grew, so did its capabilities. Just as in the case of Marvin's promotion of the unambiguous detection of continental drift, however, there were also highly useful applications of great importance to the Department of Defense. But with its growth in capability came a threat to limit the extent to which the Smithsonian could continue to gather, analyze, and publicize knowledge about Earth.

Obtaining the Expertise to Process the Network Data

In addition to building staff for the observing stations and the administrative infrastructure to manage tracking and the Moonwatch campaign, there had to be people who could collect the data and turn them into useful knowledge.[845] The key to understanding the evolution of the photographic network through the 1960s is to appreciate the constant effort to obtain and retain critical expertise.

Leland Cunningham continued to advise after he returned to Berkeley, but Whipple needed in-house talent in celestial mechanics and geodesy. Until Sputnik, specialists in celestial mechanics in the United States were hard to find.[846] Through Hynek, he found George Veis, who had just completed his doctoral studies at Ohio State University in 1958 with a very timely dissertation, "Geodetic Applications of Observations of the Moon, Artificial Satellites and Rockets." Veis arrived at the SAO in 1959, becoming "the guiding hand of the program as it evolved from satellite surveillance to a satellite geodesy program."[847] Veis's training at Ohio was perfect; since the late 1940s, Ohio State had contracted with the Air Materiel Command's Mapping and Charting Branch and developed a strong geodesy program with the gravimetric portion led by astronomer N. T. Bobrovnikov, who with other faculty fostered a training program that graduated some of the strongest contributors to the field, starting with William Kaula in 1955.[848]

Veis returned to Greece in 1961 but continued, like Cunningham, to advise, retaining a dual appointment at the SAO through the 1970s. He collaborated variously with Luigi Jacchia as well as a host of young, mathematically adept astronomers such as Yoshihide Kozai and Imre Izsák. Kozai, from Japan, and Izsák, from Hungary, refined theory that would inform satellite reduction techniques for geodetic information, primarily for the determination of the exact positions of each station, all on the same reference frame. Directed by Veis, they then reversed this procedure, using the geoid to predict satellite positions. Thus, they created a practical procedure that linked all known physical, atmospheric, and electromagnetic perturbations in an empirical scheme that allowed for rapid comparison between computed and observed orbital elements, using the residuals to improve the computed orbit. What became known as the differential orbit improvement program (DOI), crafted by Veis with the assistance of those named above and a corps of students and SAO staff, was constantly refined, improved, and applied in both directions. It was more than a single program, however; it was a system of linked programs and required a broad range of talent to set up.

Kozai had been passionate about mathematics and celestial mechanics since high school in postwar Japan, consuming Hagihara's classic *Foundations of Celestial Mechanics*, and then becoming his student at the University of Tokyo.[849] According to Kozai, "In 1945–50 Japan was isolated geodetically," and so Hagihara assigned Kozai the problem, using lunar occultations. He studied secular perturbations and soon became engaged in lunar occultation observations and both astrometric and geodetic reduction tasks. Fred Whipple stopped off in Japan on his way to the 1958 International Astronomical Union meetings in Moscow, always looking for new talent in celestial mechanics, and through Hagihara, he found Kozai. He was impressed with Kozai's spirit and promise.[850]

Kozai took a leave of absence from Tokyo to work with Henize and Hynek. After computing an improved ephemeris of Sputnik, he developed a practical theory of the motion of near-Earth artificial satellites.[851] Kozai also paid attention to using the satellite data to examine Earth's gravitational field, using insights he had gained in a study of Saturn's north-south gravitational asymmetry.

Izsák had been a staff member at the Konkoly Observatory in the early 1950s, near Budapest, Hungary, working in celestial mechanics. He emigrated to Austria and then Switzerland during the Hungarian Revolution, and from there came to the SAO. By May 1958, he had finished a study of first-order secular perturbations on an artificial satellite caused by an oblate Earth, and by early 1959, with access finally to high-speed computers, was applying radial velocity techniques to analyze simultaneous observations of satellite Doppler-shifts from three stations.

Kozai, Izsák, Veis, and others Whipple brought to the SAO joined local talent such as Sterne, Whitney, and Jacchia, as well as recent Harvard PhD students such as Don Lautman, to build a new and highly competitive workforce. As Whipple expressed his strategy in his director's report for 1958, reduction and analysis was to be a routinized and highly efficient process. But to be so, it needed a core of experienced talent able to devise computational methods that could be made routine, for both hand and machine calculation.[852] Left unsaid in 1958 was the fact that those he collected together to devise the computational methods would also be best qualified to exploit the results. Meanwhile, there had to be people who could interact with the new machines that would vastly accelerate the calculations.

Computations for Predictions

In 1963, E. Michael Gaposchkin, then supervisor of the Digital Computer Program at SAO, co-authored a report with J. R. Cherniak presenting one of the DOI elements they had been developing. They called it "SCROGE," for "Smithsonian Computations Relating Orbital Glimpses Everywhere," a highly sophisticated system for predicting satellite positions that demanded a sizable fraction of the running time of the IBM 7090 then used for satellite tracking. It could determine when and where a satellite would be visible and what tracking rates were required in order to direct the Baker-Nunns to follow it.[853] SCROGE also produced an advance schedule that was wired to all observing stations, giving a one-week ephemeris for upcoming observations. Only in this manner could the various observing teams and the central coordinating facility properly manage workflow.

SCROGE was the critical element for maintaining a normative global operation; even though it was constantly being refined, it was one of the many core functions of the satellite-tracking program that kept the Smithsonian at the front line. DOI and SCROGE were typical of the elements of what was by then a huge computational effort that fed not only the communications infrastructure that kept track of satellites, but was also providing fundamental data for a wide variety of scientific goals, especially for geodesy.

By the time of their report, they felt that SCROGE could successfully direct the Baker-Nunn cameras automatically, though with close oversight. Indeed, field personnel soon were reporting that its operation was "uncanny."[854] A year later, Gaposchkin, whose position had gone through at least two name changes in the interim, including Chief of the SAO Computations Division that managed the program, provided a comprehensive operational template for applying what was by then the third edition of the DOI program, not only to acquaint "neophytes" in its use but also to refresh those already exposed and involved within constantly changing teams hoping to apply it to a growing variety of problems. Gaposchkin introduced it as an "inevitable" product of "the early years of the space age," starting with a primer in basic celestial mechanics, followed by a step-wise "how-to" briefing on how to apply the program.[855]

The DOI could provide predictions for observed satellite positions as a function of space and time, or its function could be reversed to apply the observed data to an analysis of the earthly positions of the observing stations. Both needed a machine-readable celestial framework: The positions of the stars themselves that were recorded in the Baker-Nunn fields had to be available in some machine-readable digital format readable by DOI, with a minimum of human intervention. Anticipating this need, in 1959 Whipple asked Veis to build a digital star catalog.

Making the Data Useful: The Smithsonian Star Catalog as a Geodetic Framework

Star catalogues, orderly listings of the positions and other properties of stars in the sky, form the backbone of fundamental astronomy. Early telescopic catalogues established a fundamental reference frame fixed to Earth, using visual observations of the exact time a star crossed an observer's meridian and noting its apparent altitude. By the end of the nineteenth century, however, largely through international efforts such as the *Carte du Ciel*, the *Astronomische Gesellschaft* sponsored a general catalog based on overlapping photographic fields that provided relative positions. These two techniques existed side by side through much of the twentieth century and reconciling the two became a central occupation of many observatories. The Smithsonian's efforts contributed to the solution of the problem.

One byproduct of the IGY and its progeny, the Space Age, was international competition and cooperation. For the latter, the U.S. Naval Observatory formed a cooperative arrangement with Yale University Observatory to meet a command order from the Navy to "provide in machine-readable form as complete a catalog as possible in both the northern and southern skies."[856] Soon, the SAO put itself in the same business. Just what constituted adequate completeness versus promptness and utility for various applications distinguished each institution's efforts, their rate of progress, and the reception of their product. Here we concentrate on the Smithsonian's effort, and will limit addressing the larger question to a discussion of its reception in 1966.

Whipple and the Smithsonian made their product eponymous, a Smithsonian trait.[857] To construct its star catalog, Whipple enlisted the gifted astrometric theoretician Heinrich Eichhorn, then at Wesleyan in Connecticut, as consultant to oversee data collection from over 70 distinct catalog sources representing the dozen major efforts since the late nineteenth century to catalog the positions, proper motions, colors, and brightness of the stars. Altogether they compiled and digitized entries for some 259,000 stars, issuing them in three different formats on magnetic tape. They also issued a set of 30 by 30 inch printed star charts that included star clusters, nebulae, and galaxies. By 1966, copies were available commercially for $20.[858]

Building the catalog was a complex and sometimes frustrating job for the staff. However, it nicely illustrates how the growing SAO staff infrastructure provided access to the talent and training needed. Katharine Haramundanis's career at the SAO best characterizes this trait. We have already noted how she started in the computations group working on satellite positions and brightnesses to issue her first "Catalog of Precisely Reduced Observations" in 1962 in a series of special reports. Then, under Veis, she led a small group of measurers and computers and also became involved with the *Smithsonian Star Catalog* after reducing satellite positions to the FK4 fundamental reference frame. She did have to learn the basics of astrometry, however, and again, she found ample guidance, not only from Veis and others but especially from Heinrich Eichhorn, who she recalls was the "driving force" behind getting everything onto a single consistent framework.[859]

Haramundanis considers the *Smithsonian Star Catalog* as the key ingredient making both the geodetic work, and the eventual Celescope cataloguing work, manageable. It was hailed internationally as a "*grand service pour les astronomes et pour les observateurs de satellites artificiels.*"[860] She felt fortunate having excellent programmers available and saw her own contribution as helping to rationalize the work flow into an organizational structure that routinized the process. There was also the sheer mechanical task of getting everything onto punched cards. The Yale Zone catalogs were already available, but little else, especially for the southern hemisphere. Data were entered twice to "ensure accuracy" but still humans had to be sure that stars were consistently identified in different catalogs. That was one of Haramundanis's responsibilities, along with the constant need to be "sure that the systematic corrections were accurately applied" following Eichhorn's direction. "That was the biggest problem."[861]

Another task, highly labor intensive, was to turn the tape catalog into a printed atlas of charts. Although the tapes and printed listings were the primary goal of the project, they decided that a printed atlas would reach a larger audience. The Smithsonian asked *Sky & Telescope*'s Joseph Ashbrook to write the foreword to what became known as the *Smithsonian Astrophysical Observatory Star Atlas of Reference Stars and Nonstellar Objects*, published by MIT Press in 1969 as a companion to SAO's 1966 *Star Catalog.*[862]

Although staff typically worked closely with the scientists, such as Eichhorn and Veis, they reported through the administrative infrastructure. Haramundanis was in the Data Division, which reported through middle managers and ultimately to Carl Tillinghast. When needed, Haramundanis worked directly for scientists, such as Robert Davis, in Celescope. In this manner, she, like so many others at SAO, acquired specific sets of analytical and managerial tools that could be applied to a wide range of problems. Tellingly, at first, "I thought of them as NASA administrators because it was all . . . NASA contracts. My check had 'Smithsonian Institution' on it, but from my perspective and the reason the Smithsonian had money for these programs was because they were getting money from NASA."[863]

Still, she took personal pride in the project, and later expressed disappointment that, although she had written draft introductions to both the *Catalog* and then later to the *Atlas*, neither was utilized. Whipple did identify her explicitly in his annual report.[864] And she did present the *Catalog* to a conference of leading military and civilian institutions concerned with uniformity issues at the University of Maryland in October 1966, which convened to assess "The Construction and Use of Star Catalogues." The conference called for establishing "an optimum, practical mode of operation."[865] And as if to answer that call, Haramundanis presented the *Catalog* as a single, coherent, electronically readable body. She described the decisions and the process of collecting the original data, processing them, and converting them into machine-readable form, including magnetic tapes, and in printed form as a star chart and lists. She discussed coverage, means of reducing all the data to a single framework, assumptions, and difficulties in the process. Her straightforward presentation stimulated one of the longest and most probing discussions at the conference.[866]

The question was, could the *Catalog* provide the basis for achieving a geodetic reference frame with 1 meter accuracy? If it could, and could be trusted, then it would become the universal standard. Some in the audience questioned this. Haramundanis replied that the *Catalog* was a necessary step toward that achievement. Others worried that it lacked sufficient data from the southern hemisphere. Eichhorn, it turned out, was most pessimistic: "He thought that 1-m geodesy was just a dream."[867] On the positive side, an astronomer from Ohio State recognized that "Smithsonian accuracy has improved," owing mainly to more powerful reduction procedures.[868] What was needed now was a more powerful tracking system. Another voice, from the U.S. Coast and Geodetic Survey, endorsed the Smithsonian Catalogue as the best available.[869] This brought out more support and soon descriptors such as "breakthrough" were bandied about. Finally even Eichhorn responded, echoing Haramundanis, that it was the standard against which further improvements could be made. Still, he wondered rhetorically "whether the SAO catalogue could be regarded as a cure-all."[870]

Although she felt that this put the Smithsonian on the map, she still was concerned that the observatory was not really contributing to astronomy as much as to an allied goal: "It was a good thing to do, but we didn't make an original contribution there because we weren't making original observations."[871] Haramundanis was sensitive to the classical role of an observatory: to make new observations. Indeed, her next assignment was to do just that with Celescope data, this one with completely new observations from beyond Earth's atmosphere, and with problems far beyond solution.

Putting It Together: The Smithsonian Standard Earth

By the time of Lundquist's summer 1966 exercise exploring all the possible applications of tracking data, the Smithsonian had already issued the first edition of the *Smithsonian Standard Earth*, the first unclassified description of Earth's gravitational field incorporating transoceanic distances with a precision of 10 meters based on both geometrical and dynamical data and analyses.[872] This was a watershed publication for Whipple, because it had been his stated goal since December 1957.[873] Virtually everyone supported this central effort. In his official history of the project, *Trackers of the Skies*, Nelson Hayes (Figure 42) summarized how Veis, Kozai, and Izsák derived the geopotential from the three-dimensional geodetic triangulations of satellites, taking into account higher-order asymmetries in the field.[874] Here we look briefly at how Whipple and Lundquist represented this work over time.

Figure 42 (*L to R*) Whipple, Menzel, Leon Campbell, and Carl Tillinghast hover over Nelson Hayes signing his book, *Trackers of the Skies*, in Harvard's Phillips Auditorium, 1968. Tracking team members also visible are Jack Slowey in the background and Rosemary Sweeney McPhee, with back to the camera. *Smithsonian Institution Archives, Acc. 16-263, Box 8.*

In his *Annual Reports* to the Smithsonian through the first half of the decade, Whipple always described the work as constant, steady progress. From 1957 through 1964, he employed hopeful terms such as "add appreciably," "achieved greater accuracy," and "greater exactness," in describing the acquisition of knowledge of Earth's density distribution, Earth's atmospheric structure, and its constant variation.[875] Typically, he described progress in terms of greater technical capability in the Baker-Nunn network such as the move from crystal clock standards to atomic time standards that enabled the entire network to be "synchronized to within a few milliseconds of time." In 1961 he held out promises, whereas in 1964 his rhetoric became more encompassing and operational, arguing that "three major geophysical topics" were at the heart of the effort: "the detailed representation of the earth's gravitational field; the geometrical relation between points on the earth's surface; and the density and temperature of the upper atmosphere and their variations." These all had to be studied simultaneously in a closely coordinated effort within a single institution. And, as he promised the regents, he made it his own personal responsibility to initiate "the coordinated attack on these problems and for monitoring their interaction with national and international programs."[876]

The Research and Analysis group developed Veis's DOI, incorporating more and more parameters over time, testing each one to understand the effects of its inclusion. An important extension by Izsák, Davis, and Gaposchkin in 1964 was to account for longitudinal effects in the geopotential, applying a sophisticated form of spherical harmonics to the solution called "tesseral harmonics." Both Izsák and Kozai were adept at applying this methodology, searching for zonal harmonic terms in early satellite data. In the tradition of celestial mechanicians, they acknowledged the ancients dating back to Tisserand, but adapted the conceptual geophysical framework developed by Harold Jeffreys and, more recently, William Kaula of UCLA.[877]

Knowing that they had to deal with a wide range of possible orbits, and that some of the most radical would also be the most revealing of Earth's dynamics, Izsák also investigated the long-term perturbations on low Earth-orbit satellites whose inclinations were near a critical value (some 63.4 degrees) that caused dynamical solutions to become unstable, rendering usual methods of approximating the orbit useless and requiring exact solutions using differential analysis. Far from being an academic problem, this solution was important in assessing what the satellite's motions revealed about the overall gravitational field as a function of the semi-major axis and eccentricity of the orbit.[878]

The 1966 contribution was updated and revised in 1970 when the Smithsonian acquired its laser-tracking systems and used them to follow some six satellites. By then, new surface-gravity measurements were available from William Kaula, and there were improved station positions based on the motions of deep-space probes monitored by the Jet Propulsion Laboratory's Deep Space Network.[879] The new solution could now boast geocentric station positions to between 5 and 10 meters of accuracy and the generalized geoid was known to 3 meters. Any further refinements required more laser stations.

This series of reports of constant refinement, together with Whipple's continued industry establishing international partnerships through the 1960s, ensured that the Smithsonian remained at the center of the continuing effort to improve the geodetic model. After the geodetic flashing satellite GEOS-A was launched in November 1965, the Smithsonian became the official site for the International Central Bureau for Satellite Geodesy, which coordinated reductions worldwide, connecting the European with the American world datum.[880]

The Smithsonian's "Standard Atmosphere"

Among his earliest duties starting in the late 1940s, Jacchia helped to revise and extend the National Advisory Committee on Aeronautics's (NACA) venerable "Standard Atmosphere"—a set of tables indicating the variation of temperature, density, and pressure with altitude, essential data for the aircraft industry. After World War II, jet flight and rocketry made higher realms both important and accessible, greatly increasing both the need and the ways in which the high atmosphere could be studied and exploited. Starting in the 1950s, Jacchia continued this work, with Whipple, Frances Wright, and others at both

Harvard and the SAO, using data from meteor tracking and later satellites to explore the properties of Earth's uppermost atmosphere.[881]

Based on his realization that meteors came in two families, Whipple believed that continued refinements in his meteor data, constantly re-calibrated by in situ measurements, would ultimately produce the most reliable atmospheric standards, as well as their seasonal variations.[882] Accordingly, when satellites became a reality, there was no question that Whipple and Jacchia would extend their work to satellite heights, and both, from their experience as part of the efforts of the V-2 Rocket Panel to update the NACA Standard Atmosphere, were well aware of the importance of standardization in establishing authority for the high atmosphere.

By 1962, with SAO astronomer Jack W. Slowey, Jacchia had collected enough information from the tracks of eight artificial satellites to be able to evaluate competing models for the high atmosphere. Separating out all effects, ranging from geodetic asymmetries such as oblateness and the diurnal bulge, as well as effects caused by the pressure of solar radiation, they were able to conclude that systematic changes in the temperature profile, seen in a semiannual cycle, could be explained by a "single underlying cause—corpuscular radiation," and that variations in this flux over the 11-year solar cycle were "quite considerable."[883]

Although there was a lingering hope among staff that they were creating a "Smithsonian Standard Atmosphere," practically their efforts helped to refine national standards, such as the U.S. Standard Atmosphere, organized by the U.S. Committee on Extension to the Standard Atmosphere, which had been established in the early 1950s and continued through the 1970s with collaborations by a broad array of military and civilian agencies. SAO reports and papers through the 1970s on the subject were cited widely by workers in upper atmospheric physics and geophysics engaged in modeling and simulations of the atmosphere, as well as workers interested in the physical processes active in the ionosphere and beyond. Among the latter was a small group at Harvard led by the British theorist Alexander Dalgarno.

Dalgarno had long been interested in physical processes in the ionosphere, the formation of forbidden-line spectra, auroral excitation, and association detachment. In the late 1950s, he spent a number of summers as, in his words, a "tame theorist" at the Air Force Cambridge Research Center, and became aware of the experimental work at the Smithsonian on the upper atmosphere pursued by Whipple's growing empire.[884] By the 1960s, Dalgarno, now at Queen's University Belfast, began looking for broader challenges for his theoretical expertise, turning to astronomy. In 1967, Leo Goldberg, the new Harvard College Observatory director, appointed Dalgarno on a dual appointment funded by the Smithsonian. As Dalgarno recalled, this was an especially good opportunity to get closer to observational astronomers and astrophysical theorists, and what made the deal sweeter was that he could bring his group of postdoctoral fellows with him, again funded by the Smithsonian: "We had a very exciting beginning with what was really a quite sizeable initial group." What really impressed Dalgarno was that "Fred was in a position, an extraordinary position that he could make almost, it seemed, unlimited appointments. And the appointments he made, he could make on an almost personal basis"[885] (Figure 43).

Figure 43 Alexander Dalgarno. *Smithsonian Institution Archives, Acc. 16-263, Box 6.*

In his *Annual Reports* starting in the late 1960s, Whipple identified Dalgarno's group variously as working on "quantum mechanics of many-body systems" and "decay of ionospheric metastable helium-like atoms," and by 1970 as "atomic and molecular processes." Whipple did not try to associate their work with that of Jacchia's group, described in 1968 as "drag on artificial satellites" and in 1970 as "atmospheric densities." Although the two groups worked in parallel focusing on the same region of space, their approaches were so different that there was little if any overlap. This intellectual separation would play out at the end of the decade as Harvard and Smithsonian staff debated their institution's future. But for now we will stay on Earth and examine another outcome of the Research and Analysis Division's industry that had been brewing for years.

Relations with the Military and the Threat of Classification

In early December 1958, Charles Lundquist, then working for von Braun, represented the Army Ballistic Missile Agency (ABMA) at a geodesy conference Whipple organized in Cambridge. Although junior at the time, Lundquist had lobbied within the military for keeping such work unclassified.[886] Whipple found in Lundquist a kindred spirit, and through the early 1960s, at least until Lundquist arrived at the SAO, the two remained in synchrony over these matters. The conference Whipple had organized in 1958 was in fact the first to publically link geodesy and space, a link explicit in both Whipple's and Lundquist's view that control of the information should not be restricted.[887]

The issue of control deepened after 1959 when NASA replaced the National Science Foundation (NSF) as the nexus for satellite tracking. The general shakeup, and deepening international tensions, led people to wonder how long would it be until all civilian tracking efforts were subjected to military oversight and control, and, moreover, who would decide what institution would track which satellite.[888] The first effort at accommodation was the creation of Project ANNA in 1960, a compromise between the Army, Navy, NASA, and the Air Force to develop a flashing geodesy satellite system. Homer Newell had testified in March 1960 that NASA wanted to develop such a system "to meet the needs of all parties, both civilian and military."[889] The question remained: Who would control it—would it be civilian or military?

Whipple had long been an advisor for Air Force (USAF) planning as a member of various boards and panels, most recently the USAF Science Advisory Board's ad hoc Committee on Sensors for Space Surveillance. In April 1960, Whipple received the Exceptional Service Award of the Air Force partly for this, and he parlayed his prominence to "mount a vigorous campaign to return ANNA to civilian control." In September 1960 he advised Air Force general Donald Yates on what NASA and DOD collaboration should look like. Whipple did not have much good to say about either NASA or the Department of Defense's (DOD) track record for openness. Lack of communication breeds waste and protectiveness, he argued, and "personally, I feel that the greatest hazard to the taxpayer's money lies in the pseudo-research programs conducted under high secrecy in the name of scientific research."[890] Project ANNA was, however, placed under DOD control by the end of the year, because, given the state of international affairs, there was much concern for keeping such knowledge classified.[891] Even so, Whipple objected publically over the cancellation of a dedicated geodetic satellite for NASA and the transfer of ANNA to DOD, and privately cautioned Bradley that "the idea that we should not be in geodetic work is, of course, ridiculous. I understand how this situation came about and suggest that you relax on it until I can deal with it in my own way."[892]

In 1961, Baker-Nunn staff reported concerns about security clearances. They knew that they needed "both the efficiency of a centralized management and quick system-responses time, and most important, the complete trust, good will and cooperation of our associates in the nine foreign nations in which our stations operate as guest institutions."[893] Whipple took the lead, using ANNA as a sounding board.

In October Whipple asked the chair of the President's Science Advisory Committee (PSAC), Jerome Wiesner, to "support the unrestricted use of a geodetic satellite and encourage the establishment of an international program utilizing the satellite to augment the scientific value of the program as well as the value of international relationships."[894] He also testified before Congress, joining a powerful team including James Van Allen and George Kistiakowsky.[895] When called to testify before the House Committee on Science and Astronautics, Whipple pulled out all the stops.

Whipple testified in March 1962 before the committee's Panel on Science and Technology, usually a friendly forum led by committee chair George P. Miller of California.[896] Whipple opened with at least a half hour of illustrated remarks crafted to demonstrate the thoroughness of preparing the worldwide SAO network. He featured the Smithsonian star catalog, then whisked the congressmen to SAO stations in Japan, India, and Spain, highlighting the interdependence of the many institutions involved. He demonstrated that the Smithsonian was active and operational, worldwide, as a civilian agent of the American people. He acknowledged the radio-tracking systems of the Navy, predicting that newer satellite systems with optimized radio transmitters, such as Transit, would be highly competitive.[897]

Whipple's main point was that all these techniques were needed collectively and their products had to be accessible. To make the system comprehensive required as many stations as possible, on land and sea. Each station not only required adequate optical- and radio/radar-tracking facilities, but also precise timing to at least one-thousandth of a second. The considerable costs involved for each station, however, could be reduced someday if satellites were equipped with their own precision clocks. And if these clocks triggered some form of flashing light on the satellite, visible from his tracking stations, all the better. To make his point, Whipple insisted that such a satellite should be openly announced with predicated timings "so that people over the world would know when and where to point their telescopes." Openness was essential for Whipple, who pictured a large, coordinated international effort, in the spirit of an extended IGY, where the satellite would be operated by NASA, coordinated through the Committee on Space Research (COSPAR) of the International Council of Scientific Unions (ICSU), and would "be welcomed internationally and would focus attention on our prime role as scientific leader in the use of space for the good of mankind."[898]

Then, in a flourish, with "good for mankind" still echoing in the chamber, Whipple read from a *New York Times* article about how ANNA, about to launch, was drawing critical concerns from scientists who had been blocked from open access to its ephemeris and to the properties of its flashing beacon. Miller asked pointedly if ANNA was military or civilian, which cued Whipple to turn to his cohorts and call for discussion and debate over the question of classification. Søren Henricksen, chief of the R&D division of the Army Map Service, testified that ANNA was neither NASA nor DOD originally, but was conceived under the auspices of the IGY as a civilian program. It was handed to NASA as a completely unclassified mission. Then Colonel Peacock of Army Map confirmed that ANNA (what had been coded "Project Betty") had been completely unclassified. However, they then both argued that the Minitrack-based system proved to be extremely critical because it taught the DOD how to increase precision and explained how this criticality convinced the DOD to acquire ANNA, with its flashing lights and Doppler ranging, and the expertise of both Navy and Army SECOR technologies (Sequential Correlation of Range). "At this time it became a classified project and NASA dropped out of the picture." From this, Whipple alluded to how "internal difficulties" within the DOD have stalled the program and now they planned only one launch: "The status quo of this project is very questionable??"[899]

Everyone Whipple called upon strengthened the argument that making ANNA classified compromised the project and hence the nation's ability to create an effective geodetic system. What was needed was an unclassified system of at least five or six high-flying satellites to tie together some "14 different major world datums." Pennsylvania congressman

James Grove Fulton then asked Whipple if it was advisable to include Russia in the effort. Whipple emphatically replied yes and referred to Joint Resolution 15 of the working groups at the fourth COSPAR meeting in Florence, April 1961. Kistiakowsky seconded this, and then Whipple sarcastically stated the obvious: "Considering the radius of damage from thermonuclear weapons, it is perfectly obvious that the accuracy for targets is perfectly adequate." Fulton then reminded the committee that Russia had appealed for open international cooperation to the UN Committee on the Peaceful Uses of Outer Space. The fact was, he added, "the nation which we feared most" is calling for cooperation. This prompted Kistiakowsky to call on Roger Revelle and other geophysicists to explain why openness was essential for science: direct detection of continental drift, polar wandering, and mantle shifting were at the top of their list.[900]

Miller cannily observed that Whipple's presence among the geophysicists at the hearings indicated that this desired information went beyond geophysics, to astronomy. So he asked Whipple to comment. Whipple eagerly complied, confirming that astronomers wanted to nail down everything that moved, confirming, tracking, and analyzing the motions. Then, as he often did when speaking before congressmen, he told a personal anecdote to illustrate why astronomers were interested, as well as to confirm the USSR's interest. Whipple had once visited the remains of the Pulkova Observatory near Leningrad. It had been completely destroyed during the war. Now, in its long restoration, he reported that they had found the original "fundamental measuring point" for Russian geodetic work dating from the mid-nineteenth century, and, with the help of the International Latitude Service, had located the exact spot to within a centimeter. "I think that is about the highest accuracy that we can consider at this time."[901] And in response to Miller's friendly query, Whipple added that these were all interrelated problems. They fit together in "an amazingly complicated and amazingly exciting jigsaw puzzle which we are trying to solve now; I think we are making enormous strides in that direction."[902]

This staged campaign in March 1962, aided by other members of the committee like Joseph Karth, carried the day and ANNA was declassified one month later, its management eventually given back to NASA.[903] Homer Newell recalled the result as more of an "apartheid" arrangement. Both programs would continue but be separate: NASA would be allowed open publication. The DOD would continue ANNA, where appropriate including NASA, "But DOD would decide unilaterally on the disposition of the data."[904] Still, shifting sympathies for declassification continued, especially in the aftermath of the Cuban Missile Crisis that October. After the first ANNA launch by the Air Force in May was unsuccessful, the second launch by the Navy in October was delayed "for fear the satellite would be mistaken for a ballistic missile"[905] during the crisis.

As DOD pursued the active flasher ANNA, NASA's Beacon Explorer series employed passive corner-cube reflectors that could be monitored by ground-based laser systems. The Smithsonian was one of the earlier institutions to experiment with laser tracking, setting up its first facility at its Baker-Nunn site in Organ Pass, New Mexico, in 1964. By 1967, it was cooperating with at least four other stations across the globe, in France, Algeria, Greece, and the original NASA facility at Goddard Space Flight Center in Greenbelt, Maryland.[906] The SAO facility at Organ Pass, a cooperative venture with

General Electric's Re-entry Systems Department, mounted a ruby laser to its Baker-Nunn. A searchlight mirror nearby collected the reflection. Their first observing run from late September through mid-October 1964 caught reflections from Beacon Explorer-A's 22 mirrors 19 times.[907]

The Smithsonian continued on in various programs such as NASA's Geodetic Earth Orbiting Satellite system (GEOS), as well as networks coordinated by the U.S. Geodetic Satellite Program's Geodetic Satellite Policy Board.[908] The SAO had primary responsibility for optical tracking but networks in the Coast and Geodetic Survey, the Air Force Cambridge Research Center, and the Army Map Service provided backup.[909] Throughout the decade, the Baker-Nunns remained the aperture champs, but it was also important to have as many stations as possible. Therefore, while the Air Force started by duplicating SAO's Baker-Nunn systems in its own network, it also added transportable camera systems to many more sites.[910] SAO's strong relationship with the Air Force, lending expertise and experience and sharing data, continued through Whipple's tenure as consultant and member of the Air Force's Scientific Advisory Board. SAO staff consulted, made site visits, and provided training for staff at additional stations in Canada. As one SAO network staffer remarked, "The B.N. game is almost becoming popular."[911]

Through the mid-1960s, the Smithsonian approached NASA several times for support to maintain its network, but the Air Force was consistently more responsive. It appreciated "the invaluable assistance of Whipple's extremely competent staff of trained scientists and engineers" and regarded continued SAO operations as essential: The "urgency and the interest to national defense of this program dictates that a contract acceptable to you be negotiated at the earliest possible date."[912]

The Air Force continued to supplement SAO programs by contracting for services, from tracking to analysis. SAO continued to train AF personnel and their contractors and to respond to tracking requests and general requests from other DOD areas. SAO's Maui Baker-Nunn station faithfully tracked Air Force launches, continuously furnishing the Ninth Aerospace Defense Division (NORAD) and SPACETRACK its results (including UFO as well as satellite data).[913] There was even a suggestion that SAO manage what was, by April 1964, the five-station USAF network, so that it would, as Whipple proposed to Ripley, "operate as efficiently and effectively as the SAO network."[914] What came out of this in October 1964 was a Memorandum of Agreement (MOA) between the SAO and NORAD to allow the two networks to operate in collaboration, sharing data on a "quid pro quo" basis with all transfers kept unclassified.[915]

Amid this positive activity, however, starting in the spring of 1965, some DOD officers tried to restrict data dissemination from USAF stations. In late April, responding to a Navy request that data and analyses from USAF Baker-Nunn sites in the Pacific be classified, especially Kwajalain and Johnston Islands, Lundquist resisted, reporting that he "told him I could not agree to such a position since we may never hear from DOD."[916] The Navy responded that "certain anti-satellite systems" installed at those sites that could "intercept and destroy satellites at several hundred miles" held "peculiar importance in the overall national security effort." The SAO's refusal brought threats from the Defense

Intelligence Agency (DIA) in August to nullify parts of the Agreement, and to classify the data already provided by the Air Force under that Agreement.[917] James Bradley, acting for Ripley, quickly responded with a strong letter of protest to Eugene Zuckert, the secretary of the Air Force.

This was hardly a "cold call." Bradley had been in contact with Zuckert since May 1964, carefully paving the way for the MOA and outlining the long process leading to the agreement. Bradley reminded Zuckert of the SAO's investment and why the DIA's demands were "a very distressing turn of events." The Smithsonian would no longer be able to exploit the data and would have to cease further support of the Air Force mission.[918] Zuckert, a Kennedy appointee and a strong voice for civilian control of the military, was due to retire in September, Bradley alerted both Whipple and James Webb. So Whipple pressed Webb warning how classification would hurt NASA's image: "Instead of the big promoter of space research in the world, NASA will be looked upon as a disguised DOD agency. . . . The SAO satellite tracking system is so closely related to NASA that any suspicion falling on NASA would also fall on SAO."[919] The SAO's allegiance was to openness and to NASA, and in turn, much of NASA's good image across the seas lay in this openness, fostered by the civilian SAO network.[920]

After meetings that went as high as the President's Office of Science and Technology Policy, the SAO weathered these threats to its continuing satellite-tracking program, weakened but intact. Data exchange continued between Air Force facilities and programs, including NORAD, and, as if to protect what independence they had, Whipple, Lundquist, Veis, and other SAO staff launched a visibility campaign presenting the SAO geodesy programs to public and professional audiences in 1966. They made their point by publishing the first version of *Smithsonian Standard Earth.*

Through the 1960s, security issues haunted the program, and serious budget cutbacks started taking their toll too. In early 1968, both the Central Intelligence Agency (CIA) and the Navy expressed concerns that the SAO was entertaining extended visits by mathematicians from Poland and Czechoslovakia, as part of a National Academy exchange program. Even though they would not have access to classified data, "it would permit [them] to participate in, and assimilate the more advanced American techniques for satellite geodesy," which would include "details of computer programs, operational procedures, and even partially developed theories" not available in any other way.[921]

With classification pressure came budget cuts. After expenditures for Apollo peaked and the Vietnam War escalated, NASA cut back in almost every area.[922] By summer 1968, station staffs overseas had been cut by 50 percent, and in May 1969 NASA threatened to cut funding for SAO satellite tracking by half again in FY 1970, which would begin in barely two months, and by 100 percent thereafter.[923] Friendly NASA managers warned Whipple to prepare somehow, possibly through the Office of Space Science and Applications, then headed by John Naugle.[924] After harried discussions, it became clear that the issue was not simply funding, but continuing support for the now-outdated Baker-Nunn tracking stations. Upgrading with at least five laser stations was essential, Whipple, Lundquist, and Bradley pleaded to a NASA committee led by Naugle, who promised to find a way to respond positively.[925] Some NASA funding did continue, supplemented by direct appropriations, establishing the enhanced stations, which Whipple claimed would be a new era for geoastronomy, "the study of the earth through astronomical observations."[926] In 1969, Whipple devoted almost two-thirds of his overall SAO appropriations request to geoastronomy.[927]

In the face of these challenges, Whipple worked hard to re-invent geoastronomy, for he knew that Ripley's staff was worried what BoB would think if NASA support lapsed. Whipple had to show "who needs the data, and why."[928] Advised by Bradley over the summer, Lundquist provided the "clarity and punch" the Castle was looking for, branding "geoastronomy" in terms persuasive enough to convince Bradley to approach Hugh F. Loweth of BoB.[929]

The FY 1971 budget call was a particularly sensitive affair for the SAO and for the Smithsonian. As BoB was transformed into the Office of Management and Budget (OMB), new levels of scrutiny and shifting funding priorities in what was a tightening fiscal universe brought into question the future of the satellite-tracking program and its geoastronomy aspirations.

Up to that time, Harvard College Observatory (HCO) tolerated the satellite-tracking program mainly because it brought in substantial operating overhead. But when relations deteriorated between the SAO and HCO at the end of the decade, and NASA funding was rapidly dropping, benign indifference turned into concern for relevance. This scrutiny prompted Whipple and Lundquist to search for broadened application and a redefinition of the enterprise. But despite Lundquist's heroic efforts, and the continued production of refined versions of the Standard Earth by Gaposchkin and his staff, when Whipple retired from the SAO directorship in 1972, and the two directorships merged into one within a new Center for Astrophysics, sympathies for satellite tracking and thus a large portion of the Research and Analysis Division waned in the 1970s. But by then, however, the infrastructure that tracking had helped to create had been turned to new challenges on the ground.

Chapter 12

Ground-Based Aspirations in Southern Arizona

At the May 1979 dedication of the Multiple Mirror Telescope (MMT) at the Smithsonian's Mount Hopkins Observatory in southern Arizona, the first large-scale optical telescope to explore a new philosophy of giant telescope building, Fred Whipple thought back to the 1960s and how he and Menzel felt what the Smithsonian Astrophysical Observatory (SAO) and Harvard needed was a "*real* observatory." Both observatories had small radio telescopes and observing stations scattered widely around the world, of course, but Menzel and Whipple still felt the sting of the Oppenheimer committee's judgment a decade earlier that Harvard could not compete with the great observatories of the West. Harvard's old Boyden Station in South Africa was still alive, though managed by a consortium of European observatories, and was less than ideally accessible not only because of its physical distance, but also by "anti-apartheid sentiment in the United States."[930] Because the Smithsonian had fully divested itself of the Table Mountain facility in California by the end of 1962, it was without a substantial and flexible foothold in the West, save for Menzel's Sacramento Peak Solar Observatory in New Mexico, still under Harvard management but wholly funded by the Geophysics Research Directorate of the Air Materiel Command as a narrowly defined facility. Harvard also had a small stake in a radio facility in Texas, the "Harvard Radio Astronomy Station" at Fort Davis, Texas, supported by the National Science Foundation (NSF). It boasted an 85-foot parabolic antenna or "dish."[931]

A "real observatory" needed a really big optical telescope and had to be capable of a broad agenda. "Really big" was hardly a constant throughout the twentieth century, but it was the chief driver for ambitious observatory directors and their patrons.[932]

Big Telescopes

Encouraged by post-Sputnik sources of funding but concerned that it retain a voice in determining what it wanted, the astronomical community began to think collectively about big telescopes in the early 1960s. To ensure that it had a voice, it organized itself into what would become a continuing series of "Decadal Surveys" on the needs of astronomy. The first survey, led by Albert Whitford in 1964, called for a new generation of large, ground-based optical telescopes.[933] Some astronomers on the survey committee saw an opportunity to rectify what they believed were inequities in American astronomy, such as the "East-West split," and so sought to "break the stranglehold that the California observatories have on large areas of research." Geoffrey Burbidge, then at the University of California, San Diego, was particularly outspoken, imploring East Coast astronomers like Whipple to take this seriously, and not concentrate only in space astronomy: "[R]ather than concentrate completely on so-called space astronomy . . . Harvard itself should build a large ground-based astronomical facility in the Southwest or in some other part of the world where the climate is good."[934]

Burbidge felt that Whipple and Goldberg as well as Lyman Spitzer were devoting too much attention to "space" and were forsaking traditional lines of investigation. Burbidge darkly observed that space astronomy was mainly an East Coast enthusiasm, "which as yet does not have many achievements to its credit but is over publicized, and in the West, the rather conservative and sometimes backward way in which astronomy is carried out by conventional means using large telescopes" is still the way that "our knowledge of the Universe advances."[935]

Whipple did not see matters the same way. He did not express concern over the East-West split and because his focus was "solar system astronomy," he did not need the very largest telescopes. He admitted that cosmology was very attractive, and certainly drew in "physicists and youngsters," but nothing, he felt, would happen for many years. Although the space program of 1964 was somewhat different from his 1946 dreams, still he expressed satisfaction that "the general progress is at a level that is not unexpected."[936] Even so, Whipple mused, with Smithsonian backing, possibly it was time to obtain "some strong ground-based equipment other than the specialized equipment that we have in meteors and satellite tracking." Instead of a single large instrument, Whipple wanted to experiment: "Certainly astronomy has lagged woefully with regard to laboratory backup experimentation, and we can correct some of that now."[937]

Whipple was being both prudent and practical. In 1964, wizened by the torturous experience of building Celescope, he adopted a cautious, step-by-step procedure. After all, as with Celescope, there was no one on either the Harvard or SAO staff who had experience building really big telescopes or managing a complex remote observatory. And he was cautioned by an unfortunate experience Donald Menzel had in trying to convince his Harvard faculty that they should build a big telescope in Hawaii.

Options

In the early 1960s Menzel felt his time was running out and he wanted to make one more big splash before retirement. He had started several successful observatories in the West, but now thought of a truly "large Western Observatory." Something that could compete with the best, but optimized for his first love, solar physics. "It was no secret," Menzel wrote in his autobiography, "that NASA was then considering funding at least one or possibly two observatories and everyone was fighting for the preferred position." Working quietly with NASA program managers, he secured "a definite promise" that NASA would fund a major solar observatory in Hawaii, on Mount Kilauea, including a 90-inch telescope. He knew that the University of Hawaii and other universities were vying, but given his leading role in the discipline, he "received the commitment that Harvard would be NASA's number one choice." By the end of the year he briefed the Harvard College Observatory Council (HCOC), receiving mild support from some but not all. "Three 'theoreticians' to my great surprise . . . who didn't know which end of a telescope to look through, opposed it, bringing up objection after objection."[938] Two members of the HCOC, Charles Whitney and David Layzer, confirmed that they were skeptical of Menzel's program. Menzel had not developed an adequate management plan for such a distant station, and they worried that these plans would "subtract from our own funding."[939] Nevertheless, Menzel called for a showdown in the office of Harvard president Nathan Pusey. Pusey, according to Leo Goldberg, was shocked when Menzel arrived with a full contingent of HCOC members who were dickering over the value to Harvard of a station in Hawaii. Although he harbored sympathies for a meritocratic faculty, Pusey was not at ease with such a confrontation. "He was pretty angry about it. He just said in no uncertain terms that he couldn't support that project, that Harvard had its fingers burned many times in the past trying to operate a facility a long distance off like the one in South Africa. In effect, he turned it down just out of hand."[940] Goldberg felt that Menzel failed to win over his staff, and that Whipple certainly got the message. As Whipple told him years later: "They gave Donald a heart attack and nearly drove me out of my mind."[941]

Menzel did indeed suffer a serious heart attack in April 1965, and soon stepped down from the directorship. Goldberg (Figure 44) replaced him. Turning down an offer to run the Goddard Space Flight Center, Goldberg leveraged a full-time Harvard faculty slot so he no longer had to be supported by the Smithsonian.[942] At first, he and Whipple tried to combine

Figure 44 Leo Goldberg in the early 1960s. *Smithsonian Institution Archives, Acc. 16-263, Box 6.*

Harvard and Smithsonian interests to build something big, somewhere: in the Southwest, in Hawaii, or in South America.

In January 1966, the SAO and the Harvard College Observatory (HCO) jointly proposed building a 24-inch reflector as a starting point for establishing an optical observatory somewhere in southern Arizona. Also, stimulated by the first National Academy of Sciences Decadal Survey recommendations by the "Whitford Committee," plans were afoot to conceptualize something on the order of 100- to 120-inch aperture, with a joint expenditure between Harvard and the Smithsonian of some $3 million.[943] Menzel, now retired, kept the issue alive, offering advice and suggesting possible private funding sources, now that NASA had committed to the University of Hawaii. The options for a site remained Arizona or Hawaii.[944] Through 1967 and well into 1968 Pusey remained hostile to Hawaii; Goldberg still favored it, but confined his search to sites in Oklahoma, Texas, Colorado, and even on Kitt Peak, seeking possible alliances with a consortium of large Midwest universities.[945]

Throughout these deliberations, Whipple concentrated solely on Mount Hopkins. The site was very dark, had stable air and low winds, and was not too far from an emerging world-class center of forward-thinking optical astronomers in Tucson. And by this time, he'd already quietly built a few buildings and established some projects and liaisons there.

Searching for a Site

At the MMT dedication in 1979, Whipple also recalled how he and his staff, especially those responsible for satellite and meteor tracking, were always sensitive to observing conditions.[946] By the mid-1960s, his Organ Pass Satellite tracking station's observing conditions were rapidly deteriorating because of the growth of nearby White Sands, and the lights of the Las Cruces metropolitan area. He could not do anything about sprawl-induced sky glow, because that would be fighting local commercial interests. Nor did he want to, since he was looking for an excuse to move. He was interested in establishing stronger relationships with the growing astronomical scene in Tucson, so looked for a darker site there, one that would not only allow his tracking stations to record the faintest artificial satellites, but also provide unhindered views of the deep sky for testing out new instruments and new technologies.[947] "It was a case of opportunism, making use of the Satellite Tracking Program to get the real observatory going."[948]

Earlier site surveys for the Harvard Meteor Project had already yielded some excellent candidates that were reasonably accessible, cloud free, and very dark.[949] Since 1962, McCrosky had headed up a cooperative project with the Harvard College Observatory and MIT's Lincoln Laboratory, funded by NASA and the Air Force, to utilize high-speed sounding rockets to inject artificial probes into the high atmosphere at meteoric velocities to produce calibration data for mass-luminosity and air-density determinations. He also led the Harvard Prairie Network, so he had at hand a team of quick-response technicians attuned to remote fieldwork, and offered a ready complement to engage in additional site searching.

As had other astronomers, Whipple found the Tucson area to be the best given all constraints, but the range that the Kitt Peak National Observatory occupied was not very high, was not protected from Tucson itself, which was expected to expand, and was already quite full of telescopes. During a visit to Tucson, Whipple, along with David Latham and others, met with planetary astronomer Gerard Kuiper. Kuiper, Latham recalls, blurted out, "Mount Hopkins. That's where you want to go, Fred." Kuiper had flown over the site and found the air over Mount Hopkins to be very stable, "and that means the seeing is going to be great there."[950]

The ridge including Mount Hopkins in the Coronado National Forest was about as far from Tucson as was Kitt Peak, but it was shielded by mountains to the north, leaving the southern sky dark. Whipple, through Ripley, applied to the Forest Service in January 1966 for a use permit and, with some delicacy, notified Sen. Clinton Anderson (D-NM, 1961–1972), a Smithsonian regent, of the impending departure from his state. Anderson, then on the important Senate Committee on Aeronautical and Space Sciences, was none too happy, but accepted the shift on political terms: "I know how ruthless the people at the University of Arizona try to move everything to their State, and I am sorry to see this small operation suffer from that desire."[951]

As the SAO's contract officer, Mark Malec coordinated and executed legal land rights arrangements, not only with the U.S. Forest Service but with lingering mining claims in the area, McCrosky assigned Charles Tougas to lead a site survey for the tracking station as well as new instrumentation. Jack Coffey, SAO's personnel director, became project administrator to create a master plan.[952] Tougas had managed Organ Pass, as well as the Peru and Argentina stations, before he became field supervisor for the Prairie Network. His first step, through Malec, was to obtain permits from the Forest Service to build some 20 kilometers of narrow access roads with grades reaching up to 23 percent: "It will be no place for any observer who has a known drinking problem," Tougas once advised headquarters, worried about the dangers of driving such a road at elevations above 7,000 feet in the colder and wetter months.[953]

When the Organ Pass Baker-Nunn moved to the Mount Hopkins ridge, it was joined by a new laser-ranging system. Sites were also designated for a new 60-inch reflector and smaller instruments including an interferometric optical sky monitor. The site was also suitable for living and working facilities, including a library, electronics laboratory, and mechanical workshop. The Mount Hopkins summit, at 8,585 feet, was left for something really big in the future.

Whipple introduced the new observing station to the Smithsonian regents in 1966, and, significantly, redefined his organizational structure. No longer organized by funded projects, as he did in the first years, now he structured his reports carefully around broad problem areas (Meteors and Comets, Meteorites and Cosmic Dust, Planetary Studies, etc.), mixing them with descriptions of observing techniques (radio, optical, and flight experiments), methodology (theoretical astrophysics), and infrastructure (Central Bureaus). To justify retaining "Optical Astronomy," he claimed there were "too few telescopes and so little observing time available" to his staff so more telescopes were needed, specifically, a medium-sized telescope for spectroscopy.[954] But there was also experimentation, a "large light collector to observe Čerenkov radiation generated by gamma rays hitting the upper atmosphere." It is this last new program that best illustrates the process through which Whipple explored new opportunities to build his empire, and it as well provides insight into his methodology for building a staff capable of exploiting these opportunities.

Opening Up the High-Energy Universe

One might imagine that Whipple would build observing programs that complemented ongoing efforts in geodesy, the upper atmosphere, and solar and stellar atmospheres. But spaceflight and new sensor technologies had made other universes accessible, and he was out to conquer them—case in point: gamma-ray astrophysics.

In the late 1950s, Giovanni G. Fazio was a physics graduate student at MIT developing detectors for measuring the highest ranges of cosmic radiation. He worked within a large laboratory that fostered projects in cosmic-ray and high-energy astrophysics specifically to develop new detector technologies.[955] He then moved to the University of Rochester to build gamma-ray detector payloads for sounding rockets with support from a small NASA seed grant, and in 1962 returned to Cambridge as one of nine new physicists Whipple hired for a multitude of projects.[956] Supported by a NASA grant that Goldberg had brought to Harvard from Michigan, Fazio was first described as a "coexperimenter" on the first Orbiting Solar Observatory, searching for gamma-ray emission from solar flares. He had contributed to theoretical studies that predicted the nature of the gamma-ray spectrum expected from high-energy solar phenomena, and also lent his expertise to the prediction of gamma radiation from the galaxy. At the same time he helped to perform isotope ratio analyses to determine exposure ages of meteorites in E. L. Fireman's laboratory.[957] Typical for the physicists hired by Whipple, Fazio made himself useful wherever his expertise was of value.

In 1958 Cornell physicist Phillip Morrison excited physicists by predicting that whatever was producing the high-energy celestial radio noise known as synchrotron radiation had to have "a really hot beast in its belly."[958] Fazio, with J. B. Pollack in the Harvard Astronomy Department, extended Morrison's predictions and set limits on the sensitivity required to actually observe these sources.[959] In other words, they established how sensitive a detection system had to be to observe the suspected high-energy sources. Thus they joined in with other groups of high-energy specialists in Russia, Britain, Ireland, and the United States to develop techniques to detect gamma radiation from likely astronomical sources, such as the supernova remnant in the Crab Nebula and the beast lurking in the center of our galaxy.[960]

One way to study them was to send high-energy detectors to the top of the atmosphere and near space. But physicists had known since the late 1940s that the highest forms of radiation hitting Earth's atmosphere produced showers of charged particles with speeds greater than the velocity of light in that medium, which violated the law of special relativity. Thus they instantly had to lose energy by emitting visible photons. Nuclear physicist Pavel Čerenkov explored this mechanism in the laboratory in the 1930s and the effect has become known by his name.[961]

Fazio applied both techniques. In 1963, urged by Whipple, he proposed a gamma-ray experiment as a complement to SAO's Celescope package on the Orbiting Astronomical Observatory, which was not funded. He then joined an AVCO Space Systems Division proposal to Kitt Peak's space division for an Explorer-class gamma-ray package.[962] Finally, later in the 1960s, with Henry Helmken and NASA support, Fazio flew a series of piggy-back spark-chamber payloads on Air Force balloon flights from the Holloman Air Force Base, and with Smithsonian support collaborated with the Tata Institute of Fundamental Research for flights from Hyderabad to observe at the geomagnetic equator. Thus initiated by Fazio and his group, the SAO maintained a vigorous program employing a series of postdoctoral fellows and staff who have continued the exploration to the present time.[963]

Fazio's group also pursued ground-based optical installations to detect Čerenkov radiation resulting from high-energy air showers. First, they placed photoelectric detectors at the focus of a large ground-based parabolic mirror that during the day fed a solar furnace at nearby U.S. Army Natick Laboratories.[964] These experiments at Natick helped them refine both detector technology and the high-speed timing systems that were essential to meeting the predicted characteristics of the suspected secondary showers of visible radiation. Not surprisingly, they confirmed that there was a fundamental limitation caused by Natick's night sky brightness, caused by moisture, dust, and the lights of Boston. Mount Hopkins did not suffer these limitations.

The Čerenkov technique could be applied with available technology and did not require major funds beyond the SAO's regular congressional appropriations. Teaming up with three others, including Trevor Weekes, Fazio first employed a set of two 5-foot searchlight mirrors on the slopes of Mount Hopkins at the 4,200 foot level (Figure 45). They placed sets of commercial RCA photomultipliers at the foci of the mirrors and linked them in a fast coincidence circuit. Test observing runs between November 1967 and April 1968 helped to establish procedure, but showed that the searchlight mirrors were not large enough to actually detect optical showers from gamma-ray primaries. By then, however, the Mount Hopkins staff had built a new steerable 10-meter optical reflector at the 7,600-foot level. This new collector, variously called the "Large Optical Reflector" (LOR) or sometimes, unfortunately, the "Gamma Ray Reflector," was an array of small tessellated mirrors clustered in a circle with a common focal point. It would reduce the detectable upper limit significantly and further constrain the nature of the sources producing the radiation[965] (Figure 46).

At the MMT dedication, Whipple reflected back over a decade, recalling that he was "haunted . . . by the multiple lens system of the fly's eye."[966] But the multiple mirror idea was not new. Long attributed to Archimedes and applied by a former Jacchia colleague who built a 1.8-meter tessellated zenith telescope mirror in the early 1950s that employed 61 interlocking mirrors, it was finding favor among astronomers hungry for photons.[967] With elements based on the Natick solar collector and with cues from the innovative Hanbury-Brown-Twiss stellar optical interferometer at Narrabi, Australia, SAO staff led by Fazio's group constructed their dish from 248 adjustable hexagonal-shaped spherical mirrors mounted in a cluster on a commercial radio antenna

Figure 45 Trevor Weekes and George Reike with one of the searchlights used to search for Čerenkov radiation from gamma-ray showers. They were successful only when the collecting aperture was vastly increased with the Large Optical Reflector. *Smithsonian Institution Archives, Acc. 16-263, Box 2.*

positioner.[968] In operation it could concentrate some 85 percent of the incident light energy onto a 2.5-inch disk containing a set of photomultiplier tubes. With an effective field of 1 degree, the instrument was well designed to detect divergent Čerenkov radiation.[969] After it was completed in June 1968, Fazio and his team used it to conduct surveys over the next several years, significantly reducing the upper limits of possible gamma-ray sources.[970]

During the deliberations over the LOR design, the Program Management staff at the SAO had to negotiate with the Forest Service, which wanted to know at every step just how much of the mountain site was to be dotted with telescopes and crisscrossed with roads. They also had to satisfy local Native Americans that their instruments would not interfere with their cultural activities.[971] By late April 1967, a master plan was finally signed, and this set in place the initial scope of the project including the sizes and placement of the buildings for the LOR, the satellite-tracking facility, and observing pads for occasional projects. There would be constant modifications.

While the LOR was under construction, spaceborne detectors on Orbiting Solar Observatory (OSO) satellites and sounding rockets were showing evidence that, indeed, discrete hard x-ray and gamma-ray sources existed, and, in the Crab, the x-rays were

Figure 46 Dedication of the Large Optical Reflector in October 1968. (*L to R*) Tucson mayor James Corbitt, Whipple, and Rep. Morris K. Udall (D-AZ) in front of the 34-foot mosaic collector. *Smithsonian Institution Archives, Image # 92-1773.*

pulsing in tune with optical and radio pulses.[972] Therefore, there was every good reason to keep looking from the ground for the flashes. Deprived of a new "discovery," nevertheless the Mount Hopkins/SAO team joined the years-long search, constantly applying improved detectors and refining observing technique.

The high-energy astrophysics group greatly contributed to the breadth of SAO capabilities in the late 1960s and 1970s. It was supported on a combination of Air Force, NASA, and NSF grants supplemented and smoothed out by the Smithsonian's appropriated federal funds. Fazio, far from satisfied with available techniques, constantly improved his detectors and played a strong role in establishing the parameters of the campaign through an exhaustive survey of both theoretical predictions and established observational limits in 1967.[973] He also attracted like-minded physicists to SAO, like Weekes, who had come first as a National Research Council (NRC) postdoc in 1967 to work at Mount Hopkins, returned briefly to his base at University College, Dublin, Ireland, but soon accepted an appointment as resident scientific director at Mount Hopkins in June 1969. Weekes was directly responsible for the maintenance and continual upgrades for the 10-meter dish facility and its elaborations over the next several decades.[974]

Despite the group's early confidences that the sensitivity of their equipment was sufficient to detect predicted gamma-ray flux from the Crab, as well as their continued successes to reduce the upper limits by several orders of magnitude,[975] it was not until 1986 that Weekes succeeded, with a team of 11 physicists from collaborating institutions, to reach the flux levels of the central source and to verify its existence at a high significance level.[976]

Building a Team for Optical Astronomy

The Baker-Nunn station and the 10-meter reflector worked in the optical range of the spectrum, but also performed specialized tasks, reflecting Whipple's interests in meteor trail reconnaissance and air glow. But Whipple also viewed Mount Hopkins as a site for a "real observatory" and so, in a careful stepwise fashion, initiated steps in that direction, always within funding levels that kept the facility below visible limits in the appropriations process. Advised by Lundquist and by Tillinghast, Whipple agreed that a major part of that plan was to create an effective management infrastructure that could handle very large projects. What he needed was a test case.

He set his sights on a 60-inch photometric telescope, similar in size to Harvard's 61-inch Wyeth at the Agassiz Station and within the Smithsonian's regular appropriated budget supplemented by grants from trust-side Smithsonian Research Foundation funds; what Whipple referred to as "pin money."[977] Starting in 1966, Whipple enlisted a half dozen SAO staff in Cambridge led by Lundquist and Latham to provide detailed scientific specifications and an overall construction plan. Lundquist provided project management oversight, whereas Thomas Hoffman contributed mechanical design expertise. Latham, still a Harvard graduate student, had experience in instrument development and a specialty in stellar atmospheres, sharing talents with Stephen Strom and Robert Noyes, and also had experience with the limits of photographic spectrophotometry. Most of them had worked extensively with the 61-inch Wyeth telescope at Oak Ridge and knew its foibles, but they had tested out the many applications that this size telescope would be able to perform under observing conditions closer to the astronomer's ideal.

Together they represented broad research interests. Latham, especially, was a good example of the synergy of the Smithsonian-Harvard partnership: a Harvard graduate student supported by Smithsonian projects (Figure 47). Latham and Strom took the lead defining the problems a 60-inch could address. The group formed into a "60-inch Performance Committee," with Latham doing much of the work to define the performance specifications for the telescope. In May 1967, Whipple formed a "Scientific Organizing Committee," made up of most of the same staff, and in June, Hoffmann had drafted technical specifications for potential contractors in a process that took another year. They engaged the J. W. Fecker Division of American Optical to produce final design and fabrication drawings for the optics and the tube, and the team added Nathaniel Carleton to round out the scientific research agenda with his interest in planetary atmospheres and expertise in instrumentation. Carleton would eventually take charge of the design,

Figure 47 In addition to his role developing Mount Hopkins, Dave Latham (r) continued to utilize the local 61-inch telescope at Oak Ridge through the 1980s. Here, he is with Robert Stefanik and one of their "digital speedometers" used to obtain highly precise radial velocities. *Smithsonian Institution Archives, Acc.16-263, Box 7.*

construction, and testing of the telescope itself, in close contact with Fecker, leaving the mount, building, and dome to other staff and a growing suite of sub-contractors.[978]

After considerable deliberation, the 60-inch was designed to operate in two or three primary modes: a very fast f/2.5 primary, with secondary mirrors operating either as an f/10 perforated Cassegrain or as an f/24 Coudé. In theory, this design could address a wide variety of problems ranging from bright stellar and extended objects to faint nebulae using a suite of focal plane instruments, and the fast f/ratio meant it all could fit in a smaller dome. By the end of 1968, teams were developing all sorts of modern devices to mount to the telescope. These included a photoelectric spectrum scanner for medium-resolution studies, a slitless interference spectrometer capable of sensing faint objects in the optical infrared, and a device, adapted from a design from the University of Wisconsin, that provided ultra-high resolution in very narrow portions of the spectrum.[979]

Tillinghast managed the project, as recently promoted assistant director for management.[980] In that capacity, he developed the SAO Personnel Office, the Contracts Office, and the first formally defined Engineering Department, giving structure to and regularizing these activities. At his untimely death in July 1969, with the 60-inch near completion, he had created the means through which it could be managed efficiently and effectively. The telescope was named in his honor at its dedication on September 29, 1970, where James Bradley, Smithsonian undersecretary, recalled him as "one of a new breed of young men created by the Space Age," who respected "both the potential and the limitations of research, and who temper their technical skills with humanism."[981] Tillinghast had been especially mindful of his challenge, viewing SAO as "a striking example" of the "federal-university relationship . . . which may well become an important approach for government and university research of the future."[982]

The 60-inch telescope itself, in 1966 considered "a medium sized telescope,"[983] was hardly a technical milestone. It required far less investment than did Celescope or the satellite-tracking network. But like Celescope, it was intended as a testing ground to build an infrastructure that could then provide a stage on which far larger facilities might be based. Accordingly, the SAO elected to manage a distributed set of contractors itself, rather than depend on a general contractor.[984] In fact, it was carefully planned as one element in a larger experiment, proposed to NASA in 1969 that, in Ripley's words, was explicitly "to establish criteria for the erection of sites for future ground-based astronomy research."[985]

With the arrival, installation, and acceptance of the mirror, dome, and structure by February 1970, the telescope was in operation by the summer.[986] Nathaniel Carleton took the lead in assembling and testing the components, looking forward to placing a new high-resolution spectrometer at its Cassegrain focus for planetary studies.

Though the instrument was designed, built, and assembled in rapid order, there were glitches along the way that reminded everyone that Mount Hopkins was not yet tamed and, for some, not isolated enough. Members of the staff constantly fought against bond issues to increase street lighting in the Tucson area, and in summer 1969, observatory staff helped the local Amado Dog Track install shields on its track lights, because the 10-meter reflector could easily detect the increased brightness of the lights whenever a race was scheduled.[987]

The 60-inch Tillinghast facility quickly established itself as a capable medium-sized observatory. From October 1970 to January 1976, Fred Chaffee reported, some 55 astronomers and graduate students observed with the telescope; 18 different focal plane instruments were built, tested, and used with the facility; and the work produced three PhD theses from Harvard and two from the State University of New York at Stony Brook, which had purchased time on the instrument. Chaffee predicted that it would continue as a productive test bed as well as a programmatic instrument, for which all sorts of instruments were now in planning stages.[988]

A "Diversified Astronomical Field Activity"

Stony Brook astronomers were not alone in seeking out the instruments and observing conditions at the Mount Hopkins site. Indeed, as Whipple had carefully planned, the site became a test bed for programs developed by NASA as well as the National Center for Atmospheric Research (NCAR) and other institutions. In a "Land Withdrawal Request" the Smithsonian made to the Bureau of Land Management in May 1968 for use of its 4,474 acres, others interested in using the site under Smithsonian auspices included NASA, Harvard, NCAR, the University of Arizona, the University of Rochester, and MIT. The request even looked forward to a time when "Mt. Hopkins develops into a national facility," responding to the recommendations of the Whitford Committee and to editorials from *Science* included with the brief.[989]

NASA was already present on the site, supporting the satellite-tracking operation. By late 1968, NASA's Electronics Research Center (ERC) in Cambridge,[990] entered into an agreement with SAO to develop the site as a "diversified astronomical field activity." In the agreement, SAO permitted NASA to use the site "as an optical test, technology, measurement and operational facility." NASA would develop standardized techniques for site testing, conducting seeing experiments to corroborate those being conducted by the SAO in order to establish the site for future uses of an unspecified character. NASA's ERC as well as the Jet Propulsion Laboratory (JPL) installed stellar image monitors, an atmospheric water vapor spectral hygrometer, instruments for microthermal measurements, and a large facility for general meteorological investigations, not only to better understand the site but also to standardize techniques for evaluating other sites.[991] At times, NASA's presence equaled the Smithsonian's. In the summer of 1970, NASA installed their Mobile Laser System near the Smithsonian's satellite-tracking laser. A 40-foot trailer weighing some 20 tons and requiring some 90 kilowatts of power and five people to operate it required upgrades to the roads and the observing site. NASA thus became a substantial source of support for the growing infrastructure and operations.[992]

Trevor Weekes reported to Whipple that 1970 would be another "year of expansion" at Mount Hopkins. At least four groups were in extended residence, and reporting on all the activities took some 14 single-spaced pages. Numerous "Visiting research groups" from different NASA divisions, including Wallops Island and the Goddard Mobile Laser System group, the Huntsville meteor group, and the Goddard gamma-ray group, all were aided by the Hopkins "Observing Services Division." Weekes predicted that "as the reputation of the quality of the site becomes known, it must be expected that the number of such groups will increase." Accordingly, he recommended that they develop one or more of the knolls to be visitor friendly with facilities ready to be adapted. Some 22 SAO staff were in full time residence now; there were weekly seminars and colloquia, and public programming occurred in Amado. Weekes reported that morale was high among resident staff, "despite recent cutbacks and uncertainty regarding future programs." Probably the greatest concern, however, was the threat of increasing air and light pollution; Weekes was grateful that Whipple had already petitioned the Health, Education, and Welfare Department, supported by Congressman Udall's office, that the Santa Rita and Santa Cruz ranges be included in an initiative to declare the Phoenix-Tucson area "an air control region."[993]

Echoing Weekes, Whipple declared that the SAO now had its own major observing facility, a "multipurpose astrophysical observatory" that was very much in the spirit of Joseph Henry's vision of a "physical observatory." Most pointedly, Whipple selected Henry's rhetoric to speak directly to his Smithsonian bosses, and indeed to anyone willing to think new thoughts about big telescopes. Henry spoke of the many odd "emanations" from the "fixed stars, or are reflected from the moon and planets, or as they may be found in the aurora borealis, the zodiacal light, and in shooting stars or larger meteors," noting that each "requires peculiar instruments, and such as are not found, at present, in ordinary astronomical observatories."[994] In like manner, Whipple hoped to set the stage for, and make an institutional connection to support, the next big thing. By the time he had prepared this speech for astronomers and staff in January 1970, he knew what that next big thing would probably look like.

Aden Meinel and a New Way to Make a Bigger Telescope

In 1960, Aden Meinel (Figure 48), the builder and founding director of Kitt Peak, and the key visionary for what became the MMT, proclaimed that, in theory, there was "no limit on the size" of a mirror.[995] Practically though, the standard assumption was that suitable equatorial mountings, their housings, and their cost, increased as a cube of the aperture, making the size of the mirror the chief hurdle.[996] Meinel well knew that larger apertures were possible only if accompanied by efforts to make bigger mirrors with faster f/ratios, for economy mainly in the mounting and housing. The Tillinghast 60-inch followed this pattern precisely.

With Kitt Peak established, Meinel left in 1961 to join the staff of Steward Observatory at the University of Arizona, where he was appointed director and professor of astronomy. In 1964 he established the Optical Sciences Center at the university in part to build Steward's new 90-inch reflector, a fast primary feeding a combination f/9

Figure 48 Aden Meinel. *Courtesy AIP Emilio Segrè Visual Archives.*

Ritchey-Chrétien and a f/30 Coudé mounted on a compact equatorial fork in an unconventional dome. It saw first light in the summer of 1968 with funding from the NSF and the state of Arizona.[997]

The Optical Sciences Center was not created for astronomy but rather "to fulfill a national need for more highly trained engineers and physicists in the optical sciences."[998] With support from the Air Force and endorsements from the Optical Society of America, and with seed funding from the University, Meinel created the Center to provide agency for attracting large contracts from the Air Force's Space and Missile Systems Organization (SAMSO) and the Air Force Systems Command (AFSC). In the manner fostered by Stark Draper at MIT, it became the Air Force's pipeline for training physicists and engineers already committed to military problems.[999] His Optical Sciences Center filled a critical need for SAMSO, the center's primary sponsor, which as of November 1966 was tasked by the Air Force to provide "cost effective multipurpose satellite systems" for "enhanced world-wide early warning, surveillance and detection capabilities."[1000]

Meinel's SAMSO connections put him in touch with plans for a Manned Orbiting Laboratory (MOL).[1001] As a space platform that would be occupied by Air Force and Navy astronauts, allegedly to "test man's military usefulness in space,"[1002] MOL was developed to operate large optical reconnaissance systems under the code name DORIAN.[1003] One design called for a long f/ratio, 60-inch diameter primary mirror mounted in a cylindrical tube that was fed by a flat, full-aperture diagonal mirror looking out the side of the cylinder. The primary mirror sent a converging beam to a two-mirror optical relay and thence to a photographic film plane operated manually by astronauts. But when the President's Science

Advisory Committee (PSAC) intervened in November 1965, the system was converted to a robotic scheme whereby a film transport system fed a series of film recovery capsules, along the lines that Davis had been offered for Celescope by General Electric subcontractors in 1958. At some point between November 1965 and May 1966, the 60-inch diameter primary was enlarged to roughly 70 inches "to provide the desired . . . [redacted] . . . ground resolution."[1004]

When MOL was cancelled in June 1969, the National Reconnaissance Office (NRO) and Air Force offered some of its hardware to NASA, "within the bounds of security." Included were elements useful to the Skylab program, but also some of the mirror blanks "for various non-reconnaissance optical projects."[1005] These ended up at the Marshall Space Flight Center. But Meinel got some seven mirrors too through Harry I. Davis, deputy undersecretary of the Air Force for Research and Development.[1006] This was a direct result of an informal meeting at the Pentagon at which Meinel suggested that the Air Force develop a synthetic aperture reconnaissance telescope. He sketched it, made a model of it at home in his machine shop, and eventually presented it to Davis: "It was on the table with other models of future aircraft for the Under Secretary [of the United States] Air Force"[1007] (Figure 49). From subsequent remarks Meinel made during his campaign to build a multiple mirror telescope for the University of Arizona's observatory site on Mount Lemmon, it is clear that considerable work had taken place conducting feasibility studies before the cancellation of the program, called at one point the "ultra high resolution (UHR) photo satellite system."[1008]

"a very large, but inexpensive and lightweight, optical telescope . . ."[1009]

While he was building LOR, Whipple was thinking even larger. He asked the engineering firm engaged in developing LOR to provide conceptual studies for a "major optical reflector." Through spring 1968 they produced a half dozen options that balanced thermal and mechanical stability levels with sky coverage and tracking capabilities.[1010]

The 60-inch had been planned as a general-purpose instrument. But its larger successor would be tuned for infrared studies. Advised by Robert Noyes and others, Whipple knew that planetary and solar system studies required infrared (IR) detectors, so he directed Noyes to team up with Strom and Lawrence Mertz along with SAO managers and engineers to decide on a functional system. In March 1969, Noyes advised the Smithsonian's Research Foundation that "several converging trends" made it important for the SAO to get involved in IR work. First, many SAO astronomers were interested in problems that could be informed by infrared observations, such as star formation, the origin and presence of planetary systems, or the physics of cool stars. Second, other groups, such as at Arizona and Caltech, had by then successfully initiated IR programs and had "demonstrated that the field is very rich." Third, Mount Hopkins proved to be a good IR site. And fourth, because the SAO was already considering building a "major optical reflector

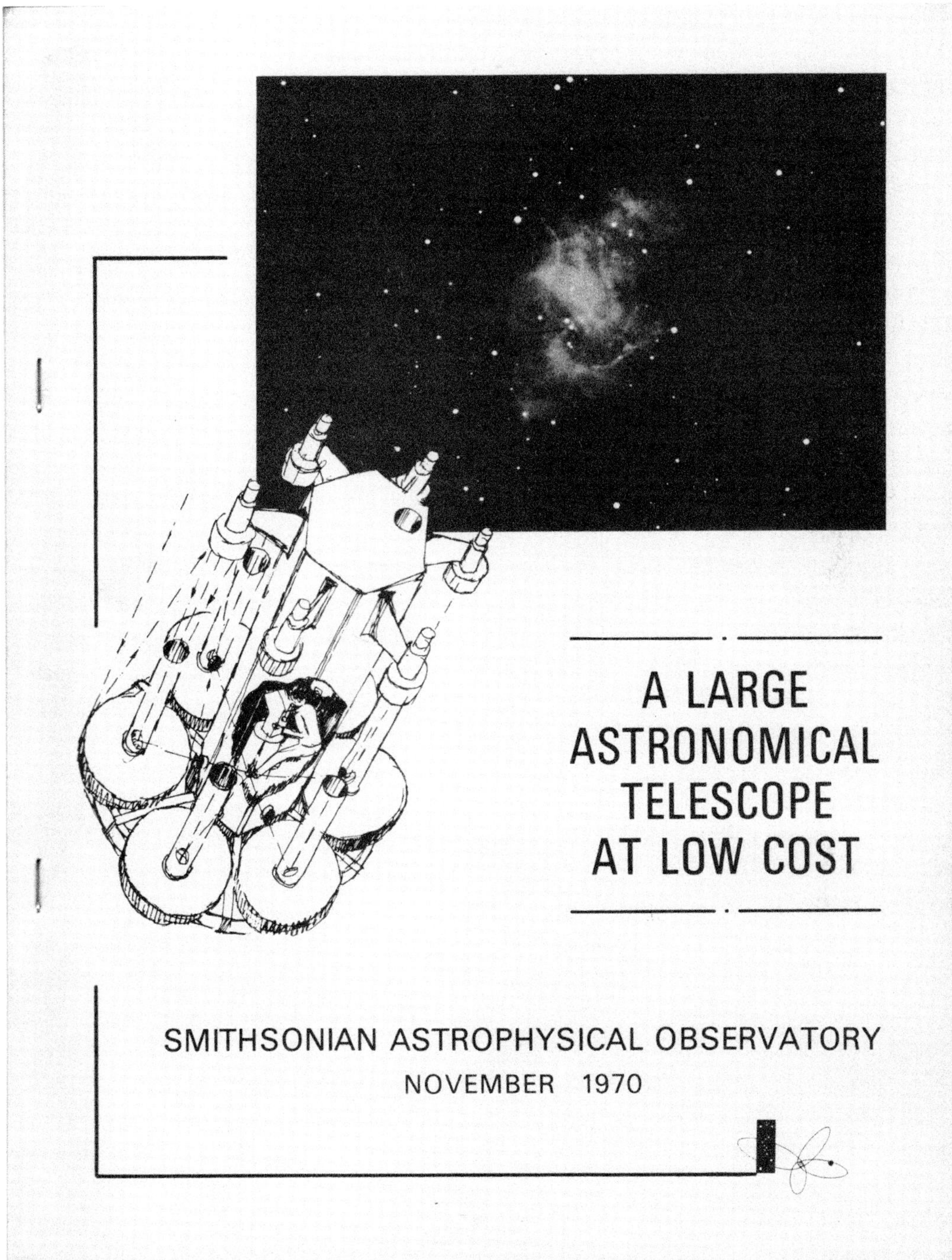

Figure 49 In early proposals for what became the Multiple Mirror Telescope, its depiction embodied Meinel's original design for a centrally supported instrument that could be folded for launch into space. Note the observer on the central axis. *Smithsonian Institution Archives, RU 468-032, Box 32.*

sometime in the future," attention to detector development was a sound and prudent step that would lead to quick scientific return with present telescopes and point out the need for larger telescopes.[1011]

Turning what Whipple and Lundquist first called a "Major Optical Reflector" (MOR) and then a "Major Optical Instrument" (MOI) into an infrared facility also relaxed optical and mechanical specifications, which led the team to consider a wide range of radical designs. After many meetings, however, going over the options the contractor provided them the previous March, they could not reach a clear decision on how to proceed.[1012] Accordingly, Lundquist split the group into subcommittees to develop detailed parameters. The scientists (Noyes, Mertz, Carleton, Schild, and Pearlman) had a month to develop clear scientific objectives. Other subcommittees were charged to prioritize options for structural engineering designs. Given the report of the first committee, all groups would report within three months to evaluate and select instrument concepts, and would then spend another half year analyzing "contending concepts." Whipple's goal was to have a proposal ready by May 1970 in time to incorporate it into the Smithsonian's fiscal '72 request to Congress.[1013]

The SAO engineers and astronomers constantly debated competing designs. Lawrence Mertz was especially attracted to an Arecibo-type design consisting of a fixed tessellated array of mirrors. It would be optimized for spectroscopy because he saw the universe in those terms.[1014] By the summer of 1968, aided by conversations with optical experts such as Meinel, he was confident that his design could deliver 1-arc-second images and thereby was only seeing-limited, like any proper ground-based telescope. The size and figure of the secondary mirror was a critical factor, however. It would be a challenge to minimize coma and other aberrations, and, as Mertz admitted, no doubt the fabrication of "such a grossly aspheric secondary is not expected to be easy."[1015]

On his frequent trips to Mount Hopkins, Whipple visited Gerard Kuiper in Tucson. At one of these meetings, Kuiper apparently discussed the plans that Meinel and Frank Low were developing for repurposing the 72-inch mirrors Meinel was to receive through Davis after the DORIAN/MOL cancellation. By early June, 1970 Whipple had also received a preprint of a paper by Meinel bound for *Applied Optics* titled "Aperture Synthesis Using Independent Telescopes," attached to a lengthy proposal by Frank Low to the "Members of AMB and ASC Infrared panels and Interested Colleagues," which explicitly described a six-element multiple telescope array for infrared astrophysics.[1016] Frank J. Low, a low-temperature experimental physicist with extensive industrial experience and expertise developing cooled solid-state infrared detectors, was attracted to astronomical research in Tucson in the 1960s and, with Meinel, began thinking about the advantages of clustered apertures for infrared studies around this time.[1017] Meinel and Low joined forces, the former providing access and optical intuition, and the latter the detector work.

Thus informed by Kuiper and most likely after he had Low's proposal to the AMB (Astronomy Missions Board) at hand along with Meinel's preprint,[1018] Whipple phoned Meinel to ask for his thoughts about building a really large telescope. Both parties vividly recalled the phone call, and that Meinel took the call just as he, Low, Kuiper, and others were meeting to contemplate how to proceed, and particularly who would fund it. They were still thinking that the optics would be of sufficient interest to the Air Force to support that part of the project. But what of the rest of the telescope and its building? As Meinel recalled: "I had no idea he was interested in it! It was so dramatic, because we were having a meeting, talking about doing something. . . . The phone rings and the secretary comes in and said, 'Its Fred Whipple.'"[1019]

Kuiper had told Whipple about the meeting, and, having absorbed Meinel's and Low's report by then, Whipple wanted to buy in. The SAO budget was still going through review, but he had already included a $500,000 line item for the telescope in SAO's fiscal '72 budget request to Congress. Because of the lengthy process of appropriations and authorizations, he could not predict "whether there are significant grounds for optimism."[1020] So his task now was to convince his own SAO study group committee to choose Meinel's design. Although Mertz resisted at first, Whipple soon prevailed.[1021]

Meinel felt that a segmented design would be difficult to produce because it required fabricating dozens of off-axis paraboloids. There also were thermal problems created by such a design that complicated its use in the infrared. A few independent mirror elements (for example, only six mirrors), however, could produce a single image field with a synthetic resolution far beyond each element, because there were ways to insert a highly sensitive optical path length adjustment system to align the wave fronts of all six systems, a process called "phasing."[1022] Phasing was the essence of "aperture synthesis" and for this reason, Meinel and Low were confident that the Department of Defense (DOD) would support the cost of building and testing the multiple telescope system. "This approach to the aperture synthesis problem appears to be quite promising, both for optical purposes, of interest to the DOD," Low argued, "and for a large ground-based infrared telescope." Furthermore, it could act as a prototype "to evaluate its potential as a high-resolution imaging system of ultra-low weight." At the least, once the telescope had proved itself to the DOD, "the major part of an extremely powerful infrared telescope will be in existence." Having the six 72-inch blanks presented a "unique opportunity," as did DOD support. "I feel it is not possible," Low concluded provocatively, "to turn down the chance to build a large infrared telescope at such low cost."[1023] Whipple conveyed Low's confidence to his committee, and everyone, with the exception of Mertz agreed.

Unlike LOR or the 60 inch, what would soon be called the MMT required congressional approval. So, coached by Bradley, Whipple's staff prepared a request to the newly created Office of Management and Budget (OMB) that had replaced the Bureau of the Budget, touting it as "a very large, but inexpensive and lightweight, optical telescope."[1024] Submitted in November, its cover included a graphic of Meinel's original concept, and the frontispiece reproduced an image of his model.[1025]

Bradley advised them to be ready to field questions that would be "shot from the hip." These would require "an equally direct and cogent answer, without having time to look around for the base line." First and foremost, they had to answer, "Why, then, prepare this?" in terms that also answered, "Why is such a large facility needed?" They answered by citing the Whitford Committee Report and the Soviet threat of building an even bigger one. They also could cite the SAO's experience with the 60 inch and LOR. Bradley advised them to dodge direct questions as to where it would be located, saying only that site comparisons are underway. Whipple was also ready with testimony to prove that this was not an impulsive interest, citing excerpts from SAO's appropriations requests from FY 1967 through 1972 that revealed a growing commitment to building a 600-inch telescope.[1026]

The Smithsonian's formal submission thus set in motion a complex series of negotiations on many fronts: between the Smithsonian and the University of Arizona, between the Smithsonian and the OMB, with both Arizona and Smithsonian with the DOD and then with the NSF. There were many issues, from shared support, to shared duties, to shared controls, and, last, to which site would be chosen for the telescope. The first three depended critically on the fourth.

Chapter 13

Building the MMT

In mid-January 1971, after they had submitted their proposal to the Office of Management and Budget (OMB) for $1.5 million over three years, Whipple and Lundquist led a Smithsonian Astrophysical Observatory (SAO) team to Tucson to work out the details. Even though Arizona felt it had secured Air Force funding, they knew it would not be enough. Arizona and the Smithsonian had to win over the National Science Foundation (NSF).[1027]

Funding aside, the biggest question was, where does it go? Mount Lemmon, northeast of Tucson, competed with Mount Hopkins. Arizona had invested heavily in Mount Lemmon; when the Department of Defense (DOD) gave over control of the area to the Forest Service, it stipulated that the astronomical work had to continue. A multi-institutional user's group managed the site funded by NASA and the recently established Universities Space Research Association (USRA) and included Arizona, Minnesota, University of California, San Diego, and AFCRC.[1028] NASA had recently withdrawn its support but Arizona was confident that the Air Force would return to support any experimentation that informed aperture synthesis.

George Low and Aden Meinel were optimistic, but Whipple had a schedule to keep to meet congressional deadlines for proposals. They all worried that the 1969 Mansfield Amendment would severely limit Air Force participation, because it prevented the Defense Department from supporting research not directly related to a "specific military function." Arizona now was just asking to support the optical experimentation. Whipple, Lundquist recorded, speaking for the Smithsonian, needed a better sense of Arizona's contributions so that the SAO could clearly "define its plans by Congressional hearing time this spring." Among many questions that he anticipated Congress would raise was where the telescope would be placed. The answer would inform the bottom line: Who would pay for what?[1029]

Whipple wrote to Robert Seamans Jr., secretary of the Air Force, promoting the joint program as a way to beat the Russians, a familiar trope in that day. Speaking informally and, evidently, with considerable familiarity, he hoped that Seamans would look kindly on supporting the Optical Sciences Center to develop "actively controlled" optics. Indeed, Whipple felt this alignment of interest more than met the strictures of the Mansfield Amendment. The DOD would merely cover costs related to aperture synthesis, whereas the Smithsonian would actually build the telescope and manage the operation: "Thus the Air Force and the DOD stand to benefit substantially from mutual experience with this telescope."[1030] Once again, Whipple applied his penchant for brinksmanship.

Seamans, though sympathetic, did not approve. The Air Force was interested in synthetic aperture technologies, but not in telescope development.[1031] Whether or not he felt limited by the Mansfield Amendment, or by shifting R&D priorities within the Air Force, such as the abandonment of Manned Orbiting Laboratory (MOL) and larger-aperture "Ultra-High Resolution" (UHR) systems in favor of smaller more rapid response "near-real time" systems, one thing was clear: Seamans put the project on notice that they needed to search for other resources.[1032] Without DOD, Arizona had to approach NASA and the NSF.

The pressure was on. Congressional hearings were due to start 19 March 1971 with testimony before the Senate and on 5 April in the House.[1033] A few days before the Senate hearings, Whipple tried to assure Bradley that, even without DOD support, the budget justifications already crafted stated the case as well as it was known at the moment, and no special pleading should be required. He emphasized that the cost of this instrument was one-fifth that of a conventional telescope of comparable power, that their push into the infrared (IR) was absolutely at the forefront of the science, and, most important, that to take full advantage of the technology, site selection was all important. Mount Hopkins, Whipple advised Bradley, was better shielded from Tucson than the nearer Mount Lemmon. Nevertheless, there would be a seven-month site survey. In the end, Whipple emphasized, Congress needed to know that the Multiple Mirror Telescope (MMT) would be the third-largest telescope and the most effective IR telescope in the world; its success would lead to "far greater telescopes than could be considered practical following conventional designs."[1034]

Whipple was just as forceful in his testimony before Senator Alan Bible (D-NV). Bible did not want technical details, just something he could understand about what the thing would be used for. Whipple painted a portrait of "great gas clouds," where stars are being born: "You might call them nurseries, or the laying-in hospitals for new planetary systems like our own." Whipple sprinkled his remarks with wonder and spectacle, but was brought back to Earth when Bible asked whether the Smithsonian "had the primary function" for astronomy in the nation.

A loaded question to be sure, and Whipple said the right thing: "No." Bible kept pushing, asking how many observatories in the United States were supported by the government. Whipple recited the list, making it as short as possible by citing only those that were wholly supported: the Navy supported the Naval Observatory, NSF supported Kitt Peak and the National Radio Astronomy Observatory (NRAO), and the Air Force, the Sacramento Peak Observatory. Satisfied, Bible admitted that he once thought about being an astronomer "until I found they weren't going to be paid." Whipple seized the opportunity: "I became interested in the same way, and never worried about being. But I have been." Bible had the last word: "I hope you have been," he laughed. "I understand you are preeminent in your field. It is nice to talk with you."[1035] Whipple came away confident of congressional support.

On the Arizona front, with DOD funding fading, Ray Weymann, director of Arizona's Steward Observatory and a highly cited theorist interested in a wide range of problems from cosmic background radiation to the evolution of stars, focused his campaign on

Arizona politicians. Writing to Morris K. Udall in May 1971, he introduced the "joint enterprise" with rhetoric similar to Whipple's, again citing the ever-present Russian challenge. Weymann also spoke of the economy of the design and seconded Whipple's appeal to Congress, indicating that he had already discussed it with Washington State representative Julia Butler Hansen, a prominent member of the Appropriations Committee and chair of its Interior and Related Agencies Subcommittee. Weymann's strategy was to push hard for the Smithsonian's appropriation, and then once it was approved by Congress, the state of Arizona then would propose to NSF and possibly to NASA.[1036]

In the next few weeks and months, other senior astronomers at Arizona, including Bart Bok, who had returned from Australia in 1966 to head the Arizona astronomy department, Gerard Kuiper, Meinel, and Low, petitioned Senator Barry Goldwater on behalf of the Smithsonian's appeal. By July, Whipple could proudly report to Meinel that "the hoped for good news is now a reality." The House approved an appropriation for some $400,000 for the first year, with the same or more in the following two years. Whipple was confident that the Senate would soon agree, because Bradley was leading the charge.[1037]

On the day Whipple testified before Congress, Ripley, prepared by Bradley, wrote to William D. McElroy, the director of the NSF, to appeal on behalf of Arizona's pending proposal. Pointing to Whipple's strong track record of "initiating other projects involving unconventional optical systems" from the Super-Schmidts and the Baker-Nunns to the 60-inch Tillinghast reflector, Ripley asked for a personal audience with McElroy to establish closer ties and to campaign for his Arizona compatriots.[1038] McElroy, however, took over two weeks to respond coldly and bluntly that the NSF was the "lead Federal agency for ground-based astronomy." He also did not want to discuss the matter directly, pointing to the appropriate NSF officers in the astronomy section, specifically Robert Fleischer, the astronomy section head.[1039]

This was not the reaction Whipple and Ripley had hoped for, though by 1971 they certainly should have expected it. Whipple was then also campaigning, aided and abetted by Ripley, for a new national facility devoted to radio astronomy that was in direct competition with NSF's NRAO and its Very Large Array (VLA) facilities. But the question in Washington was who controlled the nation's science? Also moving through Congress were bills H.R. 3820 and S. 4453 of the 91st Congress, "To establish a Department of Science and Technology and to transfer certain agencies and functions to such Department." This reorganization would bring all science functions of the government under one department, including NSF, NASA, the Atomic Energy Commission (AEC), DOD, and those parts of the Smithsonian primarily concerned with research, such as the SAO.[1040] McElroy's declaration, and Bible's inquiries, reflect this very volatile context.

Who Is Responsible for Astronomy? NSF vs. Smithsonian

Weymann had already met with Fleischer earlier in February and received a negative and puzzling response. Prompted by informal suggestions from members of a newly constituted Second Decadal Survey, that the astronomical community might be more receptive to the radical new telescope design if they proposed to build a small prototype, Weymann asked Fleischer if he would be receptive to such a proposal. Fleischer said he might be, because he doubted that the Smithsonian would get all of what it was asking from Congress. In fact, Fleischer added darkly, Arizona was probably better off going it alone in this project. He also predicted that it will require a "long and hard campaign" to persuade the astronomical community of the need for this project, and without community approval, NSF would not support it.[1041] Fleischer had concerns about the feasibility of the unproven radical design, but what was really bothering him was any cooperative arrangement wherein NSF would support the goals of the Smithsonian Institution. Fleischer felt that acquiescing to such a commitment was a threat to NSF's hegemony.[1042]

Weymann's negotiations over the next several years with Fleischer and NSF would not lead to support, nor would negotiations with NASA be successful for Arizona's share of the cost of the MMT. NASA had also expressed an interest in aperture synthesis, but after McElroy's declaration, both Arizona and SAO planners worried that NSF would be miffed at the further competition. But first there was the question of feasibility.

Meinel was annoyed by the repeated calls to prove the concept with a prototype. Speaking from experience beyond the classified barrier in February 1972, Meinel told Fleischer that "astronomers do not know of the extensive work done in the past decade." From his standpoint, "It hardly seems necessary to duplicate these experiments when most of it is declassified at this date. I can assure you that the design we have evolved for the SAO/UA project has been arrived at in light of previous work." Further testing was just not needed, Meinel asserted: "When one faces engineering decisions one learns much concerning the options open at the decision point."[1043]

The point was, Meinel and Whipple regarded the MMT *as* the prototype. In all the designs to that date, the focal plane was internal to the mirror cluster and had to be large enough to include a chamber for the observer. Meinel took care to include the impression that the observer actually had to be in the chamber, admitting much later, "Well, I had to cover up the spaciness by putting a human in it . . . [i]n order to get it published."[1044] Such a design could not really be prototyped, nor was there reason to do so, as Meinel well knew, because the ultimate design for the ground-based version would have an external truss, not a central core.[1045]

Fleischer, a 1951 Harvard PhD with eclectic interests in stellar and solar astronomy, joined NSF in 1962. He generally took a "strong approach" to his job; his "belief in himself served him well in many ways, but caused him grief from time to time."[1046] He was especially adamant that NSF be recognized as the principal federal agency for ground-based astronomy, in spring 1969 calling for all federal administrators who supported astronomy in some way to agree to that fact.[1047] So Fleischer vigorously defended McElroy's assertion of NSF hegemony over ground-based astronomy during the negotiations with Arizona and Whipple. Looking at the strategies employed by Arizona and the SAO will help to better understand NSF's concerns and assertions.

Over summer 1971, as the SAO, Smithsonian Institution, and Arizona developed a formal Memorandum of Understanding (MOU) to clarify shared responsibilities and a workable management structure, they also debated how to approach NASA without annoying NSF, given that NASA was indeed interested in supporting mission-related, ground-based facilities in Hawaii and Arizona. Meinel had wanted to approach NASA because, not surprisingly, he saw the MMT as a step in the "development of an engineering model of a large space telescope." But SAO staff felt that Meinel's enthusiasms had to be controlled.[1048] They knew that NASA was already involved because a Smithsonian staffer found out that NASA's Marshall Space Flight Center had initiated a Phase A definition study of an optical telescope assembly for a Large Space Telescope.[1049] Meinel knew this from his negotiations with NASA program managers over their share of the DORIAN 72-inch mirrors. Meinel had secured flat blanks, whereas at least one of the six samples at Marshall had already been slumped and figured to f/2.2 for testing to see if the design could still hold a sufficiently precise optical figure.[1050] Evidently NASA was invested.

In mid-August 1971, the Senate restored the Smithsonian's original proposal at the $500,000/year level for FY '72. Whipple, chairing a new "MMT Steering Committee," put out an all-staff call to both organizations for final input to the scientific and technical requirements for the MMT, due as soon as possible but within 10 days. The SAO's new assistant director for management, John Gregory, would coordinate the responses. Latham, now project scientist, and Low, the program scientist, would set priorities.[1051]

The group debated the proper f/ratio for the primaries through November, balancing costs against fears that too much slumping would damage the "egg-crate matrix" substrate or the thin front plates of the quartz mirrors.[1052] By November, after tests at Corning, they decided on f/3 because the cost savings on the building would overcome the increased costs of optical fabrication and would provide a nominally larger field of view.[1053] They all came together in Cambridge in mid-October to decide on the MOU, confirm the f/ratio, and establish all the major design elements, including the dome, the control systems, and the alignment and guiding systems. Although not an immediate requirement, the design had to be compatible with an eventual phased-array capability or an active optics control system.[1054]

By October, Latham had completed a preliminary comparative survey of the two possible observing sites: Mount Lemmon or Mount Hopkins. In the visible range, Hopkins's southern sky was 50 percent darker than Lemmon, and in selected "clear" spectral bands, the advantage was even larger. There were no measurable differences in water vapor absorption between the two sites. But at the moment, no one wanted to go any further to state an actual choice until the steering committee had a chance, over the next few months, to thoroughly review all the findings with all parties.[1055]

Whither Harvard?

Only SAO and Arizona people were identified among those attending the Cambridge meeting. The minutes, however, were distributed to 29 SAO staff and at least 8 Harvard staff. Leo Goldberg did not attend, though he had been invited. Whipple, of course, knew that Harvard had different interests. Led by Goldberg, Harvard astronomers were now more interested in a new conventional telescope, of some 90-inch aperture, sited in Chile and built in collaboration with Yale and MIT. Goldberg certainly knew what the SAO was planning. He, Menzel, and Whipple had been conferring since October 1967 on strategies for funding large observatories and Whipple reminded them repeatedly of his goals for Mount Hopkins.[1056] By 1968, Goldberg had shifted his vision to Arizona as well, and was contemplating an alliance with the University of Texas for a cooperative telescope on Kitt Peak or even in Oklahoma.[1057] In May, Goldberg strongly endorsed the SAO's "Land Withdrawal Request," which explicitly included the intention to develop the site into a "national facility" featuring a "really large telescope," and predicted that Harvard would "soon become a full-fledged partner in this enterprise."[1058] Goldberg was also on Frank Low's distribution list in June 1970, proposing the "Six-Element Infrared Telescope," and Whipple copied him in February 1971 after his first large planning meeting with the Arizona astronomers.[1059]

No evidence has been found, however, that Goldberg paid much attention to the MMT initiative until Robert Bartnik, Whipple's assistant director for management, sent him a draft of the MOU in June 1971, along with an explicit invitation to the October meeting. Whipple wanted Harvard participation, especially if Harvard shared some of the costs. He hoped to put a statement in the MOU indicating that the SAO would assist the University of Arizona's efforts to raise funds "possibly in cooperation with the Harvard College Observatory."[1060]

Goldberg stonewalled for over a month, claiming that he did not want to comment on the matter "since on the face of it the agreement seems not to concern the Harvard College Observatory." Goldberg also objected to any "collaborative effort with another university. [It] is simply not in keeping with the intent of the original Harvard-Smithsonian agreement, either the written form or the gentleman's version." Whipple was going ahead one way or another, Goldberg believed, and "this is certainly not an acceptable basis on which I could recommend Harvard's participation in such a costly long-range project." This, Goldberg admitted, copying Dean John Dunlop, as well as Whipple's senior staff, was his personal opinion and did not reflect the opinion of the Harvard Observatory Council.[1061]

Bartnik advised Whipple that Harvard indeed cooperated with other universities such as a new infrared ballooning program with the University of Arizona (Figure 50) as well as a joint program between SAO, Harvard, the University of Texas, and Bell Labs to construct a millimeter wave telescope.[1062] Goldberg, in fact, possibly unknown to Bartnik, was also courting Texas, as well as MIT and Harvard in various ventures, proposing relationships that were similar to what the SAO wanted to do with Arizona.

Goldberg's objections went far deeper than the MMT project. Ever since Pusey's rejection of Menzel's proposal for a major solar telescope in Hawaii, due in large part to his divided faculty, Goldberg felt that Whipple "decided that he just wasn't going to have any more to do with that Observatory Council." This annoyed Goldberg, who as Menzel's successor felt he should have some oversight on Whipple's activities: "Really," Goldberg felt, "he didn't have that power of decision." Goldberg viewed the agreement between Harvard and the Smithsonian as cooperative, which to him meant that Whipple "couldn't unilaterally decide that he wasn't going to have anything more to do with Harvard. But he was going to go ahead and do what he wanted to do and use the facilities, use the students, and so forth."[1063]

Figure 50 Robert Noyes, Giovanni Fazio, and Doug Kleinmann with their balloon-borne 40-inch infrared telescope payload, initiated in1971 and seen here nearing the first of four launches in the mid-1970s. *Smithsonian Institution Archives SIA Acc. 16-263, Box 4. Courtesy of Rick Stafford,* Harvard Gazette.

Goldberg defined "cooperation" in terms of informed and shared control. Just as he acted to broaden membership in the Harvard Council, widening it beyond senior professors to non-faculty and fellows and students, he expected the cooperative agreement with the SAO to extend to mutual governance. What annoyed Goldberg no end was the fact that, whereas Whipple enjoyed an oversight role as a member of the Harvard Council, there was no complementary oversight for the SAO. There was no advisory circle of senior staff, no visiting committees. So in effect, the SAO, essentially Whipple alone, had a voice in Harvard appointments, whereas the Harvard "department was never consulted on new appointments to the Smithsonian. There was such an asymmetry in the situation"[1064] Goldberg's annoyance extended to a deep concern over shared responsibilities for undergraduate teaching and graduate training.

Whither NSF?

Goldberg's refusal to cooperate did not noticeably influence the deliberative process. Fleischer's refusal did. In November 1971, Fleischer demanded clarification of many issues, ranging from what other federal grants Arizona had at the time to how the MMT compared with other large federally supported telescopes. Fleischer was particularly incensed that the ongoing inquiries by the Second Decadal Survey committee were pinching the NSF.[1065] Weymann dodged Fleischer's questions, emphasizing that the MMT was not just another large telescope, but would be beneficial to understanding techniques for active optical alignment and control for telescopes on the ground and in space. It was moreover, an "essential first step" for any plans to build a 600-inch class telescope. He also reminded Fleischer that senators Barry Goldwater and Paul Fannin, as well as Congressman Udall, "strongly supported and endorsed this item in the SAO Congressional request."[1066] Whipple also weighed in, making much the same points but adding emphatically that the support Arizona was asking for was not duplicative of what the Smithsonian was providing. Arizona was asking the NSF only to support the critical element in the design of a new generation of telescopes, the optics, and not the mundane side, which was being carried by the Smithsonian's appeal to Congress.[1067]

Fleischer would not buy any of this, and engaged the matter mainly by making accusations. He was annoyed that the Smithsonian was "acting as a Government Agency," given it was requesting appropriations of Congress. And it was "irresponsible" to presume that the NSF would support Arizona's share of the project. As had Goldberg, Fleischer scolded Whipple for not keeping him sufficiently in the loop even though he admitted that he was well aware of the initial approach Whipple had made to OMB in 1970. But, Fleischer felt, that initial approach had led everyone, including the members of the second Decadal Survey, to believe that the MMT would be supported solely by the Smithsonian and the DOD and that competitive NSF funds would not be requested. Thus Fleischer was surprised when Arizona came to the NSF asking to support "their share." He stated bluntly that any additional funds had to be raised by Arizona from its own resources, and that the mere fact of the MOU "could not be a forcing element to funding by the Foundation."

"I definitely react against any piecemeal attack on the Foundation's scarce resources," Fleischer cried, "using agreements to which the Foundation has not been a party as basis for pressure to acquiesce." With that remark, Fleischer abruptly signed off, "I look forward to your reply."[1068]

Whipple's first draft replying to Fleischer was just as contentious. Certainly there was no "irresponsibility" pursuing what he believed "to be a true breakthrough" in telescope design. The MMT, Whipple predicted, would be the prototype "for all telescopes larger than the 6-meter USSR instrument" whose "huge cast mirror" was already a "dinosaur" with no more evolutionary potential. Then, as a challenge if not a direct slap, Whipple lauded the "forward-looking attitude of the OMB and of the Congress" as they have now saved years in the march to truly large aperture telescopes. Indeed, he had been "highly responsible" to astronomy, to the astronomical community, to the government, and "to the National Science Foundation for relieving it of a considerable financial load." He ended his draft by challenging the NSF to get on board.[1069]

What he actually sent was more prudent. Whipple admitted that the process had happened quickly and serendipitously. For the Smithsonian, it was the sudden appearance of the six mirror blanks, "and the keen interest of dedicated scientists in five research centers," as well as the now soon-to-be-expected appearance of a report by the Second Decadal Survey Committee that endorsed the MMT. Fleischer had to consider how the MMT would help the NSF "determine whether the concept should form the basis for a really large national instrument." Accordingly, Whipple invited him to reconsider, and join in providing "modest support" for "making use of existing light weight mirrors rather than letting them go to waste."[1070]

Weymann was dismayed with Fleischer's harsh assertions. Both he and Meinel believed their proposal was in step with the Decadal Survey recommendations, and so made some small revisions and resubmitted it to Fleischer, confident that a competent technical review would support them.[1071] Fleischer glanced at this new proposal but still fumed at the Smithsonian's actions as he saw them: compromising the NSF's centrality, which, he believed, was essential for a long-term national plan.[1072]

Weymann and Meinel were partly vindicated when the Second Decadal Report appeared in June as "Astronomy and Astrophysics for the 1970s." It did suggest "prototype tests" before one in the 150-inch to 200-inch class "should be built." But the Second Decadal Survey Committee also had warmed considerably to funding essentially what Arizona was asking for, seeing the MMT as the intermediate prototype for a truly huge instrument.[1073] SAO staffers were elated with this news: "It shows we are in the current mainstream of thinking in terms of national priorities." The MMT was the second-highest priority in the committee's highest-priority category, and at the time the MMT was the only project identified that was well into the hardware stage: "Indeed, without SAO/SI's initiative in seizing the opportunity to build the MMT," Bartnik claimed, ". . . this vital national need would have still remained only an idea."[1074] Thus bolstered, the Arizona team modified its NSF proposal, negotiated again with Fleischer over budget, and added a feasibility study, despite Meinel's attempt to assure Fleischer than none was needed. By this time, Meinel, Low, and Weymann had also given up on DOD. Meinel had, in fact, made his final pitch to DOD back in December 1971, calling his appeal the "moment of truth" to join in to build this "landmark in optical technology."[1075]

Fleischer didn't budge even after the revised Arizona proposal met all of his conditions, and NSF leadership remained mute.[1076] In the end. John P. Schaefer, the University of Arizona president, formally objected to the NSF director, accusing Fleischer of being "completely unfair. . . . Recent direct and indirect communication from Dr. Fleischer now leaves a real taste of ashes."[1077] Arizona eventually secured state support for partial funding, amounting to some $1.5 million. Most of the balance was made up by Smithsonian supplemental appropriations, some internal grants, and private donations, but there were still serious cost overruns by late 1975, owing mainly to a delayed construction schedule caused by the tight budget and NSF's refusals. Ripley finally stepped in, campaigning for, and receiving, additional funds through the Smithsonian regents.[1078] Appeals like his eventually allowed for the completion of the telescope. But in the process, the Smithsonian indeed, as Fleischer always believed, became the primary institutional home for the project, especially after it was finally decided to place it on Mount Hopkins and not Mount Lemmon.

Where to Put It?

Kuiper always called the project the "Mt. Lemmon infrared facility." It was his singular goal. Beyond providing a new large infrared telescope in short order and at low cost, Kuiper campaigned that "it will demonstrate the soundness and economy of the MMT concept" as a test case for "a National IR/Optical instrument in the 400"–600" class."[1079]

Of the two sites, Lemmon had better infrastructure but Hopkins was darker, and would likely remain darker in the future.[1080] Yet a thorough test had to be made. In early 1971, David Latham led a team conducting simultaneous comparative site surveys. He designed sky photometers and collected other standard tools to record the seeing and night sky brightness simultaneously, as well as complete meteorological profiles and radio frequency interference levels, all using new pulse counting techniques for automatic recording. Organizing a small staff of observers at both sites from both SAO and Arizona personnel, he conducted site testing through 1971 as the two institutions sought funding and negotiated responsibilities.[1081]

Whipple well knew that there were factors beyond site quality; both Low and Kuiper were wedded to Lemmon. Kuiper, however, had many projects pending and Meinel's enthusiasms constantly wandered; at the moment, he was excited by solar-power generation. Meinel still expected that his laboratory would build the mirrors, the active optics, and the electronics.[1082] At the end of the summer, the two institutions formed a "Joint Multiple Mirror Telescope Facility Committee" to oversee all matters, including site selection.[1083]

In October, to no one's surprise, Latham presented his findings favoring Hopkins. They also knew that the Forest Service was reassessing the area near the Lemmon site because of local community recreation requests, and was considering a total ban on all new construction. Kuiper was not happy with this; he applauded any effort to preserve dark skies, but not at the expense of building a new telescope.[1084]

The final showdown came in Tucson in January 1972, at Kuiper's Lunar and Planetary Laboratory on the Arizona campus.[1085] All other matters were dealt with first, including a decision to slump and figure to f/2.7 and to build an integrated, box-like structure to house the telescope, where the optics were supported by an exoskeletal structure rather than a central column (Figure 51). Then they debated Latham's report.[1086] Kuiper and Low still preferred Lemmon for its facilities and existing infrastructure. Weymann at one point rose, and to some surprise and amusement, as consummate theorist, presented time and distance statistics from road trips he had made from the university campus to both locations. It was some 42.9 miles and 69 minutes to get to Lemmon, and 62.2 miles and 101 minutes to Hopkins. "He stated that he felt the difference was not significant." Whipple, tiring of the endless discussion, called for a vote, placing all four SAO votes for Hopkins. Then Meinel rose and seconded Whipple, stunning the room to silence. He and others agreed that the Hopkins site would strengthen "SAO's fund seeking position." Weymann, however, pleaded for a 48-hour extension because Provost A. B. Weaver had become "visibly upset" during

Figure 51 The completed MMT at the summit of Mount Hopkins, in its characteristic rotating cubicle dome. *Smithsonian Institution Archives, Image # 98-015.*

the discussion and left the meeting early. Bartnik stayed in Tucson to clear matters up with Weaver, whose main objection was that Meinel had bolted. But Weaver also demanded a full cost accounting before the decision for Hopkins could be finalized. With that promised, only Kuiper remained wedded to Mount Lemmon. This was the end of the matter.[1087]

Latham continued site testing on Hopkins through much of 1972 before his team decided which of various knolls and ridges was the absolute best. Meanwhile, the Steering Committee continued to lobby federal, state, and local sources. In June 1972, it prepared a brochure titled "Impact of the MMT on American Astronomy," in which they emphasized it would not only serve two major astronomical research centers, but also be available to all astronomers at least one-sixth of the time. "Of equal importance," the report once again concluded, echoing Kuiper, the MMT was "indispensable" as a steppingstone to a truly giant telescope.[1088]

A Bittersweet Reality

The decision for Mount Hopkins came as a blow to Arizona. By September, Low left the Steering Committee, Kuiper backed off, and Weaver was relieved by a new vice president for research who, a good year after the decision, told Whipple that the relative operating costs were still a very sensitive issue. Claiming they made the original agreement to site at Hopkins assuming it would cost no more than at Lemmon, he added, "This decision, of course, cost us two of the key initial figures in the development and left some scars."[1089]

Whipple remained steadfast in his devotion to the observatory. These were uneasy times for him, not only because of his deepening rift with Goldberg, but also because of the attention he was attracting for his many exploits: greater scrutiny from the central Smithsonian, the Office of Management and Budget, NASA, NSF, and even the American Astronomical Society. Indeed, the growing rift between Goldberg and Whipple would soon bring internal and external scrutiny by both Harvard and the Smithsonian. And this would lead, by the summer of 1973, ultimately to Goldberg's departure to head Kitt Peak, Whipple's retirement from the directorship, and the creation of a new institutional entity, the Center for Astrophysics (CfA). Whipple, however, never retreated from Mount Hopkins. As late as March 1973, he was still taking the lead responding to concerns from Arizona over governance and shared costs. Whipple always assured Arizona that once operating and support costs were clarified, along with "growth" costs, the remaining "user's fees" for the two sites would be comparable. But soon, growing cost overruns required that Whipple engage new champions, finding them in the new CfA director, George Field, and the new Steward Observatory director, Peter Strittmatter. He also relied on old friends and colleagues such as Jesse Greenstein (Figure 52), who had chaired the Second Decadal Survey and had many ties with Whipple and Harvard as a loyal member of Harvard's Board of Overseers.

Greenstein, one of the most influential names in the profession, convened yet another review board, this time friendly, to support the MMT. They concluded that it was an

Figure 52 Jesse Greenstein. *Courtesy of the Archives, California Institute of Technology.*

"important step," was very "realistic" and being "carried out responsibly" and, in sum, "little except money is lacking in the conceptual design and the operating plan."[1090] This newest Greenstein report was meant for the NSF and it was the key that secured Ripley's attention and willingness to marshal the Smithsonian's regents as well as sympathetic Arizona congressmen to support the MMT. In 1975, Ripley used a key conclusion from the Greenstein report, that further delays would create even more cost overruns, to convince all parties, with the exception of NSF, to fully fund the project to completion.[1091] In the end, delays did produce serious cost overruns due to inflation. There was some local negative media, but the Smithsonian mounted a campaign to counter it.[1092] By June 1978, with first light in sight, the SAO media staff had done their job, and in two more years, Walter Sullivan of the *New York Times* proclaimed that, with the "new revolutionary astronomical instrument" now nearing completion, the "New Telescope Is Considered a Breakthrough: Third Most Powerful Telescope."[1093]

At its opening in May 1979 the MMT was administered as a department of Steward Observatory, with support from Arizona, the Smithsonian, and, later in the year, finally by the National Science Foundation for focal-plane instrumentation. The MMT thus became another one of the growing number of "users" on the Mount Hopkins campus. Through much of 1980 and 1981, to be expected for so radical a design, scheduled astronomical observing at the MMT was a small fraction of the operating schedule, but it gained a definitely increasing fraction of the time over continued construction, engineering and maintenance that brought it into full operation by 1982, when, in May, it was rededicated as the Fred Lawrence Whipple Observatory.[1094]

The Mount Hopkins facility was actually not Whipple's most ambitious mission at the time, in terms of cost and scope. But it was his largest and most visible institutional investment. The potentially far larger campaign, which he and Ripley had finally abandoned in 1971, promised to establish the Smithsonian as a major player in radio astronomy as well. But it had proved to be too much and too politically dangerous to take on at a time when the national institutions of science were being scrutinized for relevance by a less-than-friendly administration in Washington, D.C.

Chapter 14

National Facilities for Astronomy and the Question of Limits

Harvard College Observatory (HCO), largely through the interests and energies of Bart Bok, was one of the first institutions to develop radio astronomy in the United States. In 1937 Bok encouraged Jesse Greenstein and Whipple to explore the physical processes giving rise to celestial radio noise detected in the 1930s by Karl Jansky and then by Grote Reber. After the war, Harvard physicists got involved. Harold I. ("Doc") Ewen and his advisor Edward M. Purcell observationally verified the 1944 Dutch prediction of the 21-centimeter spectral line of hydrogen. Sticking a copper-lined pyramidal plywood horn out of a fourth-floor window of the Harvard Physics Department, they detected the line in March 1951.[1095]

Ewen and Purcell's detection energized Harlow Shapley, who encouraged Bok to mount "an adventure in radio astronomy" by starting a group and installing a modified 24-foot radar antenna at the Harvard observing station at Oak Ridge, recently renamed the Agassiz station. Bok also initiated a graduate training program in a cooperative venture with Ewen.[1096] Bok felt that he had "missed the boat" in the optical competition to discern the spiral structure of the Milky Way. He knew that radio techniques would provide a far more powerful mapping tool.[1097] Bok, Ewen, and their students (Figure 53) quickly developed the radio facility at the Agassiz station using Harvard patronage and then support from the Office of Naval Research and the National Science Foundation. By 1954, as Menzel championed the idea of a regional consortium for radio astronomy, Bok and Ewen proudly reported the measurements their growing band of graduate students were taking on 21-centimeter radiation at Agassiz, producing the first radio-astronomy PhDs.[1098] And by 1956, just as Bok was leaving Harvard to direct the Commonwealth Observatory in Canberra, Australia, his students opened the equatorially mounted 60-foot George R. Agassiz Radio Telescope, supported by the National Science Foundation (NSF) and the Research Corporation. It was then among the largest steerable telescopes in the world capable of observing the universe at 21 centimeters.[1099] But not for long.

By the early 1960s, whatever lead Harvard had enjoyed had vanished, as Bok's students, such as Heeschen and Drake, had left to build national facilities supported by the Department of Defense (DOD) and the NSF, at a time when there seemed to be no funding limits for radio astronomy. The issue for any one institution was only what specific design strategies they should choose. At first, with seemingly limitless funding, the question was

Figure 53 Harold Ewen (*far left, back*) and Bart Bok (*far right, front*) and their radio astronomy students in the mid-1950s. (*L to R*) Frank Drake, William E. Howard, Mary Connelly, A. Edward Lilley, T. K. Menon, and D. S. Heeschen. *From the Papers of David S. Heeschen, NRAO/AUI Archives.*

only who would do what. But once funding limits started to appear in the mid- to late 1960s, the question became, Which design was best?

Those few astronomers and physicists left at Harvard who were devoted to radio astronomy had even deeper concerns. Save for radio meteor studies, few astronomers at both Harvard and the Smithsonian defined themselves by instrument specialty. When Bernard Lovell, a prominent British radio astronomer, addressed a combined audience in a packed Phillips Library in April 1964, discussing how his huge 250-foot diameter Jodrell Bank radio telescope coordinated with Smithsonian Astrophysical Observatory (SAO) optical studies of flare stars with the Baker-Nunn network, the *SAO News* blithely quipped, "Since his talk mainly concerned radio astronomy, not *all* SAOers present understood *everything* he said."[1100]

One Harvard professor, Edward Lilley, listened closely as Lovell envisioned giant radio telescopes in the future. Through Lilley's efforts, by the late 1960s, the Smithsonian had come to lead what was called the North East Radio Observatory Corporation (NEROC), a multi-institution campaign to establish a gigantic, fully steerable and tracking radio-radar astronomy dish. This chapter traces the campaign for NEROC's radio-radar telescope and how the campaign reflected not only the astronomical goals of astronomers such as Lilley but also the institutional goals Whipple harbored, aided and abetted after 1964 by Ripley. Once again, Whipple's goals meshed with Ripley's to restore the Smithsonian as a national resource for science.

But Ripley and Whipple's ambitions coincided with a resurgence of a conflict that first arose in the 1950s when the possibility of federally funded national observatories for radio and optical astronomy posed a threat to traditional institutions and power structures.[1101] In 1956, the main issue was whether the observatory would be managed by a regional elite group of universities, AUI (Associated Universities, Inc.) including Harvard and MIT, or by a fully representative body of universities.[1102] Now, a decade later, the stakes were higher, in sum and scale; the conduct and organization of science at national levels was very much in flux, especially after there were serious setbacks for science funding at the end of the 1960s.[1103] It is within this context that this chapter explores why the Smithsonian's fully steerable 440-foot, radome-enclosed radio-radar telescope was never built, and examines what the effort revealed about limits to growth at the Smithsonian.

Ed Lilley's Vision

In 1963, A. E. "Ed" Lilley felt that Harvard's radio facilities were hopelessly outclassed. He doubted that NSF would help, given that it was devoted to building the National Radio Astronomy Observatory (NRAO) in Green Bank, West Virginia. So he approached Leo Goldberg and Donald Menzel with a proposition.

After training at Harvard, Lilley received his PhD in 1956 for a well-cited study of the relationship of interstellar gas and dust as revealed by 21-centimeter radio observations with Harvard's new 24-foot parabolic dish. Lilley then joined the Naval Research Laboratory (NRL) in Washington, D.C., where radio astronomy was heavily supported. There he had the use of NRL's 50-foot dish, and came into contact with people talking about building dishes more than 10 times that diameter. By 1959, Lilley had returned to Harvard on the faculty, and by 1961 was using Harvard's new 60-foot dish to measure atmospheric temperatures on Venus using radar techniques in support of planned Mariner space probe missions.[1104] He was also active in spaceborne microwave radiometry, collaborating with radio astronomers at MIT and JPL and others in the northeast region on related projects, including Mariner 2 observations of Venus in December 1962, which confirmed Earth-based radio estimates that the Venusian surface is "extremely hot, dry, and hostile to terrestrial forms of life."[1105]

Lilley proposed a cooperative regional effort in radio and radar astronomy between MIT, Harvard, and the Smithsonian Institution. Goldberg was generally sympathetic to the idea, but both he and Menzel knew that NSF was not possible. And as for the military, the Office of Naval Research (ONR) had "single handedly started radio astronomy at Cal Tech, at Michigan, at Berkeley (Hat Creek), [and] at Illinois" so was not likely to engage in anything much larger.[1106]

Lilley envisioned something much larger: a collaboration that would leverage MIT's Air Force–funded Lincoln Laboratory facility in Lexington, Massachusetts.[1107] Lilley's work on planetary radar closely matched interests at Lincoln, which with Air Force support focused on air defense systems. It was just then building an ultra-precise, fully enclosed 120-foot radar-radio telescope on Haystack Hill in Tyngsborough, Massachusetts. The site already boasted an 84-foot dish on nearby Millstone Hill, in operation since 1957, as well as a 60-foot microwave dish.[1108] Lilley hoped for increased access to Haystack's huge facilities, and worked to encourage them to build even larger facilities. Collaboration was the key, gained from his experience with the Mariner 2 program and his work with MIT's M. L. Meeks on microwave studies of Earth's atmosphere.[1109]

In March 1963, backed by Menzel and Goldberg, Lilley brought together local radio and radar specialists to search for common ground. It was a rather open-ended affair and some walked away with misimpressions. Edward Purcell and Menzel thought Lilley wanted to convince NSF to move NRAO headquarters to Cambridge. Menzel even wrote a draft proposal to that effect. But for Purcell, as well as Lilley, the idea was to establish a regional collaboration. It would work only if "working scientists can agree on instrument."[1110]

But what kind of instrument? Meeks was heavily preoccupied building the 120 foot dish at Haystack and was not responsive, and neither Menzel nor Goldberg saw it as a mission. So in September, Lilley appealed to Fred Whipple to take the lead in the program, because MIT remained reluctant to move beyond its plans for Haystack. Lilley argued that Harvard and the Smithsonian had to make a decision while funding existed, and what was required was at least a fully steerable 300-foot dish, larger than Bernard Lovell's Jodrell Bank at Manchester and more versatile than a new 300-foot transit telescope at the National Radio Astronomy Observatory.[1111] Whipple liked thinking on this scale, and with his encouragement Lilley managed to bring Goldberg and Menzel back together on several occasions and got them to agree, at least, to allow him to approach MIT for access to Haystack, as well as to campaign for an even larger facility.[1112]

But what kind of facility? The National Academy's first decadal survey in November 1963, headed by Albert Whitford, did not favor any one design or strategy.[1113] Survey committee members gave almost equal weight to upgrading present university-based radio astronomy facilities, large single-dish facilities, and interferometric arrays. There was growing sympathy for the latter, as they were the key to high resolution. The Europeans had shown particular ingenuity in developing radio interferometers, but not in combination with a large collecting area. As *Science* reported, the Whitford Committee, as it came to be called, concluded that "complex arrays are not the complete answer; . . . fully steerable single paraboloids of the largest feasible aperture" were also necessary.[1114]

But who could afford to provide the "complete answer"? In the mid-1960s, equivocal statements such as this one, reflecting the traditional astronomical community, were not yet problematic as long as the nation could afford multiple lines of attack. Parallel efforts would not soon exceed available resources, when multiple lines included competing programs in optical, radio, and space astronomy. Thus in 1964, the Whitford Committee concluded that over the next decade, among several options, all high priority, would be the construction of "Two fully steerable 300-foot paraboloids." And highest priority was for a national facility consisting of a "very high-resolution array with great collecting area" consisting of many distinct, large, steerable dishes set miles apart.[1115]

Whipple accepted the challenge posed by the Whitford Committee's conclusions. It would be a perfect growth opportunity in a highly competitive area. In late fall 1963, after receiving support from members of the Whitford Committee, Whipple provided Lilley with two SAO-supported staff positions and a planning budget of $30,000.[1116] Lilley immediately asked Harvard's president, Nathan Pusey, to match it, but Pusey refused. Goldberg, even more critical, viewed Lilley's zeal as morally questionable: spending money to raise money. Goldberg's concern was not an issue for Menzel nor especially for Whipple; both constantly played brinksmanship games with complementary funding schemes. Whipple urged full speed ahead; in mid-December Whipple and Charles Lundquist arranged for a meeting with NASA's local office to determine interest there. NASA, however, was not yet in the business of supporting radar astronomy and so was reluctant to support Harvard directly.[1117]

Whipple had been in contact with Lincoln Laboratory since 1956, developing means to derive meteor orbits by radar to complement his ongoing photographic efforts. Widely separated radar stations could track the ion column produced by a meteor as it cut through the atmosphere, triangulating to yield the velocity vector of the meteor. When observed in quantity, this yielded information about shower radiants and orbits.[1118] By the end of the 1950s, the SAO was utilizing both Lincoln Laboratory facilities and radio meteor networks in Havana, Illinois, and in Texas supported by the National Bureau of Standards. Led by Gerald Hawkins, the radio meteor studies lasted through the early 1960s, supported as well by the Air Force, the National Bureau of Standards, and the National Science Foundation.[1119]

Whipple found in Lilley a kindred spirit, certainly someone who shared his view of the value of stronger ties with Lincoln Laboratory. Like Whipple, Lilley was familiar with multi-institutional collaborative efforts and with multiple funding sources. He had been supported by Harvard, typically through private sources such as the Sloan Foundation and the Research Corporation. And Lilley had done his homework, contacting experts who knew how to develop large facilities such as electron accelerators, as well as managers of large optical observatories such as the Mount Wilson-Palomar complex. From all of them, he found strong confirmation that he needed planning support: Bruce Rule of Palomar, in distinction to Goldberg's disinclination, stated bluntly that "monies spent in the proposal will be the best you will ever spend."[1120]

By December, Whipple committed more staff and advised Lilley to create an administrative structure to handle all the details. He and Menzel also re-established contact with MIT physicists Charles Townes and Bruno Rossi to explore three options: buying into Haystack's new 120 foot, building a new 210 foot, and planning for a 500-foot dish. The most Menzel would support was a known size, such as a 210 footer based on the Parkes design in Australia. MIT, on the other hand, was bolder and more expansive. By June 1964, with its 120 foot nearing completion and promising revolutionary precision, MIT physicists were not interested in anything less than a 300 footer as the next step.[1121] Whipple continued to back Lilley with logistical, financial, and strategic support, and joined Lilley on several occasions searching for support from NASA and from the Sloan Foundation. Once Ripley was named to succeed Carmichael, Whipple and Lilley led a contingent to Litchfield, Connecticut, in August 1964 to gain his approval. Lilley now had an institutional platform.

That platform was shaped in various ways. In early 1965, amid continuing questions about NSF supporting individual research at the Smithsonian, the Bureau of the Budget (BoB) finally clarified its position regarding institutional efforts. NSF could support "service functions such as surveys and the sorting and records of specimens" and the "development and use of special instrumentation."[1122] This, Whipple saw, was how the Smithsonian could compete. The key was to establish a consortium, but instead of a regional facility, it would assume national scope and in essence provide a service function.

Just at this time, in what was a typical Bradley initiative, contemplating the inevitable U.S. Government Accountability Office (GAO) audits, he invited congressional staffers to visit the SAO to get a chance to appreciate its strengths and importance to American science, but also to sound them out. In one visit, George Evans, clerk of the Subcommittee on Appropriations that oversaw the Smithsonian, asked some pretty pointed questions about radio astronomy. As Bradley reported back to Ripley, "Mr. Evans mentioned the Sugar Grove fiasco" where the Naval Research Laboratory "expended some $85 million in a colossal failure to produce a large astronomical antenna."[1123] Whipple and Lilley were well aware of this "misadventure" and, reported Bradley, mounted a capable response disavowing any similarity between Sugar Grove and the SAO's intended facility. Evans, though satisfied, still warned them that "other agencies were highly competitive with Smithsonian for appropriations." This interaction sensitized Whipple and Lilley to the challenges ahead. But it also encouraged them because, as part of a larger strategy, Bradley had reported that Ripley's staff decided that the SAO was to be a test case for the Smithsonian overall in that "new authorizing legislation would give the Astrophysical Observatory the added stature of Congressional approval of its programs."[1124]

CAMROC and NEROC

In summer 1965, Lundquist brought together Harvard (Lilley), the SAO (Whipple, Lundquist, and Purcell), and MIT (Jerome Wiesner, Phillip Morrison, J. Orlen, and Alan Barrett) to create the Cambridge Radio Observatory Committee, or CAMROC. Well-known scientists and administrators, Wiesner especially had chaired the President's Science Advisory Committee (PSAC) during the Kennedy administration, was then dean of the School of Science at MIT, and was soon to become provost. Along with Morrison, he was well known in Washington circles and soon reported that NSF would consider a planning grant from CAMROC. Lilley now had his platform and a path.[1125]

Lilley prepared and submitted proposals to NSF in July 1966, asking for more than $1 million. By then, CAMROC decided that it should be administered by an independent corporation, and so transformed itself into the Northeast Radio Observatory Corporation (NEROC).[1126] With NSF support, NEROC was incorporated by the middle of 1967 and

Lilley became executive officer, under the guidance of Wiesner and Purcell, chair and vice-chair, respectively.

NEROC was established just when the astronomical community was growing more sensitive once again to continuing tensions between government support for national facilities and traditional lines of support for university-based research. One reaction had been the creation of private consortia, but there were also heated debates over what the most effective technologies were for new facilities. To deal with the former, at the urging of Harvard's highly influential Harvey Brooks, a former member of PSAC and the National Science Board, and now chair of the National Academy of Sciences (NAS) Committee on Science and Public Policy (COSPUP), NSF called for an update to the Whitford Report which had led to the Second Decadal Survey led by Jesse Greenstein that had deliberated over the Multiple Mirror Telescope (MMT).[1127] But to face the latter question about the most effective designs, NSF created another review panel for radio astronomy led by the innovative physicist Robert Dicke of Princeton. His Ad Hoc Panel for Large Radio Astronomy Facilities held hearings in Washington, and in August 1967 issued a report suggesting that NEROC-type facilities, as Lilley darkly summarized, be "deferred until more is known of Arecibo-type spherical dish and [the Committee] hopes that NEROC group will assist in making comparisons between . . . these two types of antennas."[1128]

The Dicke Panel had been more concerned with setting priorities based on technical feasibility than on cost, in spite of the fact that the Johnson administration was imposing "increasingly strict limits on non-Vietnam spending." So the two it endorsed were the least challenging and most practical. First was a $3 million upgrade to the Arecibo antenna to improve its reflecting surface to collect radiation as short as 10 centimeters. Next, they voted to add seven more 40-foot elements to an interferometric array operated by Caltech at Owens Valley, at a cost of some $15 million. The Dicke Panel felt these were both prudent steps toward the largest proposed facilities. But those, the panel concluded, had to be postponed until proved feasible. Thus, both NEROC's proposal for a 135-meter, fully steerable parabolic dish enclosed in a radome, costing some $27.8 million, and NRAO's Very Large Array, an interferometer with thirty-six 25-meter dishes arranged along three 21-kilometer arms, to cost some $52 million, were deferred.[1129]

Wiesner was not happy. He had hoped the Dicke Panel would be decisive in favor of one design. Their indecision, Wiesner charged, would cause American radio astronomy to fall further behind Europe and Australia.[1130] The Dicke Panel did feel that both projects should proceed in planning and that all designs be considered. They liked the Haystack 120-foot instrument and were partial to the radome-enclosed design for large, single parabolic dishes. The goal in all of this was to reach full-sky coverage and arc-second resolution over as broad a frequency range as possible and as deep as possible. Their report created a flurry of meetings between astronomers and engineers to assess merits of spherical dishes, large steerable paraboloids, and distributed arrays. Stimulated by Wiesner's objections to NSF, the group proceeded with a lengthy study that was issued in March 1968, after a $900,000 request from NEROC to NSF was cut first to $600,000 and then to $500,000 in May.[1131] These were austere times, to be sure, and these cuts severely slowed progress and increased tensions.

Ripley's Activism

In November 1967, Ripley happily reported to Whipple that a recent GAO audit of SAO operations, indeed the one Bradley knew was coming, had been "appraised most highly by the Bureau." There had been concerns about the SAO's meteoric growth and lack of clear authorizing legislation. Ripley was relieved, but cautioned Whipple that BoB was still concerned that any further expansion be solidly based on external funding, and new starts had to be avoided.[1132] So whither NEROC? This is the likely reason why the SAO became Ripley's test case.

Ripley was deeply concerned with the ongoing question of the Smithsonian's continuing viability as a national resource for science. Throughout its history, and especially after World War II, when Congress debated placing all statutory agencies under the General Services Administration, the Smithsonian vigorously defended its independent status, and its pipeline to congressional appropriations. Now the NSF's drive to establish national centers for scientific research, altering the infrastructure for collective research and centralized resources, was a new threat compromising the Smithsonian's independence. Ripley feared on the one hand that, without a national mandate, the Smithsonian would soon outgrow its ability to attract sufficient appropriations from Congress, and on the other, was ineligible for support from the new civilian federal funding mechanisms for distributing support for science.

As we've seen, this was a growing concern dating from Ripley's first years as Secretary, when Congressman Albert Thomas inserted language into the record that effectively blocked researchers from any agency or independent entity receiving federal appropriations. This blocked the Smithsonian from eligibility for competitive research grants from the NSF "for research funds on a peer basis."[1133] As Ripley described it to Leland J. Haworth, NSF director, in March 1968, the Smithsonian had succeeded in securing modest increases in appropriations to offset the loss of NSF support, but they were just treading water. Accordingly, the Smithsonian had no choice but to push hard to increase appropriations for all its "infrastructure and management and collections needs."[1134] Ripley's strategy was to reassert the Smithsonian's traditional role in American science, as the birthplace of many services such as the Weather Bureau and the Bureau of Fisheries, and as the "godmother" of aeronautical research. As a "universal institution" for science, in Hunter Dupree's words, Ripley reminded Haworth that "a salient fact is that the Smithsonian is the principal national agent for the husbandry of scientific collections. [The] entire scientific community has a stake."[1135]

Ripley knew that since 1966 NSF had supported "national programs" proposed by other agencies, whereby NSF was the sole funding source. But the Smithsonian was not part of this adjustment. The Smithsonian now asked the NSF to reconsider, offering a partnership: The Smithsonian would "push for increased direct appropriations for its entire infrastructure for the management of collections and facilities, but NSF had to recognize that the Smithsonian could not achieve its mission on appropriations alone, given its history." Allowing Smithsonian staff to continue to communicate its research professionally and to make Smithsonian's unique resources as well known and accessible as possible on a peer-reviewed competitive basis, NSF would in effect be acknowledging the Smithsonian's role in securing the health of the scientific infrastructure of the nation. Ripley's appeal also implied that the Smithsonian had the right to secure direct appropriations for national facilities available to one and all."[1136]

The Smithsonian Takes the Initiative

Ripley wanted to restore the Smithsonian as a central institution for science in the United States, and NEROC's telescope became one of his test cases. He knew he had to make one more pitch to NSF, and that it would in all likelihood fail, clearing his path to Congress via the congressional members on the Board of Regents. Accordingly, in the summer of 1968, Bradley told Lilley to organize a meeting in Cambridge to develop a proposal to NSF quickly. Assuming that would fail, they would send that proposal through the regents to the Office of Science and Technology and the Bureau of the Budget. This meant that they had to propose NEROC as a national facility, in order to meet the needs of and gain the support of astrophysicists around the country. A national character was considered necessary for a major federal appropriation.[1137] Time was very tight, given the budget process, so they also decided to mount a phone campaign alerting scientists and administrators within NEROC's domain of the change in scope.[1138]

At a lunch at the MIT Faculty Club, Bradley pressed the NEROC leaders for facts and figures he might use as briefing materials that argued why the Smithsonian should be the "authorized agency."[1139] Bradley knew there were several deep problems with this course of action. First, in reply to his questions, astronomers feared "that a national designation would tend to lead to creation of a staff-in-residence, which would duplicate or displace the staff of NEROC." There was also the question of how NSF might react to a competing national facility. Next came issues of governance, which Bradley insisted would have to be at the discretion of the Smithsonian. Wiesner agreed with this, but continued to resist the implied national scope the Smithsonian demanded. Only Whipple was totally positive, predicting "that radio astronomers from all parts of the country would expressly be welcomed to share in the use of the telescope."[1140]

After their meeting, Lilley polled members of NEROC while Wiesner and Bradley reached out to science policy figures such as Hugh Loweth in BoB and Don Hornig, director of the Office of Science and Technology, as well as Haworth.[1141] By mid-September Bradley provided Ripley a general strategy schematic, and where problems might lie. They had to find a way to surround the NSF, but knew that NSF was committed to a large interferometric array, the "Very Large Array," rather than a single "big dish."[1142] Loweth advised that NSF had authority for only one new large radio facility, and at that time, even the Navy was cutting back on its radio astronomy programs, so he wondered why the Smithsonian felt it was the right place. There was no alternative but to proceed, Bradley advised Ripley privately. Standing still would soon "subordinate Smithsonian's traditional pursuits in astrophysical research to determinations and funding by NSF."[1143]

Bradley decided that Ripley's strongest suit was to emphasize that the Smithsonian had been in astrophysics for more than 70 years, now had some 500 people involved in the effort, and was already working with a "strong company of associated universities." He felt that Loweth's "unfavorable reaction" was fairly predictable but hardly definitive, and in sum encouraged Ripley that the time was right to engage the regents, get clearance from the BoB and, if necessary, the White House, and then, led by friendly congressional regents, march on to Congress.[1144]

Bradley knew they had to act quickly before debates over a national policy for the organization of science created an immovable logjam.[1145] Brooks's earlier urging for a new "super Whitford Committee" had created the Second Decadal Survey that Greenstein chaired and then the Dicke Panel, both of which, as Lilley feared, became roadblocks to a quick decision.[1146] A third reason for haste was that, as Bradley well knew, Ripley had just won a long battle over continued responsibility for the Science Information Exchange, resisting calls by BoB and NSF that it relinquish the long-held national responsibility to another coordinating body, possibly the NSF, in light of management policies set in place by NSF in 1963.[1147] There was no question that the Smithsonian had to act fast. Ripley, always provocative, wanted to hold a national meeting on the matter.

Whipple stayed in the background, making personal appeals to as many astronomers as he could muster by telegram, by phone, and by direct contact. He had not been an early advocate of a national meeting, preferring a quieter, personal recruiting scheme, but Bradley overruled him.[1148] Bradley knew they had to gain a broad base of visible support from astronomers in as many congressional districts as possible, and especially from astronomers who had visible roles in the competing projects. Through these efforts, orchestrated mainly by Lilley, Bradley, and Wiesner, the beginnings of a coalition started to form, but it remained tenuous because of the continued frustrations that astronomers were experiencing getting anything big approved.

Accordingly, they engaged Frank Drake, the Cornell astronomer and Arecibo director who harbored resentment toward NSF when it initially failed to fund improvements to his dish.[1149] Drake had also advised on the NEROC design and was in close contact with Harvard, MIT, and Lincoln astronomers. He agreed that a national meeting would break the perceived deadlock caused by the Dicke Panel. It was also "a real opportunity" to gain the attention of the Smithsonian's Board of Regents: "Meeting and action should be national and called by Smithsonian."[1150]

Ripley's Call to Washington to Debate the Needs of Radio Astronomy

Alerted by NEROC partisans in Cambridge, the *Boston Herald* issued a strongly worded statement that the "U.S. Falls behind Europe in Radio Astronomy." It quoted Geoffrey Burbidge, always ready to lend a hand, who ascribed the failure not so much to bleak budget prospects, but to the "regional attitudes [that] are diverting radio astronomers from combining for the common good." Burbidge was a member of the second Decadal Survey headed by Greenstein, and used it as his bully pulpit: "The high energy physicists banded together and decided [what] they wanted. . . . If radio astronomers could have done the same thing they would have had a large new telescope four years ago."[1151]

In very short order, with a haste that caused some confusion because not all NEROC members got the word, Ripley invited 35 radio astronomers, including NSF and NASA representatives, to attend a two-day conference in late November to debate the needs of radio astronomy in America. Bradley asked NRAO astronomer John Findlay to chair the meeting and orchestrated the proceedings to avoid comparing competing designs. The purpose of the meeting was not to compare the two, but to discuss the filled dish "exclusively."

This strategy worked. The group recognized an "urgent need" for a fully steerable, large aperture dish as a national facility, left site selection for another time, and, most of all, announced that it "strongly supports" the plan to have the Smithsonian submit a proposal to "appropriate agencies of the Federal Government." And finally, mostly directed to NSF, the group agreed that the present NEROC design was "close in size and general specifications to a feasible optimum design, and that it should be used as the basis for final design work for a Smithsonian telescope."[1152]

Findlay, a senior NRAO radio astronomer who had directed Arecibo in 1965–1966 and was then on the NAS Space Science Board, played a key role in the deliberations. He had been a strong advocate for synthetic aperture designs, but was regarded as a broad-minded technical expert who could see multiple sides of an argument. He vigorously called for a combination of technologies, including a study for NRAO on "the largest feasible steerable telescope."[1153]

After the meeting, Ripley asked Findlay to lead a new advisory group named STAG (Smithsonian Telescope Advisory Group) to implement these conclusions. STAG membership included leaders in radio astronomy across the country, with only minority representation by NEROC members. This placed the Smithsonian center stage, but deeply worried Weisner, who suspected that NEROC itself was being "squeezed out."[1154] Findlay authored the proposal for Ripley, while Whipple did what he could to address Wiesner's concerns. But what dominated everyone's attention was the drafting and formal submission of the legislation for review by the Smithsonian's Board of Regents, which was due to meet in regular session within days.

Ripley's Strategy

Ripley sold the $30 million project to the Board of Regents by outlining where it fit into the Smithsonian's longer-term plans and identity. Laying out examples of SAO programs that enjoyed national standing, he went further, citing other parts of the Smithsonian as examples of national standing, such as the Woodrow Wilson International Center for Scholars, and the "sponsorship and management of national-scale cooperative projects: like

STRI [the Smithsonian Tropical Research Institute in Panama], the Chesapeake Bay Center for Field Biology," and other endeavors. Along these lines, in 1969 Ripley would rename the Smithsonian's Museum Bureaus, emphasizing their "National" status.

Ripley's key strategy was to make the regents aware of the fact that the SAO was actually on thin ice, if anyone in Congress cared to notice. Whipple had noticed, of course, knowing that the Act of August 22, 1949, was the sole legislative act authorizing the observatory's existence: a means of performing "solar observations at high altitudes."[1155] But "if the Regents approve and the Congress ultimately enacts this proposed legislation, it will stand as an unprecedented legislative recognition of the research capability of the Smithsonian Astrophysical Observatory."[1156] The regents had to act to redefine SAO, and act they did. With the regents' quick approval, within a month, over the Christmas holiday, the Smithsonian's congressional regents, including senators Anderson, Fulbright, and Scott, introduced authorization bill S. 705 with the promise that "under the administration of the Smithsonian, the telescope will therefore be a truly national facility serving national goals."[1157]

Ripley's strategy of subjugating NEROC to Smithsonian leadership through STAG was not taken lightly by the rest of the NEROC members. Some demanded that, henceforth, there had to be legal agreements and a clear plan for overall management. Lilley pleaded with Whipple on several occasions to assure Wiesner that NEROC's interests were being served. Whipple said he would do so, but he did just the opposite. As the regents were approving the submission of the legislation in mid-January, Whipple asserted at a STAG meeting at Lincoln Laboratory on 3 January that STAG reported to the Smithsonian, and not to NEROC.[1158]

In reaction, NEROC members created a counter "Salvage group" that first met in a member's home, and then more formally at Lincoln Laboratory.[1159] Members worried that "if SAO developed an in-house staff for the telescope, it could compete with the [national community] which the observatory is built to serve." Irwin Shapiro, then an MIT professor of geology and geophysics who had been a prominent Lincoln Laboratory staff member through the 1960s, applying techniques in planetary radar at Haystack, sought an alternate route: the "formation of a national consortium of universities which would work strongly with SI." For Shapiro, STAG could well act as a core group, but it had to be expandable if more institutions wished to join in. Lilley disagreed, fearing that "the in-house/out-house staff problem might arise again. [1160] Lilley knew that a strong "in-house" capability was required for instrument development to keep the facility at the cutting edge. But on the other hand, a strong "in-house" faculty of research astronomers, which would be expected if STAG ran the show, would look suspicious to any outside funder, or to community-based leaders, who were ostensibly looking out for the "have-nots"—in the case of Kitt Peak, it had been university-based astronomers in the East and the Midwest who had little or no access to large optical telescopes.[1161]

Why the sensitivity? Some of the members may well have recalled Whipple's efforts in 1955 to centralize the process of establishing a national optical observatory in the Smithsonian. We have seen that Whipple had been named chair of a "Needs Committee" to outline what support astronomy needed to solve the major problems it was then facing. But then Whipple parlayed this responsibility into visibility for the new SAO by publishing the resulting papers, *New Horizons in Astronomy*. By the end of 1955, Whipple even wanted to convert his ad hoc committee into a working committee for a National Astronomical Observatory. He and Menzel, who was then the president of the American Astronomical Society (AAS), tried to orchestrate a plan whereby the Smithsonian would apply to NSF for enabling funds to continue the deliberations. Their effort was not appreciated by the NSF advisory panel for astronomy or by West Coast astronomers.[1162]

Moreover, Ripley's proposal would consume the entire 1967 Dicke Panel budget. To make matters more convoluted, MIT was now certain that the Air Force would no longer support Haystack and had thrown in with NEROC completely.[1163] This further separated NEROC from the Smithsonian's takeover in what MIT called the "Cambridge end-run." The way Whipple viewed the situation, as he confided privately to Lilley and Shapiro, was that Wiesner's and NEROC's objections were merely a contest over "power."[1164] The project did indeed end up as a Smithsonian-led proposal for a 440-foot dish in a 550-foot radome (Figure 54). Lilley and Whipple, of course, couldn't pull something this big out of the hat with Congress alone. But the Smithsonian regents could, urged on by S. Dillon Ripley.

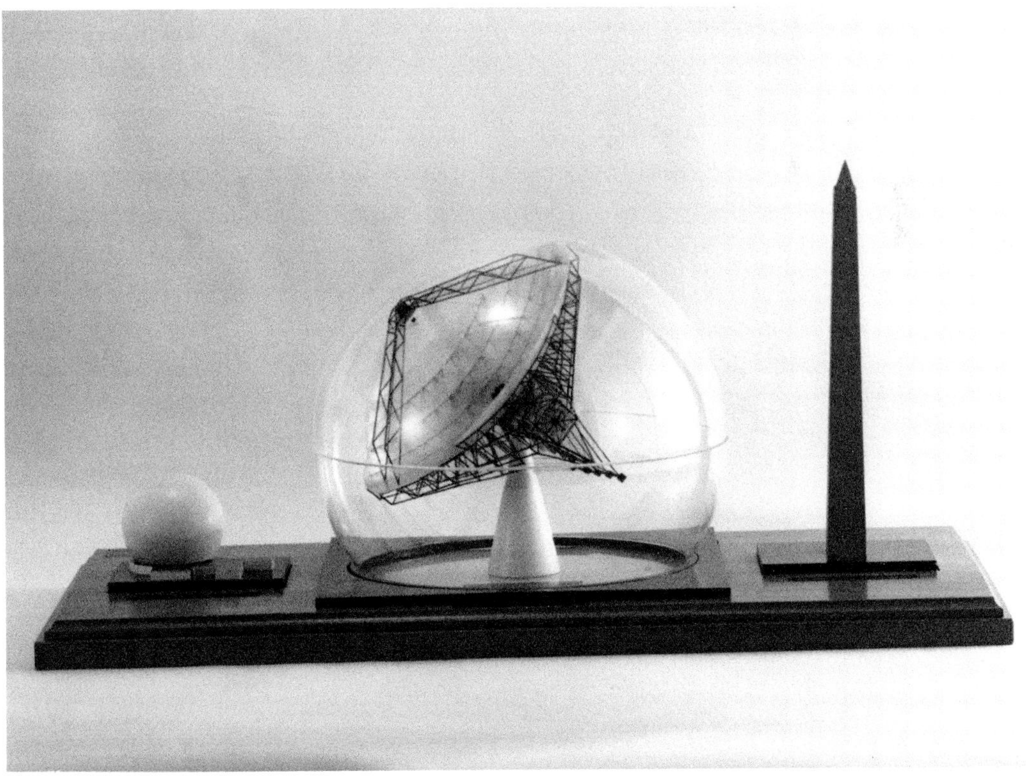

Figure 54 Display model of NEROC compared with an existing radio facility. The Washington Monument was included for scale. *Smithsonian Institution Archives, Acc. 16-263, Box 4.*

Deliberations across the Board

The year 1969 could well be remembered as one of uncertainty in radio astronomy. As the Smithsonian submitted its proposal to BoB and courtesy copies were sent to NSF and elsewhere, as members of STAG and NEROC debated, and as the second decadal survey chaired by Caltech's Jesse Greenstein was continuing to deliberate, NSF reconvened the Dicke Panel.[1165]

Through the spring months, Bradley took over the lead from Whipple in dealing with Wiesner and managed to repair some of the damage to NEROC/Smithsonian solidarity. At least as long as Bradley said the Smithsonian would consider a multi-institutional consortium, based possibly on the Smithsonian's successful Woodrow Wilson International Center for Scholars, which included an independent board of trustees, Wiesner agreed not to sabotage the Smithsonian bill if and when it reached Congress. Simultaneously, Lilley worked with Shapiro and others to create a management structure that was workable, but still vague enough to accommodate both the Smithsonian and NEROC visions.[1166] Flow charts and block diagrams went back and forth between Lilley and Whipple through March.

The Wilson Center's board of trustees model resembled a consortium, and as Bradley admitted (to Whipple at least) the Wilson board is "indeed a strong Board. . . . In all candor, it is stronger than I would like to see." Of course, Bradley knew the Wilson Center was lodged physically within the Smithsonian and was managed "perfectly well" by the secretary and board of regents. But NEROC would not be local, so Bradley advised Whipple to call it something that would appeal to Wiesner: a "Committee of Cooperating Users for the telescope." It would not have permanent members, there would be no ad hoc congressional nor presidential appointees. All managerial functions would be defined by the scientific board (STAG, in Bradley's mind), and Bradley advised Whipple to select a director "to plan, perform, and submit for review the matters to be decided by the committee" as quickly as possible.[1167]

As Bradley coached Whipple on how to satisfy Wiesner, Sidney Galler, the Smithsonian's assistant secretary for science, courted BoB on various Smithsonian initiatives, including the radio-radar telescope, and found deep problems. John Donald "Jack" Young, who had been at NASA in the early 1960s and then had moved to BoB, was opposed; it "completely distorts the 'SI fiscal picture.'" He also feared this would harm future relations with NSF. Young also strongly advised that the Smithsonian delay the legislation, at least until the incoming directors of NSF and the Office of Science and Technology (OST) for the Nixon administration (William McElroy and Lee DuBridge, respectively) were settled, and the Smithsonian did due diligence seeking out NSF's position. Young warned Galler that the BoB was preparing these comments formally, but Galler hoped BoB would hold off at least until NSF's position was known.[1168]

Nevertheless, the Smithsonian pushed ahead. The reconvened "second Dicke Panel" now felt that the preliminary design studies for NEROC and the NRAO proposal for what became the Very Large Array (VLA) were sufficiently mature to recommend they be implemented with "equal urgency." Both had to be "nationally available" to astronomers. Cost was not an issue for the panel, but it did note that the VLA had a slight advantage because it could be built in stages, whereas the radio-radar dish was all up.[1169] Whipple alerted Lilley to be ready to travel to Washington on a moment's notice through late July and early August.[1170]

Bradley put the final touches on the briefing to Congress that he thought was set for September 1969, only two months hence. Ripley would lead off, expressing the "national need for an instrument of this kind" and then paint the Smithsonian as the institution to make it happen. Then Whipple would explain the nature of radio-radar astronomy and how it would contribute overall to science. He would also have a prepared rebuttal to any BoB objections (budget restrictions, program priorities, pending studies, question of "science organization"), but he would hold off using it, Bradley advised, until the question period to avoid a defensive or negative tone in the opening presentation.[1171] Bradley also had Findlay and Purcell testify after Ripley and Whipple addressed the subcommittee. And if the congressmen wished, Lilley would summarize.

What made life even more tense was that NSF's Haworth had stepped down, and the views of the new director, William D. McElroy, confirmed only in mid-July, were not known. Ripley and Bradley soon secured a hearing with McElroy on an unrelated matter, finding him generally friendly and receptive.[1172] It was not so, however, with BoB. As it turned out, both the House and Senate hearings were postponed to beyond July 1970, because the parties were far from a clear resolution; specifically, BoB was charging that it would "not be prudent to initiate this ultimately very costly project at this time." The second Dicke Panel report and the Greenstein Committee results were not expected until sometime in the spring of 1970.[1173]

Waiting for Greenstein

Harvey Brooks had campaigned long and hard to rectify the deficiencies of the first Decadal Survey, negotiating ground rules with BoB's Loweth and NSF's Haworth. Brooks felt that optical astronomy had to share equally with radio and space astronomy. Astronomers wanted this, as did NASA, and thus the Greenstein Committee was charged with encouraging NSF "to study astronomy in cooperation with NASA."[1174]

In January 1970, Greenstein rationalized that the second survey was necessary only five years after the first led by Whitford because "certain fields had been omitted and should now be included." These included space astronomy, solar physics, and radio astronomy. The first Dicke Panel report on radio astronomy facilities as well as an Astronomy Missions Board report to NASA demanded a more balanced review. There was also a recent survey in the physics community that impacted astronomical priorities, so there was every reason to make the new effort.[1175] The Greenstein Committee included names familiar in this story, such as Leo Goldberg on the central committee. Dave Heeschen, Bernard Burke, Geoffrey Burbidge, and George Field were on a subcommittee for radio astronomy. The SAO's Robert Noyes and Giovanni Fazio were also panelists.

The radio panel, headed by Heeschen, took center stage. Throughout the fall, Harvey Brooks pushed vigorously to get the panel to select a single project, keenly sensitive to the negative assessment from BoB that astronomers could not make up their minds on anything. Brooks also played a familiar card, warning that if the United States did not act fast, it would fall further behind the Europeans and British.[1176] Brooks's highly influential COSPUP had already endorsed the VLA. But in a preliminary report in December, Heeschen's panel endorsed a little bit of everything, calling for the Arecibo upgrade and recommending major funds for the radio-radar dish and the VLA. Burbidge was the only member of the panel to agree with Brooks at first; the rest of the panel resisted.

The debate continued through the spring and fall of 1970 in parallel with hearings in the Senate and House. Heeschen expressed "mixed feelings" that the VLA was NSF's choice. And Burke, an advocate at first for the NEROC design, worried, with Brooks, that if the large dishes in Europe proved successful, they would be the "'kiss of death' to the big dish."[1177] In the end, the Greenstein Committee's conclusions, in spring 1971, reflected Brooks's push for a single instrument decision, and the NSF's push for VLA. The case for the big dish was desperate. Just how desperate requires looking at a larger landscape.

The Nixon Years

Several significant national trends heightened tensions during these years, all precipitated by tightening budgets and shifting national priorities. One was the Mansfield Amendment, which, though repealed in 1971, effectively dampened military support for basic research.[1178] Its effect was widespread, especially on NSF to pick up the slack but as well on numerous large initiatives like NEROC to show relevance to military needs. By summer 1970, in response to these pressures, but related as well to continuing campus tensions about military sponsored research at MIT, the Air Force transferred Lincoln Laboratory's Haystack facility to NEROC and MIT, with support from both NSF and NASA.[1179]

The second shift was Nixon's reorganization of the BoB into the Office of Management and Budget, which eventually institutionalized "management by objective." Nixon's order required OMB's centralized clearance "of all department, agency, and commission budgets *before* they were submitted to Congress."[1180] These shifts were part of a general tightening of control and closer oversight over the planning and management of scientific research through a "master plan for science policy."[1181] Edward E. David, who had replaced Lee DuBridge as Nixon's science advisor, warned that "things are going to change."[1182]

Smithsonian Testimony

Things were changing already for the Smithsonian. Bradley's strategy for the big dish was finally put to the test in July 1970 when it was presented to Rep. Frank Thompson Jr.'s House Subcommittee on Library and Memorials within a larger submission dealing with the full Smithsonian. Bradley asked Thompson (D-NJ), well known to be a friend of the arts but critical of the Smithsonian's growth in the 1950s, to call these hearings to respond to growing criticisms of Smithsonian operations in recent years, especially the concern that many Smithsonian programs were not covered, or had expanded greatly beyond, their authorized mandates. Publically, Thompson's charge was to "conduct hearings into the entire Smithsonian-from basement to attic."[1183] Congress had not examined the Smithsonian for more than a century, at a time when the Institution "was a mere fraction of its present size" and critics were asking why.[1184] The radio telescope would be the last subject on the docket.[1185]

During some seven days of testimony from Ripley and staff, much of the attention was given over to a public controversy over the siting and extent of the Joseph H. Hirshhorn Museum and Sculpture Garden, the public versus private nature of the Smithsonian, and the nature of its financial management and employee relations. Finally, the question of the Smithsonian's bid for a "Proposed Radio-Radar Telescope, Report on H.R. 13024" was raised.[1186] On 29 July, the sixth of the seven days of hearings, Ripley, accompanied by Whipple, Findlay, and others, testified for the telescope. Ripley provided the atmospherics, claiming that the Smithsonian, once again, found itself "on an incipient bandwagon long before the smoke is cleared and the bandwagon itself has come into view." This was certainly, to his mind, the case for astrophysics, in which the Smithsonian had devoted its energies long before it was popular in America. He then appealed for the radio-radar telescope, describing it as a "truly national facility serving national goals" under the administration of the Smithsonian. It would be national insofar as the Smithsonian will establish an Advisory Committee, "a pattern which is very familiar to the Smithsonian, as we have a wide variety of commissions, committees and boards to advise us on so many of our disciplinary concerns."[1187]

Thompson interrupted Ripley frequently with comments, quips, and questions. Asking early on about costs, Ripley deferred to Whipple, who said that with delays they would escalate, and had already done so since the initial budgets were developed in 1967. To this, Thompson wisecracked: "Maybe if we could get this in the Defense budget it would be all right, but then I would be against it, at least, as it stands now."[1188] But now, turning to Whipple, Thompson asked why such an expenditure was needed by astronomers.

Whipple called upon the wonders of the universe, using superlatives for how modern technological prowess needed to be harnessed to understand them. Citing the "unbelievable power" of quasars, radio sources, and pulsars, he telegraphed a clear message that such power might someday be understood, and even generated locally to solve humanity's energy needs. The big dish would be the key to unlocking these mysteries: It would detect the weakest radio signals from the farthest distances. It could well aid the nation's space program by acting as a receiver for transmitters aboard deep-space probes, or it might even be the means by which life could be detected elsewhere in the universe. When Thompson asked whether there were military applications, Whipple responded: "No, it is not contemplated for defense use, but one never knows when such an instrument might be of value to the national defense."[1189]

Promises flowed endlessly; Whipple predicted that sooner or later some government agency would want to fund the instrument, so they were now asking only for support to ensure that their construction package be ready. Thompson cut in again, asking whether other government agencies had similar plans. This was Whipple's cue to introduce Findlay, who testified that nothing like it was being proposed by any other government agency.

Thompson and his committee well knew about other proposals, but these were for arrays, upgrades, and smaller dishes. So they asked Whipple, why the 440 footer? On and on the questions went circling the same subjects and themes. After a committee member asked whether the different designs promoted by the competing interests were complementary or competitive, Findlay responded they were complementary and that they had made this clear to the BoB. Thompson then snorted: "It is very hard sometimes to get them to understand."[1190]

Indeed, BoB had also asked, why the Smithsonian? Galler had answers at the ready from Lilley and Whipple and he had implicit NSF approval through Haworth and McElroy. With this, Galler put it to rest, stating that the matter had been raised and dealt with: "There is no incompatibility, no competition. Dr. McElroy made it very clear that he endorsed the Smithsonian's efforts for the dish. That is all I would like to state."[1191] In the end, Thompson thanked the testifiers, and reiterated his view, expressed to his colleagues on the bench, that this proposal "fits well into the general activities, current and future, of the Smithsonian."[1192] With that comment, Whipple was excused and Ripley and Galler went on to the next subject, the Radiation Biology Laboratory, and, as it turned out, how it fit into national planning and coordination compared with, once again, the NSF.

Whipple stayed in the room in case there might be more questions. One of the five congressmen, Indiana's John Brademas, indeed had more questions for him relating to land acquisition, design, and construction costs. But Brademas pressed on, sermonizing more than questioning. These hearings were his opportunity to air what he felt was a most egregious state of affairs in the nation, only hinted at in these hearings "in very gentle tones" by "a most distinguished scientist in the person of Dr. Whipple." He observed that scientists across the nation were "in far more trenchant and indeed savage tones indicting the present administration for failing to invest adequately in basic research." Citing a recent issue of *Science*, Brademas warned that scientists had to organize: "They have nothing to lose but their laboratories."[1193]

Both Thompson and Whipple warily endorsed Brademas's sermon. But this chorus prompted Iowa representative Frederick Schwengel to rise and speak from "the other side of the aisle." He cautioned his "scientist friends" to "start setting priorities . . . and not just lay the whole thing in our lap or the White House's lap." Thompson readily admitted that his own criticism was in part political, but Whipple did not skip a beat, facing Schwengel:

> Also as a farm boy from Iowa—I spent the first 13 years of my life on a farm, a real dirt farmer in Iowa. . . . [Schwengel: 'You have a fortunate background'] . . . I have much the same point of view that you do, and I would like to see us get as much as we can out of every dollar of the taxpayer's money. That is one of the reasons that this

> design has been so very carefully engineered to be sure we are getting the very best product, as well made as possible, at the cheapest cost, to give us the science which we so desperately need.[1194]

To drive his point home in terms Schwengel could relate to, Whipple admitted that "well, we are digging in very poor soil for funds these days, I must say." This was Thompson's cue to reassure Ripley and Whipple that though BoB's assessment was important, it was not determinative. "I'm glad to know you are from Iowa." Thompson added, "I gather Mr. Schwengel is very sympathetic to this idea now." To which Schwengel added, "This fact has raised the stature of the entire Smithsonian Institution."[1195]

Ripley expressed satisfaction with the well-orchestrated Thompson hearings. He reported Smithsonian-wide that the hearings were positive, offering as examples compliments by Thompson and Brademas. If Whipple's performance was typical of the testimony delivered by Smithsonian staff, Ripley had good reason to be proud, but he had been warned.[1196]

Beyond the Thompson Hearings: "Is SAO Needed?"

During the long process of transforming BoB into the OMB, OMB's oversight of the Smithsonian was now in its General Government and Management Division, not its Science Division. This caused deep concern within Ripley's circle, because even under Science, relations were far from comfortable. Just at that time, for instance, Charles Lundquist was preparing a "sales' document" for Whipple on the MMT to send to the new OMB, in addition to their continuing push for NEROC.[1197]

Even before the transition, BoB budget analysts in the Science Division had questioned the scale of SAO's activities. Analysts such as Loweth and Young viewed the SAO as an anomaly and not an organic unit within the Smithsonian. More seriously, BoB felt that national support for astrophysical research should reside solely with NSF. "To the contrary," Bradley heatedly advised Ripley in January 1970, these were only the analysts' opinions, and not the expressed feelings of the decision makers such as Haworth. Bradley promised to realign the OMB staffers, even some who thought that SAO be given wholly to Harvard. "Someone," Bradley charged, "is incredibly disregarding the statutory basis for SAO."[1198] Just as Ripley had advised the regents in November 1968 about the weakness of the statutory basis for the SAO, Bradley felt it was absolutely essential now that the Smithsonian fight for the big dish. It would, at the very least, validate the SAO's national status.

Despite Bradley's assurances, more trouble was brewing. In the early summer of 1970, just before the transition, Kenneth Schwartz, an analyst in the BoB Science Division, was assigned to focus specifically on the SAO. To prepare his report, Schwartz consulted with the NSF solar astronomer Gerald F. Anderson, who was responsible for NSF's array of national observatories. The initial draft was not for circulation until Schwartz's boss, Hugh Loweth, sent it upstream for comment and approval in late August. By then, however, the transfer of OMB cognizance over the Smithsonian had changed to the Government Division and a new team of analysts had been formed there to examine the Smithsonian overall. Initially, the principal examiner was OMB staffer Dean Anderson, but within a week, the OMB staff was further reorganized, and Anderson would now assist another staffer, Ted Lutz, in the investigation.

Part of Anderson's responsibility was SAO oversight, so he was in contact by the end of the year with Raymond Watts, Whipple's administrator. Anderson had Schwartz's report and spoke of it generally with Watts, but took several months to send a copy to John Jameson, then Ripley's director of the Office of Programming and Budget. [1199]

Schwartz's "SAO Issue Paper" viewed the growth of the SAO in the 1960s "like many other Smithsonian activities," as "generally unstructured."[1200] It reviewed how, since 1955, Whipple had built up what had originally been a traditional department of the Smithsonian and took it in many new directions with significant NSF and NASA funding, while, at the same time, integrating itself with the Harvard Observatory. Whipple's "policies worked well" in the first half of the 1960s, but in the past few years, cuts in NASA support and complications and competition with NSF had led to a slower rate of growth. Any increased appropriations for the SAO could well constrain the Smithsonian overall, so it was "particularly appropriate at this time to raise questions regarding the future of SAO." Bluntly, the "SAO Issue Paper" asked: "Is SAO needed?"[1201]

After belaboring pros and cons, the curt answer was: SAO's role "is hazy." Its "unique and special capabilities" were not so much institutional; they lodged within individuals who might well take jobs elsewhere and so were not institutionally defined. Satellite tracking was, of course, a unique function and it was not likely that it would survive if SAO were to be "dissolved." Asking what would it take to keep SAO alive, the Paper reviewed its major assets: a large computer center funded by NASA, and of course, the Mount Hopkins complex. The computer was contingent on the satellite work and was not a real Smithsonian asset. Therefore the real costs of continuing SAO, to the Smithsonian, were in positions and salaries. If all the instrumentation were to be deeded to NSF or to AURA, then Smithsonian astronomers could still qualify for observing time.

The "SAO Issue Paper" also asked how invested the Smithsonian was in SAO. Ripley, the paper noted, was making "rapid movement" toward the arts and humanities. And in the sciences, the Smithsonian's real push was now in the environmental and biological realm, not in the physical sciences represented by the SAO. Bradley was the SAO's main champion on the Mall; certainly he was "the best informed of the Washington staff on its operations."[1202] So if the future of SAO was indeed in the secretary's hands, and if he continued to shift away from the physical sciences, then it might well be better to transfer SAO to NSF in one of these ways:

1. "Turn SAO into a national center" [funded through NSF and managed by a consortium of universities as are Kitt Peak and NRAO].
2. "Offer the management of SAO to Harvard" [very much as Cornell's operation of Arecibo].
3. "Meld Mt. Hopkins facilities into the Kitt Peak operation and discontinue funding of SAO as a separate entity."

4. "Meld the bulk of SAO personnel and facilities" [into the Association of Universities for Research in Astronomy (AURA), which, Schwartz mistakenly thought, operated both Kitt Peak (yes) and NRAO (no, operated by AUI)].

The "SAO Issue Paper" concluded that the fourth option was the best. It recognized that there were areas of SAO activity, such as satellite tracking and geodesy, that did not fit AURA's interests. And it was these very activities in which the SAO was unique, or at least was ahead of other civilian institutions, such as NASA.[1203]

The "SAO Issue Paper" concluded that there seemed to be no "convenient home for SAO." It was an oddball for the Smithsonian and belonged more logically under NSF directly. But for the transfer to work, none of the four options came without problems, although the fourth was "the most promising of the lot." It would place SAO within NSF's sphere and provide a home for SAO staff at Harvard, which conveniently was a member of the nine-member AURA consortium.[1204]

Then came a fifth option: Retain SAO as a bureau of the Smithsonian "but reduce its level of support," which would result in staff cuts. This would retain the undesirable "preserve" of funding for astronomy within the Smithsonian and would not aid the "centralization of basic research funding in NSF." Even OMB realized that this would not work. So in sum, the "SAO Issue Paper" favored the fourth option and recommended further discussions with NSF senior staff regarding such a shift. And if NSF proved to be receptive, then the OMB views should be shared with SAO staff "to elicit their views" in open meetings between the SAO, Smithsonian, NSF, and AURA staff. Everyone would be free to "air differences of opinion."[1205]

By late February, OMB's Dean Anderson was planning to visit the SAO. To prepare his respondents, he sent Schwartz's "SAO Issue Paper" to Jameson, "to provide a basis for discussion." Indeed, Anderson admitted, "The memo is frank in places—hopefully not embarrassing," adding that "it wasn't intended for SI-SAO review; just in-house musings." But most of all, he emphasized: "It doesn't necessarily reflect the current views of OMB, General Government Division, or me!"[1206]

Jameson quickly relayed Anderson's radioactive leak to Watts in March 1971, urging him to take it seriously. He well knew that OMB was still in a fluid state, and the statements in the "SAO Issue Paper" may now be "only interesting historical documents" because OMB had recently shown some mercy in the budgeting for FY 1972. But, he warned, these questions would no doubt arise sooner or later. And he was right.[1207]

What is striking about this document, and its distribution, is that it reveals how the existence of the SAO, at the scale to which Whipple had raised it, was a challenge for those tasked to set priorities for allocating federally funded resources, setting institutional priorities and structures, and devising means to manage and control them. As he transmitted his draft to Jameson, Anderson explicitly distanced himself, not specifying which of the options he favored. One thing was clear: These were the issues that needed airing. They indeed must have made for sobering reading, and could well have played a role in Bradley and Ripley's ultimate decision to drop the big dish.

Preparing for Senate Hearings in the Midst of Chaos

Although the big dish was now front and center, the Thompson Subcommittee had been more concerned about the Hirshhorn matter, especially the possibility that its sculpture garden would split the National Mall at 7th Street.[1208] The Senate would not be similarly distracted, or benign. At about the same time, there seemed to be a thaw in NSF/Smithsonian Institution (SI) relations, as newly appointed assistant secretary for science David Challinor reported to Ripley in February 1971 that "with the arrival of Bill McElroy at NSF the anti-SI forces there seem to be retrenching." Challinor cautioned, however, that "obviously, we are in a sensitive time period where we all must be sure that we have a uniform approach. I know all agree that it would be tragic to muff this opportunity."[1209]

The big dish was reintroduced 30 March 1971, in the Senate by Anderson, Fulbright, and Hugh Scott, and in the House by Thompson and Lucien Nedzi. All save Thompson and Nedzi were regents. This time, Bradley and Ripley emphasized that the Thompson hearings unanimously voted to report their recommendation to Congress, and they hoped that the new Congress would agree.[1210] But they well knew that the OMB draft was lurking.

Later in April, however, radio astronomer Bernard Burke warned Lilley that Jesse Greenstein was now sympathetic to Harvey Brooks's continuing concerns over the lack of unity among astronomers. He now favored the VLA over the big dish, and, worst of all, felt that the Smithsonian's continued push was impeding progress of the VLA, as it too was being discussed in Congress as part of NSF's request.[1211] Greenstein was especially important to SAO at that time, because he was also deeply involved in deliberations brewing at Harvard over the future relationship between Harvard and the Smithsonian.[1212] So in May, as congressmen were becoming aware of Greenstein's support for the VLA and were asking strong questions, and with the OMB "SAO Issue Paper" in the background, Bradley conferred again with Loweth at OMB and then with Findlay and Whipple. It was clear now, Bradley advised Ripley, that support for the big dish was eroding, as was support for the SAO overall. Bradley also reported that Harvey Brooks had urged Whipple to postpone "our bid for authorization of the big dish for this year."[1213] Any evidence of disagreement on priorities, Brooks felt, was likely to be used as an excuse to delay funding anything. "If we don't get the VLA this year, then so far as I am concerned the game is dirty pool rather than scientific statesmanship, and everybody better start pulling his own political wires. But let's try statesmanship for one more year." Knowing what else was lurking, Whipple could not resist Brooks's logic, and so complied. Brooks immediately and publically hailed Whipple's decision as a "very statesmanlike action on the part of the Smithsonian," which made it possible for the radio astronomy community to stand behind a single facility in the coming fiscal year.[1214]

Ripley now seized the diplomatic high ground, writing Anderson, Thompson, Fulbright, and Scott that "we feel for the good of the scientific fraternity in general that we should abide by this new committee's ordering of priorities." And in a statement prepared some months later by Bartnik regarding the Thompson hearings themselves, Ripley candidly admitted that "this exercise in introspection, albeit somewhat hurried, was really

Figure 55 Edward Lilley. *Smithsonian Institution Archives, Image # SIA2017-090561.*

healthy for us all."[1215] Of course, he was referring to all of the issues raised at the hearings and was, in effect, answering the OMB; all the Smithsonian programs were to survive, live long, and prosper, but at the expense of the big dish.

There was plenty of mopping up to do, with the regents, Congress, and members of NEROC. Lilley (Figure 55) scrambled to save what was left of NEROC at Haystack, managing to convince sympathetic administrators at MIT and in the Air Force to keep it going on what was, in effect, a survival budget. As part of this campaign, the Haystack facility was transferred totally to MIT in the name of NEROC, and soon some NSF funding was secured.[1216] In his notification to NEROC members, as well as to NSF and NASA, Whipple stated only the facts and offered no comment other than his hope that this action of stepping back "will be a contribution to developments in radio and radar astronomy."[1217]

Conclusions and Analysis

The VLA was indeed included in NSF's authorization bill that year, and the appropriation was approved in early 1972 for construction to start in 1973.[1218] But the big dish was never revived, at least not in the form proposed in 1969–1970. NEROC, confined to Haystack, continued on for a number of years. There were also sporadic thoughts at the Smithsonian to revive the 440-foot project, but after 1972, the demands of radio astronomy had gone from the centimeter range to the millimeter region, necessitating much more mechanical precision.[1219]

Ripley initially saw the radio telescope facility as an opportunity, but the nature of that opportunity changed with changing conditions. It may have begun as a means to assert a national role in astronomy and to validate the very existence of the Smithsonian Astrophysical Observatory, but it ended, from the fear of further exposure to critics at NSF and OMB, as a convenient expendable to protect the continued health of the Smithsonian Institution overall.

Whipple, like Ripley an institution builder, knew when to step back. Just as in his motivations for Project Celescope, Whipple's priority was not so much in the science, but in building an infrastructure for science. Both acted to retain and strengthen an institutional structure, and both faced numerous challenges to more central parts of their respective institutions. But they also knew, given OMB's scrutiny, that there were limits.

When asked about the fate of the radio-radar telescope in 1976, Whipple recalled only that it was "stopped by the national committees because they wanted the large array instead. Well, that was a disappointment."[1220] But for Bradley, when asked two years after the fact, upon his retirement as undersecretary, it was just a fact of life in Washington—he conflated the delay caused by the need to present it to a new Congress with what was about to happen to Whipple: "In the meantime, the directorship of the Astrophysical Observatory was changed, and the former director became a senior research astrophysicist. A new director was placed in charge of both the [Harvard] College Observatory and the Smithsonian Observatory, with different emphasis on the various aspects of their work." For Bradley, saving the Hirshhorn was the critical and most painful issue. What was going on up at Cambridge was, he felt after the fact, quite natural in the constantly changing and increasingly competitive landscape of American science.[1221]

Throughout the campaign, Whipple had faced numerous challenges to his empire. Celescope was finally launched, but there were deep concerns over the quality of the data being returned, so he had to marshal forces to save that program. Mount Hopkins was proceeding more or less smoothly, but was another area of tension with NSF and the University of Arizona over the site for the new instrument. The SAO was also in the midst of struggling to upgrade the tracking program to remain competitive. These were hardly distractions; indeed, they all were close to Whipple's core interests. But none was as close as an internal conflict with Leo Goldberg that had been brewing for some years. As Whipple campaigned with Ripley, in early April 1972, Goldberg abruptly took a leave of absence as director of the Harvard College Observatory to assume the post of director at Kitt Peak. His departure was a shock. It highlighted the severely deteriorated relations not only with Whipple, but as well with Harvard-Smithsonian affairs. How this factored into the whole picture will be covered in the next chapter, but it is important to realize that this fight was going on in parallel to the one for the big dish.

Chapter 15

Growing Fault Lines in Cambridge

Brian Marsden, whom Whipple had hired away from Yale in 1965 for the satellite-tracking program to strengthen expertise in celestial mechanics, remembered sitting on Whipple's office floor sometime in the late 1960s, the two of them wading through a mass of scattered project status files. Marsden recalled that Whipple paused at one point, pondered the many projects before him, and then, looking at Brian mused, "What kind of monster had [I] built?" In the way Whipple asked it, Marsden knew he did not want an answer; "[T]hings were getting a bit shaky" at the time, Marsden knew, given the difficulties Whipple was having with Leo Goldberg over governance and planning, and over what Goldberg saw was a serious imbalance in teaching loads between Smithsonian Astrophysical Observatory (SAO) and Harvard College Observatory (HCO) staff. For Goldberg it was a four-headed monster, which George Field later clarified were most likely "Harvard, Smithsonian, NSF, and NASA."[1222] With the Vietnam War escalating and funding sources tightening with NASA cutting back, threatening to cut SAO's satellite tracking up to 50 percent in 1969, this was a time of retrenchment. The Office of Management and Budget's (OMB) "SAO Issue Paper" had raised many questions: Could the Smithsonian replenish the support that NASA had provided, or would some other institutional relationship be more desirable? That was the OMB's message, and to some extent that would also be the question posed by both Harvard and Smithsonian staff.

In the March–April 1970 edition of *SAO News* Fred Whipple tried to assure staff that though there had been cuts and there would be more, they were trying to minimize the damage, negotiating with NASA officials "to clarify the role of the SAO in the space program—both present and future." Satellite tracking, the largest employer of SAO's rank and file, would be hit, requiring a reduction in staff. Essential services were still intact, "though cut to a workable minimum." One bright spot, Whipple observed, was cooperation with HCO to expand the facilities at 60 Garden Street.[1223]

Indeed, new and enlarged facilities had been made possible by large bequests from Richard Perkin and John G. Wolbach, as well as a grant from NSF. They helped, but didn't erase growing tensions between the SAO and HCO, tensions that led to a deep institutional restructuring. Many of the issues were not so different from what other large-scale academic enterprises at Stanford, Michigan, and MIT faced in trying to balance traditional university policies and procedures with the autonomy of the individual scholar within industrial-scale governmental and military research enterprises that defined research in terms of institutional mission or national needs.[1224] For Harvard and Goldberg, however, it was also institutional interdependence. And for Whipple and the SAO, it was institutional autonomy and heterogeneity. People on both sides of the divided house viewed it as a clash of personalities. Menzel and Whipple had close personal and social relations, close enough that Whipple's daughters gleefully called Menzel "Donald Duck" at department parties when he quacked at them.[1225] But not so Leo Goldberg; there was a philosophical distance and a clash that led to his resignation, to Whipple stepping down, and to a consolidation called the Center for Astrophysics.

Leo Goldberg

From his liberal upbringing, training, and nature, Leo Goldberg believed in egalitarian, government-supported national facilities that could compete with the established elite institutions in astronomy.[1226] By the early 1960s, he was among the most powerful voices in the profession. He brought his many passions from Michigan to Harvard to build a vigorous program in solar astrophysics, taking a leading role in NASA's Orbiting Solar Observatory series. From Goldberg's many campaigns at the national and the international levels, such as helping to instigate the Decadal Surveys by the National Research Council, and his humanist-inspired campaigns in his capacity as president of the International Astronomical Union, it was clear that overall he felt strongly that astronomers had to speak for themselves in setting priorities for funding large projects.[1227]

When news spread of Goldberg's impending move to Harvard, Jesse Greenstein wrote to congratulate him on returning "to the scenes of your youth." Greenstein hoped that Goldberg would stimulate a renaissance in solar and stellar astrophysics at Harvard and restore the reputation of the place.[1228] Goldberg responded that he looked forward to establishing new associations with ground-based solar facilities and resolved to "do whatever I can to help make Harvard a strong center for all kinds of astronomy."[1229] Goldberg, however, also agreed with Whipple that continuing the Smithsonian's traditional program of routine solar monitoring from the ground had to be abandoned in favor of putting a bolometer on a satellite, adding that the Smithsonian was better off spending money on a theoretical group interpreting data collected either at Kitt Peak or from space.[1230]

Indeed, in the first years, regarded generally as a Harvard professor but supported equally by the Smithsonian, Goldberg worked with Whipple to build a strong solar and stellar atmospheres group extending from theory to the laboratory, and to space.[1231] In March 1962, after Whipple's repeated efforts, Goldberg was converted to senior federal career status as a physicist at the SAO. Whipple had the slots available because he was unable to retain Sterne or Hynek after Harvard rejected them. Goldberg was more than acceptable to Harvard. As a Smithsonian administrator explained the situation, "The only reason a man of his [Goldberg's] caliber is available at all is that as a member of the staff of Harvard University, he is familiar with and attracted to our program which permits a high degree of utilization of his knowledge in a most rewarding way."[1232]

Following Menzel's failed efforts to establish a large solar telescope in Hawaii, but very much in sympathy with him, and in tune with Greenstein's admonitions, Goldberg pushed to build a large telescope for Harvard in some suitable climate. His efforts with Yale and

MIT and then again in Hawaii led to naught with both staff and Harvard administration.[1233] It must have frustrated Goldberg no end that by then the University of Michigan had established an experimental solar radiometry station on Maui.[1234]

Meanwhile, Goldberg's programs in space solar physics were burgeoning; not as large as Whipple's Celescope and initially almost as frustrating, but ultimately highly successful. He built a powerful shock tube laboratory where gases could be heated to mimic the characteristics of the solar photosphere, chromosphere and even corona.[1235] He led staff, including Hubert Reeves and W. H. Parkinson, to design and build instruments for the Orbiting Solar Observatory (OSO) series as well as an Advanced Orbiting Solar Observatory. After the flights of OSO IV and VI in the late 1960s, his group was "flooded with data."[1236]

Soon after taking the Harvard College Observatory directorship in 1966, and turning down simultaneous offers from Yale and then from NASA to direct the Goddard Space Flight Center, Goldberg also assumed the chairmanship of the Department of Astronomy and became intensely concerned with graduate education. He and Whipple were able to expand office and laboratory space, but Goldberg failed to establish long-term plans for a major ground-based optical facility or expand the Harvard academic faculty and, most frustrating, compel those with tenure to teach.[1237]

Growing Concerns

In his 1967–68 annual report, Goldberg listed some nine Harvard faculty at the senior, or tenured level, a few visiting professors, and an unspecified number of junior faculty and lecturers. There were 22 undergraduate "concentrators" in astronomy and 54 graduate students. Some 13 PhDs were awarded during the period.[1238] He left unsaid the fact that most of the classes and theses were managed by SAO staff, not Harvard faculty. This situation had been festering for some time, but Goldberg only started complaining about it, in surviving records, at least, in late 1970 when faced with a deeper and more systemic problem: Harvard's lack of oversight of Smithsonian appointments.

Goldberg complained to Harvard's dean John Dunlop in November 1970 that there was a lack of parity between Harvard and the Smithsonian. This was especially worrisome because so many SAO staff taught: "The disadvantage of this arrangement is that we have become so dependent on Smithsonian personnel without having a voice in their selection." Harvard's teaching standards have suffered as a result, Goldberg implied, which was intolerable for a man who "had zero tolerance for second-class work or thinking." Most of the Smithsonian staff who were teaching, Goldberg added, were better described "as highly trained specialists lacking the breadth that is ordinarily expected of a Harvard professor."[1239] Goldberg was evidently between a rock and a hard place, and grew deeply frustrated, annoyed, and even fearful that the SAO would absorb the Harvard College Observatory.[1240] The way Goldberg viewed the problem:

> . . . there was no joint planning of any kind. Menzel and Whipple had developed a kind of informal mutually supportive relationship, in which each did pretty much

as he pleased, but with the understanding that the other would back him up. They socialized together, and outwardly they got along very well.[1241]

Menzel had, after all, created the space for Whipple's empire, and the two enjoyed more than 30 years of close cooperation that began as a mentorship. Menzel knew and was comfortable with Whipple's style, which could, indeed, frustrate those around him. As Charles Lundquist observed, Whipple was a firm believer in organizational independence. At times, he felt seriously hampered by Smithsonian oversight, as well as NASA oversight. The more voluble members of his staff sometimes reflected his disdain for bureaucracy; pondering what should be placed in a time capsule for a new Smithsonian building at Harvard, Charles Whitney suggested including a memo from an especially annoying high-level Smithsonian official, or even the official himself![1242] And of course Whipple chafed under the micromanagement of administrators such as Graf and Keddy. It was just at this time that Bradley appeared on the scene, which saved the day.

Tensions existed well before Goldberg took over. Whipple had failed to secure Harvard appointments for his most senior SAO staff, and those who stayed, such as Charles Whitney, clearly recall the "Harvard snobbism . . . a certain disdain of the Smithsonian and anybody who had anything to do with it."[1243] In 1959, some staff attempted to establish a Smithsonian-Harvard Observatories Round Table to facilitate communication, but it proved to be ineffectual and short-lived. When Goldberg took over, he tried to expand the Harvard Council to include Smithsonian staff, including Andrea Dupree and Eugene Avrett, but again this proved to be ineffectual because they could not speak for the Smithsonian unless Whipple led the way.[1244]

Whipple's own staff sensed his distant nature. Whitney, one of his project leaders in the SAO's atmospheres group and so quite close to the man, recalls feeling uncomfortable with the "looseness of the structure." Aside from Whipple, Lundquist, Tillinghast, and leaders in the satellite tracking areas, the structure of the scientific problem groups was constantly in flux, depending on the needs of the problem area, the funding, and expected deliverables to satisfy funders and patrons. Although there was not much migration between the larger areas, for example, solar and stellar physics as distinct from planetary and meteoritical studies, within each of the groups, as problems changed or were redefined, resources were adjusted to suit immediate needs.[1245] The issue at hand was resource allocation, which everyone "left pretty much in Fred's hands." There were no organizational meetings "where various project leaders could discuss the importance of their work in the face of other project leaders. There was no give-and-take on the question of allocation." In June 1967, several SAO staff members happened to discuss the matter during a meeting of the American Astronomical Society. The discussion continued through informal "BYOB" lunch meetings in Cambridge, led by Whitney and close colleagues in the atmospheres group, including Avrett, Wolfgang Kalkoffen, Myron Lecar, and Owen Gingerich. The meetings continued until Whipple learned of them and of their intention to suggest the creation of a Smithsonian Observatory Council. As Whitney recalled: "Fred found that very threatening: 'What does this mean? What are you doing here? You seem to be attacking me and my administration of the observatory.

I want you to stop it and get out of that and stop having these meetings,' and he called me names actually."[1246]

Smithsonian staff generally enjoyed and flourished but Whipple's distance from his staff and his penchant to make decisions without broad, inclusive deliberation were worrisome. Lundquist speculated that Whipple "just didn't want to spend time on things like that."[1247] As long as a staff member's research did not require large-scale investment, and fit within the needs of the institution, they were quite free to do what they wanted. Dave Latham recalled a definite feeling of autonomy, but it had to be under the radar, or under the condition that "if you write proposals to NASA and they give you money, you get to do it."[1248]

In parallel to tensions caused by the teaching situation, there was also a lack of coordination committing to major new observational programs and facilities. Goldberg knew what Whipple was doing in southern Arizona.[1249] Through 1970 and 1971, Whipple stepped up the development of Mount Hopkins, dedicating the 60-inch telescope to Carl Tillinghast and pushing ever harder for the MMT. He did these things using only the Smithsonian's annual appropriations, in full view of anyone who cared to pay attention. But what Whipple did not do, according to Goldberg and staff members in both organizations, was to adequately engage Harvard to their satisfaction in developing Mount Hopkins. He kept negotiations away from the Harvard Observatory Council until an agreement had been hammered out between Arizona and the Smithsonian. When the news broke, Whipple and Goldberg met with an assistant secretary of the Smithsonian to try and straighten out the situation, but it was clear to Goldberg that Whipple was not of a mind to subject himself or his observatory staff to the whims of the Harvard Observatory Council.[1250]

Indeed, Whipple made it clear that he wished to distance himself and the SAO from Harvard. Writing to an inquirer in March 1970 who had asked about the state of university research, he responded openly that universities were in a time of "transition and [it was] a difficult period for them." Because there was such social pressure at that time for universities, even the elites, to educate "practically everybody," they would be drained of all other responsibilities, and "research efforts will probably move toward institutes and away from universities." Too clumsy and too burdened with tradition, and facing a society where science and technology would be evermore essential, Whipple would do almost anything to avoid Harvard's oversight.[1251]

A Call for Long-Range Planning

Matters came to a head in April 1970, when the latest HCO Visiting Committee convened at the observatory. This committee well knew what Whipple was doing on the SAO side. The Smithsonian seemed without limits with plans for the North East Radio Observatory Corporation (NEROC) and the Multiple Mirror Telescope (MMT). But what was Harvard doing? This latest committee included Dillon Ripley as well as National Bureau of Standards director Louis Branscomb and H. Irving Pratt, chairman of the Board of Trustees of the Pratt Institute. But astronomers took the lead. Jesse

Greenstein confronted Goldberg, asking him what he was doing about long-range planning and stability. Specifically, if NASA were to disappear, or NSF were to be downsized, what was Harvard astronomy doing to prepare? Greenstein had collected comments from graduate students who worried that HCO did "not engage in major research programs in some of the frontier fields of astronomical research." They pointed to hot areas such as x-ray astronomy, which was "presently restricted to a very few government-supported laboratories, not one of which is connected with a department of astronomy at a university."[1252]

Goldberg was pessimistic. He spoke a bit about his joint planning with MIT for the Chile venture, but left little hope that it would happen in the near future.[1253] Naturally, the committee wanted to know what Goldberg was planning to do about this situation. Goldberg took the charge to heart, cancelled a planned sabbatical, and contemplated ways to include the Smithsonian Astrophysical Observatory to leverage programs that would strengthen Harvard's offerings in astronomy.[1254]

During his visit, Ripley privately asked Goldberg frank questions about the Smithsonian relationship, and then wrote Whipple suggesting that it would be a good time to "have a joint meeting on common problems in research and teaching shared by Harvard and SAO." Ripley worried about declining budgets and reported that the recent dedication of the Mount Hopkins station had raised concerns over why Harvard was not more involved: "Surely there are areas where we can consolidate our resources for mutual benefit?" Ripley felt it would be a good time to think of new ways to build Whipple's empires; the Smithsonian had just gained a significant new advocate: former NASA administrator James Webb was now a Smithsonian regent and had "many contacts and special knowledge."[1255] Whipple responded that Harvard and the Smithsonian were already cooperating, sharing shop facilities and computer facilities. They were also cooperating in their radio astronomy initiative, and their talents in stellar atmospheres had been completely integrated. Conversely, Whipple was open to discussing "mutual problems of SAO and HCO," adding that the only areas Harvard was not interested in are "our general meteor programs, the satellite-tracking program and the Celescope program."[1256] Whipple copied Goldberg in his response to Ripley and Goldberg readily agreed that SAO and HCO were cooperating scientifically. However, Goldberg pointed out, what had worked smoothly for the original players (Carmichael, Menzel, and Whipple), operating cordially and informally under "no set of formal ground rules governing the relationship," was not the way of the future. With NASA funding on the wane and OMB deliberating over who controlled science, the stakes were now just too high.

In the wake of the Visiting Committee's and Greenstein's call for long-range planning, Goldberg felt it was a good time to make changes in the ground rules, so he continued to push Whipple and Ripley. Goldberg now demanded that "we do more than merely cooperate in scientific research and that we try to work toward single, joint programs in those areas in which both institutions are doing major research." Goldberg admitted that Harvard had no interest in the SAO core programs such as satellite tracking, and he countered that the SAO had little interest in his solar physics programs. But there were areas of mutual interest: in optical and radio astronomy, and in stellar atmospheres. He therefore called for more "joint planning" in appointments of staff and planning of major facilities and "in the shaping of long-range strategy generally."[1257]

All this was brewing while Ripley's team was preparing for the Thompson Subcommittee hearings on the Smithsonian's growth. And it was there that Whipple pushed to situate SAO as a "unique" federal agency able to respond to the needs of mission-oriented agencies such as NASA, as well as be engaged in truly basic research, pursing astrophysics autonomously. Terms Whipple had employed, such as "the national astrophysical observatory" certainly did not calm Goldberg's nerves. It would only exacerbate what he already saw as a critical imbalance in governance: While Harvard astronomy engaged in shared governance, and was subject to annual oversight, the Smithsonian had neither.[1258] Whipple was of course on the Harvard Council, but there was no counterpart for the Smithsonian.

This is what Goldberg wanted to change. Ripley, according to Goldberg, was in favor of the change, but Whipple strongly advised against it, trying to keep the Smithsonian free of, and hence not subject to, Harvard's planning.[1259] Whipple would not tolerate what he perceived as the intransigence of Harvard senior faculty, slowing down, if not stalling completely, any initiatives for growth. His refusal to compromise also raised questions in Washington.

A Fact-Finding Mission from Washington

In mid-June, knowing that OMB's Dean Anderson was planning to visit SAO as part of his ongoing investigation, Ripley directed John Jameson to accompany him to assess the questions that OMB was raising and the problems that Goldberg had charged. They picked a time when Whipple was out of town so they could tour laboratories freely and speak with both "bench scientists and administrative personnel." They wanted to see how real the programs were as described in Whipple's recent SAO brochure *Astronomy for the Seventies*, prepared for the Thompson hearings. They looked for evidence of "Earth as a Planet," "the Solar System," "energetic phenomena in the universe" and reported back to Ripley that this "programmatic split is a convenient way of classifying crosscutting aspects of various projects, but the reality of SAO organization lies in the projects themselves and in their funding."[1260]

Making sense of the SAO seemed to be the point of the visit, but there were deeper concerns raised by OMB: assessing limits to growth. Whipple's testimony for the Thompson hearings did not set any limits to SAO growth. His only allusion to limits was made to Harvard's Dean Bundy in an oddly framed promise that the SAO would not grow beyond the scale of HCO, a claim Goldberg felt was hollow and was really just a form of Whipple's brinksmanship rhetoric.[1261] To test this assertion, Jameson and Anderson compared the relative contributions of SAO and HCO with various initiatives and found that, for optical astronomy, at least, Harvard funding was negligible compared with NASA's funding of Smithsonian activities. In fact, they found SAO staff very worried about continued NASA funding, given that both satellite tracking and Celescope were in their last days. SAO staff still hoped for a big new NASA start, possibly in gamma-ray astronomy. The Smithsonian

Institution was barely able to support core infrastructure with overhead from NASA funding, and Harvard was not even in the picture. So continuing NASA support was key: "Without it, SAO as an organization would collapse."[1262]

Anderson and Jameson gave good marks for the valiant efforts trying to save Celescope's data, especially as a good example of collegial cooperation between the two observatories.[1263] They also complimented SAO for finding ways to remain competitive, obtaining major NASA support for satellite tracking and for its biggest product to date, *Smithsonian Standard Earth*. But they also listened to several SAO staff who thought that the SAO had "lost its lead" in several areas. Because NASA was noticeably drawing to itself more and more of the infrastructure for space activities, staff feared that the overall importance of optical tracking was on the decline, especially because it had become "more methodical than innovative." This was not necessarily bad, as it fit in with Smithsonian tradition. But was tradition defensible in the face of increasing competition and deteriorating support for science overall in the nation? Indeed, "Especially in past years of heavy NASA funding," Anderson and Jameson observed, "Smithsonian/Washington has tended to view the SAO as a 'rich cousin.'" But now, what of the future? In their report, reviewing "Impressions of the Management and Operation of SAO," they concluded that the "SAO is largely an autonomous operation with little input from the Smithsonian's Washington staff" and, evidently, little regard for it.[1264]

These "impressions" reached the heart of Ripley's concerns, raising questions that would, in fact, foreshadow those asked by OMB later that year. The OMB report would state that it felt that the SAO continued to deserve federal funding in some form. But, as Dean Anderson observed, there was "a real issue as to whether SAO should retain a special 'preserve' of funding as is now provided by the Smithsonian at about $2 million per year." If NASA funding were to continue to decline, could the Smithsonian make up the deficit? The answer: probably not. And if not, Anderson concluded, it was probably time to "more clearly draw the boundaries of SAO activities—especially in regard to the purchase of major new equipment and facilities."[1265] Anderson's assessment was in Whipple's hands by March 1971, but probably came as no surprise, given what was happening even closer to home.

Goldberg's Resignation

Whipple and Goldberg met with Ripley two weeks after Anderson and Jameson's fact-finding trip in June. In November, deeply frustrated, Goldberg reported to Dunlop that not much was accomplished there because Whipple expressed nothing but satisfaction with the present arrangement, and Ripley said little, taking "comfort in silence." In the ensuing four months, Goldberg claimed, there had been no substantial discussion of the issues, even after repeated attempts. "Unfortunately, Professor Whipple and I seem to be far apart, both in our understanding of how well the Harvard-Smithsonian relationship is working and how it ought to be modified."[1266] Goldberg was convinced that if things did not change, astronomy at Harvard "will be in danger of losing its independence and of becoming an appendage of the SAO." Even though both observatories were suffering from "drastic budgetary limitations" at the moment, he darkly knew that Whipple would find a way to revive his programs and "I expect the numbers to skyrocket."[1267] Overall, Goldberg remained deeply frustrated by the imbalances he perceived in teaching and oversight.

But Goldberg also thought he finally had an ace: an offer to accept the directorship of the Kitt Peak National Observatory. Informing both President Pusey and Dean Dunlop of the offer, he pushed for a showdown with Whipple in Dunlop's office. There he boasted that colleagues around the country had been urging him to take the job, bluntly stating that Harvard was "not exactly an administrator's paradise." Its only advantage and "greatest asset," Goldberg asserted, was "its incomparable student body." He claimed he had to make a decision soon, in a matter of days.[1268]

The job offer gave Goldberg the leverage he wanted to deal with Whipple, and he surely applied it.[1269] A day after he announced the offer, he devoured background materials on the history of the SAO-HCO arrangement to prepare for the showdown with Whipple and Dunlop. By the time they met in November 1970, Goldberg had found what he needed to argue that what he was proposing for the Harvard-Smithsonian arrangement was in fact implied in the original defining documents: parity between the two institutions.[1270] Goldberg felt the problem was "clearly and unmistakably expressed by Fred when he said that he would refuse to engage in any kind of programmatic discussions with the Harvard Observatory Council." His position, Goldberg asserted, was "in clear violation of paragraph 4 of the Menzel-Carmichael agreement."[1271]

Whipple's intransigence, Goldberg charged, was stimulated by Menzel's failure some five years earlier to secure the Harvard Observatory Council's approval of his plans for an observatory in Hawaii. Goldberg admitted that the council may well have exhibited "bad judgment," but it was merely responding to what it thought were "badly-prepared proposals." Goldberg added that he felt it was "shockingly unfair to accuse the Council, by its actions" of having brought about Menzel's heart attack. He was annoyed that Whipple repeatedly used this argument over the past five years as an excuse for not cooperating, especially now with the Visiting Committee's request for better long-range planning for HCO. Revealing his bias against building big telescopes for the sake of building big telescopes, a dig at the MMT and quite possibly the radio-radar telescope, Goldberg confirmed that he planned to resign both as chair of the Department of Astronomy and as director of the observatory on 1 July 1971. Thus he gave credence to rumors that had been flying, well before any papers were signed. Goldberg's intent was to bring action, and it did.[1272]

Within a few weeks, both Pusey and Ripley called for yet another blue-ribbon "review committee" to examine the Harvard-SAO relationship. Once again Greenstein would take the reins, leading what Ripley called a "*super nova* constellation of astronomers." There was some haggling over the membership of the ad hoc committee; the Smithsonian feared that Harvard would fill the chairs with astrophysicists. Lundquist and Whipple both ardently campaigned with Ripley's office to add representatives from planetary astronomy and geophysics and space physics: "classical, academically oriented astronomers . . . have little understanding of the special organizational problems confronting non-university research laboratories such as the SAO."[1273] In this manner the selection process not only became a test case for inclusiveness, but also clearly identified the disparate and competing communities and cultures that SAO now embraced.

When Harvard senior astronomers learned of Goldberg's decision, and the impending inspection by the new Greenstein committee, set for late January 1971, they did not remain silent. David Layzer wrote bluntly that Goldberg's departure was a "heavy blow to the University." Not mincing words, he traced the problems back to the Shapley administration, and how in his wake, Menzel's ascendency and the arrival of the SAO caused Bok to leave, which to Layzer was a blow of similar magnitude for Harvard astronomy. Layzer traced how the graduate program suffered after Bok's departure, and how others such as Thomas Gold found the Harvard-Smithsonian arrangement untenable and also left: "One of the reasons for his departure was concern over the growing ascendancy of the Smithsonian, which he used to describe as a huge digestive apparatus indiscriminately gobbling up large contracts and excreting small ones."[1274]

Layzer, a former Menzel student in atomic physics who had been a postdoctoral fellow at Michigan under Goldberg and who had returned to Harvard in the 1950s, outlined how Goldberg had tried to take corrective steps once he succeeded Menzel: "For the first time in my twenty-seven year long acquaintance with the Observatory, all the professors are permanently on speaking terms with one another." Obviously overwrought given the circumstances, Layzer saw Whipple as the root of all problems and Goldberg as the thwarted savior of Harvard astronomy.[1275] Layzer copied Goldberg, who responded in kind. They and others would be called to testify as "witnesses" in some several days of questioning staff and administrators in what Greenstein later referred to as "an emergency in the Astronomy Department" at Harvard.[1276]

"An Emergency in the Astronomy Department"

Greenstein's committee met on 22 and 23 January in Cambridge with Dunlop and Bradley. Their goal, according to Pusey and Ripley's December letter of invitation, was to evaluate the situation and what needed to be done to correct deficiencies.[1277] Any thoughts that this was a pro forma event were killed with Goldberg's announcement and letters of complaint, which were distributed to the members of the committee. As Greenstein described it, the committee's deliberations were "made in a somewhat more difficult situation than was contained in the original charge."[1278]

The meeting opened with Goldberg's letter, then Whipple spoke and answered questions, and this led to a general discussion. The committee then conducted brief interviews with a dozen witnesses, from senior staff to graduate students. Greenstein felt there was "frank and full" conversation, and that the committee quickly realized that there was not enough time to hear all aspects of the issue. So they left out budgetary matters, there were no tours of facilities, no formal or celebratory dinners. This was serious business.[1279]

Committee members included Greenstein, Kuiper, Purcell, Ed Salpeter of Cornell, John Simpson of Chicago, and Spitzer. Whipple's plea for a more broadly based panel did not happen, although Simpson and Spitzer were heavily involved in space research, as was Kuiper. Whipple and Goldberg were on call during the two days of hearings. Layzer recalled specifically that Ed Purcell seemed "somewhat distressed by the meeting." Indeed, the atmosphere was tense.[1280]

Greenstein kept detailed notes during the hearings and his liner notes tell the tale: "Should boundaries be sharpened?" and should both HCO and SAO "be replaced by Astrophysics Center"?[1281] He also recorded that Whipple felt that everything was "OK," giving all credit to Lundquist. Whipple also brought up the student problem saying it was distorted, and that the Harvard Council under Goldberg had grown until "it is too large and unwieldy." Layzer, who Greenstein noted was supported by the SAO, was the most critical, reflecting a persistent theme in the hallways: "Half of the SAO appointments were irrelevant to astrophysics," and, while they were "not at issue," they still absorbed resources and set priorities. Alex Dalgarno, the distinguished atmospheric physicist who held a dual appointment funded by the Smithsonian, was more moderated and diplomatic, saying there were good things resulting from the association of the two institutions. But these benefits, he added, were "unplanned & accidental." He also observed that because hirings were not bilaterally coordinated, some research programs were planned independently and "often duplicated." His most negative observation was that not all SAO staff were up to par.[1282] Although in retrospect Dalgarno recalled that Whipple ran the place "like a kingdom," at the time he moderated his views as a witness, suggesting that fingers could be pointed in both directions.[1283]

Most of the witnesses, other than Whipple, agreed with comments by Associate Professor I. John Danziger, who was hired away from Caltech by Goldberg in 1967, that the "relation must be reviewed." But not all of them felt, with Danziger, that the SAO was "far too large" relative to the HCO. On the SAO side, John Wood testified that his opportunities were excellent because of the infrastructure and flexibility SAO provided. He added that protracted "policy discussions are paternalistic & protective." Under Whipple, Wood claimed, "supervision ends after appointment."[1284] Ed Lilley reflected similar views, strongly supporting Whipple's objection to HCOC oversight: "Council has been a talk factory, wanting unanimity."[1285] Lilley here reflected his failed appeals to Harvard to support his radio-radar dreams.

SAO solar astronomer Eugene Avrett thought the relationship worked for him; data from OSO were readily available no matter what side of the house he was from. But he wished the two directors could be more cordial. Edmond Reeves, a senior research associate at Harvard in Goldberg's solar group, complemented Avrett's remarks, citing examples of cooperation and expressing appreciation for the abundant computer time from his SAO colleagues. He voiced the common view that "working level people find means to get together." But then, when asked if he was involved with Mount Hopkins, Reeves responded that "Whipple ran away with it.[1286]

The MMT was a sore point. But Danziger countered Goldberg's claims, saying that, indeed, Goldberg was always informed, because he asked Danziger to attend the planning meetings. Danziger, though, had other gripes: "Only Goldberg and Dalgarno care about teaching" and Harvard's "excessive dependence upon Smithsonian for app[ointments]" was a real problem.[1287]

Lundquist tried to clarify how Whipple managed the some 50 projects at the SAO that were presently funded or somehow related to outside agencies. He explained that Whipple "reshuffles people as project leads require" in support of a wide range of projects, including Harvard projects like Goldberg's OSO. His strategy was to demonstrate all the support the SAO was providing Harvard, real and in kind. He was not shy about expressing his frustration with Harvard's project management, or, more pointedly, lack thereof. As Greenstein recorded his testimony: "Cannot find any consensus at HCO. No one to talk to—no administrative manager for science." Lundquist felt that the working groups were well integrated because the workers found ways to manage their projects. However, on the Harvard side, "some mechanism for consensus" was needed. There had to be some clear mechanism for decision making so that priority decisions can be made. People had then only to learn to "live with system."[1288]

Indeed, but which system? The committee looked for systemic causes for the problems. Did the physical isolation of the HCO-SAO complex, almost a mile up Garden Street beyond Harvard Square, make the staff too "mentally isolated" from campus? Did the SAO's presence make Harvard staff feel isolated from academic affairs? Was there a better way to organize the staff? The committee discussed options for changing the relationship, ranging from terminating the arrangement completely, to continuing with two directors but creating a strong advisory group drawn from both institutions. They also explored ways to strengthen the Harvard department to address hot areas in astrophysics. Many names were raised to take over key positions or to succeed Goldberg. This bouncing around finally stabilized on the single major question: In effect, was the $2 million annual federal appropriation for SAO enough for Harvard, given the fact that the HCO lacked an academic endowment but got back a fraction of the Harvard overhead from its large NASA grants? It was clear that the budget situation was complex but it was also clear that the HCO was aided substantially by the SAO. Before that question was answered, another was raised: Has "the potential of this combination . . . been fully realized?" The sense of the committee was, on both points, no.[1289]

Greenstein, clearly pained by the proceedings, jotted, "Collaboration must cause serious emotional problems for both top administrators."[1290] His draft summary, written in late February, noted the "strong difference" between Goldberg and Whipple, resulting in almost complete lack of communication between them. Maybe, Greenstein reflected, it was time to combine the directorships. But how could this be done if the two institutions had to remain distinct, yet cooperate? SAO wanted unity, and HCO diversity, as reflected in the two directors:

> Whipple operates part of the SAO like an efficient government laboratory which seems quite independent of Harvard and without much impact on it. He feels, with some emotion, that he should not compromise his authority on important decisions regarding SAO. He expressly said he would feel unbearably constrained by the need to submit appointments or major policy decisions to a council or committee. He feels that the indecision of the HCO Council shows the failure of the committee approach.[1291]

Goldberg also possessed a "strong personality" and "in spite of a devotion and outstanding ability in research is also strongly interested in teaching." Their clash, Greenstein wanted to believe, was not so much personal animosity, but caused by the fundamental differences between the two organizations. The SAO's research effort was bifurcated: "one part centered on the interests of the Director" which he described as largely "special tasks for NASA." The other half included broad aspects of astrophysics, and it was in this half they found "a close and successful collaboration with HCO personnel and students." However, it was just in those areas of overlap where "the most serious questions of principle arose, representing the area of conflict between the Directors of the two institutions. Perhaps this was inevitable."[1292]

Greenstein's draft analyses were tortured and indecisive. His early drafts had produced strong pushback from some of the members of the committee. Everyone seemed to agree that there had to be a clearer institutional separation while retaining collaboration at the working level. This was considered the norm among those, such as Spitzer, who were familiar with the complexities of associating industrial-scale laboratories with universities.[1293] At one point, someone suggested establishing a Joint Advisory Committee, four representatives from each organization to advise both directors. "The Joint Advisory Committee should exclude from its interests the area of direct active scientific responsibility of Professor Whipple," one draft suggested, something that clearly would not fly with Goldberg. Its purpose was to promote coherent "science-policy-planning."[1294]

At the very top of the legal yellow note pad Greenstein had used to keep track at the January 1971 meeting was the telling query, "Age of retirement at Smithsonian?" Harvard's retirement age was effectively 65 in spite of the Age Discrimination in Employment Act of 1967, and James Bradley confirmed that Smithsonian high administrators typically retired at that age as a matter of good form, though they retained their grade and could serve in various capacities after that time. Bradley also reported that Whipple had once confided to him, "When you go, I go" but, Bradley added, he could stay until age 70, some six years hence.[1295] Greenstein did some checking as he prepared the report for Ripley and Pusey, and in fact suggested they seek a lower mandatory age limit and that it be made the same for the directors of both institutions.

Greenstein took well over a month to revise the draft, adding numerous comments and suggestions from members of the committee. Spitzer worried that an early draft seemed too hard on SAO and Whipple.[1296] He did some editing he hoped would reduce the appearance of blaming it all on Whipple. Kuiper was adamant that "it would be unfair and not in the national interest to suggest a replacement for Fred Whipple." For Kuiper, the man who headed an Institute defined that Institute, reflecting on his experience inspecting the Kaiser-Wilhelm-Gesellschaft in Germany right after World War II, "the directors of these Institutes were irreplaceable because the Institutes reflected the particular insights and drives of the directors." The Dutch-American Kuiper then added: "The U.S. has been slow in learning from this German experience." Kuiper indeed had struggled to build just this kind of institute at the University of Arizona in Tucson, his Lunar and Planetary Institute, and was sure that Fred felt the same about his in Cambridge.[1297]

Before modifications were made, both Pusey and Ripley had seen a leaked rough draft, and this annoyed Spitzer no end. Spitzer's anger centered on "certain derogatory implications that might be read into this draft." It was Bradley who leaked them, Greenstein replied darkly to Spitzer, "Apparently any business with Smithsonian may otherwise become public knowledge."[1298] Greenstein hastily prepared the final draft, telling Dunlop that it was the only one to believe.

Meanwhile, Goldberg announced formally that he was taking a two-year leave of absence from Harvard to head Kitt Peak. He would definitely step down as HCO director, but he held on to his professorship for the time. Clearly, neither the Greenstein committee, nor Pusey and Ripley, were ready to meet Goldberg's demands. But equally likely, he relished the new challenge. One way or another, his departure sparked a whole new round—in fact, several rounds—of committees, oversight hearings, and search committees that kept all the same issues in the air for months. Goldberg did not leave empty handed. He told Greenstein that his "decision to leave Harvard" should not be viewed as a "negative one." In fact, the shock of his leaving had finally loosened up the purse strings at the university; Dean Dunlop had just assured him that he would increase the faculty slots in astronomy, and that Ed Purcell would lead yet another interdepartmental committee to search for candidates, including a new director and at least two additional senior professors.[1299]

On 1 March, Greenstein submitted his final report, recommending strongly that Alex Dalgarno assume the HCO directorship in an acting capacity, and announcing that George Field topped the list of candidates that the new Purcell search committee should consider, along with other prominent names. Harvard, above all, needed a strong leader "interested in the Harvard problem." But on the other side of the matter, the Smithsonian also had to think about new leadership. "Given the age of the present Director," it was time, Greenstein felt, to plan for Whipple's successor. The new director had to be a Harvard professor, and had to agree to oversight by a Joint Advisory Committee. For the moment, Greenstein advised, the Joint Advisory Committee should not oversee any of Whipple's areas of "direct active scientific responsibility," just those of joint concern with Harvard.[1300]

Although there had been questions about their administrative reality, no one questioned Whipple's personal areas of "direct active scientific responsibility." No one attempted to publically enumerate them as they were clearly stated in Whipple's own blueprint.[1301] More than half of his statement highlighted satellite-tracking data, and linked Mount Hopkins activities to the measurement of atmospheric and environmental conditions. The SAO's stellar atmospheres activities were summarized in two short sentences relating to OSO data, whereas several paragraphs were devoted to planetary studies and work on the "smaller bodies in the solar system" including a full paragraph devoted to laboratory facilities for "mineralogy and isotopic composition of meteorites and lunar rocks." He also acknowledged the SAO's international services, such as the "IAU Astronomical Telegraph [*sic*] Service," the Central Bureau for Satellite Geodesy, and the "Smithsonian Center for Short-Lived Phenomena."[1302] Whipple's focus on the areas in which the SAO provided unique products and services confirmed Kuiper's arguments for the criticality of Whipple's leadership. Greenstein's final report expressed this view, dividing the SAO universe in half, describing one half as Whipple's interests, and the other as "a broad range of astrophysics."[1303]

This distinction did not please Whipple, nor would he accept the specter of retirement. Later in May, after additional rounds of meetings by the Overseers Committee to Visit the Department of Astronomy that still involved Greenstein, and the efforts of Purcell's search committee, where many of the same issues rose again, Whipple wrote very personally to Greenstein: "The Grim Reaper is taking my old friends at such a fast rate that I certainly do not wish to lose any because of misunderstandings or lack of communication." He wanted to believe that Greenstein had tried to be "fair and unbiased in judging situations," so he wrote a very long letter to state his case. At the top, the letter read, "My real concern is the shock of your refusal to believe that I was responsible as Director for the good aspects of the SAO outside the field of my own interest."[1304]

Whipple was very proud of the stellar atmospheres group, which he knew Greenstein cited as "perhaps the best in the world." Whipple felt he deserved full credit for this, because it came about in "consequence of a long-range planning and effort on my part," recalling his feelings at the outset of the Celescope program that the "space program" would return a flood of astrophysical data that had to be analyzed. Personally, Whipple added, somewhat disingenuously considering the topic of his doctoral thesis and his various attempts to break into astrophysics in the 1930s, that he claimed to have found stellar atmospheres "boring," but he did encourage Charles Whitney "and backed him to the limit in setting up the group which you now see." Whipple added that he had hoped that Whitney would have been his successor as SAO director someday, and also regretted that Stephen Strom left because he was not respected by Goldberg and Harvard. But overall Whipple felt that it was because of his own "foresight and action" that the group was brought into existence and remained strong.[1305]

Whipple also raised Lilley's frustrations, claiming that Goldberg did nothing to support Lilley and so that was when he had stepped in, bringing the resources of the Smithsonian to the challenge. "However disgruntled Leo may feel at the result of his lack of action and my initiative, I take the credit for doing something constructive, useful and with considerable potential for the future."[1306]

Conversely, Whipple openly acknowledged that the "Dalgarno group" owed its life to Goldberg. But, in fact the SAO paid all the bills, and "the dollar cost is great." This was Whipple's main gripe: The SAO had paid the bills not only for Dalgarno, but for Goldberg as well until 1966, and in fact for all of the new Harvard professorships in recent years. He had built up an institutional infrastructure that made it possible to attract first-rate scientists and provide a stage to plan on a scale unprecedented for university-based astronomy. "I think, Jesse, that you grossly underrate my ability to engineer large projects."[1307]

For examples, Whipple summoned his universe of projects such as the Super-Schmidts, the Baker-Nunns, and even Celescope, which he asserted "has attained its original goals but was expanded by NASA and became so distorted from our original concept that I have never been happy about it." Next came the SAO's ongoing campaign to build a major telescope at Mount Hopkins. Whipple was proud of their progress thus far, with "no encouragement whatsoever from Leo and mostly with our interior funds." He had great confidence in the MMT concept, and felt it was "too bad HCO shows no inclination to join us." This was much the same, he felt, for NEROC: "It falls into the category of a large national facility and I merely try to do what I can towards its fulfillment."[1308]

Greenstein gingerly took Whipple's points one by one. He agreed that they were losing too many friends these days "to enjoy losing more by unfortunate differences of opinion." But, he added, "It is hard to mediate between friends when a situation arises with attitudes so frozen that conciliatory gestures would appear merely spineless." Asserting that "there is no one who can step into your shoes," he advised Whipple to craft a solid succession plan. "I most certainly do not underrate your ability to engineer large projects," Greenstein continued. "Anybody who can live through being involved with a space program as you have, has both self-control and scientific leadership." But from the testimony they collected, Greenstein also delineated Whipple's blind spots. He was "distant" as a director and was "difficult to access," protected by an administrative assistant who took too much responsibility and authority, assuming too much "power" himself. Greenstein also reported that people suggested he create formal divisions led by strong associate directors to lessen this negative impression.[1309]

Greenstein also worried what all this fuss would look like to a new Harvard president, as Pusey was retiring. What would he think of these committee reports and critical letters in the file? "The letters put on record by friends and foes of both sides will do astronomy no good in the University, and I wish that some of them had been much more restrained." But it was this deep polarization that had to be mended: There was too much gossip, not enough communication, too much "inbreeding" within competing cliques. In sum, Greenstein sighed with relief that his terms on Harvard's Board of Overseers and on its various Visiting Committees were at an end. He had more than enough travail, "but remain concerned about the future of my Institution of 34 years ago."[1310]

Menzel watched all of this, but waited until Dunlop asked him specifically for his views. Up front, he thought that Goldberg's "accusations" about the SAO and its director were just not correct. Yes, he felt, there were "deficiencies," but the responsibility lay with Harvard: "I know from long past experience that there is no real problem communicating with Professor Whipple." Menzel and Whipple certainly had had their own disagreements in the past but, as Menzel recalled poignantly, they had sat down and worked out their differences. Goldberg, Menzel felt, was "at least as responsible as [Whipple] for any breakdown of communication. . . . [S]ome of the sarcastic letters I have seen from him to Professor Whipple are not calculated, in my opinion, to help the situation."[1311]

Menzel agreed with Whipple's negative opinion of the Harvard Council. The director had to set research agendas, so there had to be two directors. He also liked the idea of a joint advisory committee for scientific and administrative policy. It would consist of the two directors, the chair of the department of astronomy, and senior staff up to six or eight people. Goldberg and Whipple were, of course, "among my very best and oldest friends. They both were students of mine." So, Menzel pleaded, "Let us ignore and forget the hurts, the jealousies, and the irritation." Somehow the relationship with the Smithsonian had to be preserved. Writing these thoughts, Menzel wearily admitted, "has taken a lot out of me."[1312]

Committee members offered many names to succeed Goldberg. Some were physicists who had experience in government and industrial laboratories. Most concern centered on the institutional arrangements they represented that posed options for future cooperation between the SAO and Harvard. When Greenstein wrote to Goldberg to congratulate him on accepting the Kitt Peak position, putting him in "direct competition with us dinosaurs of Pasadena," Greenstein also articulated the problem: that a great opportunity had been somewhat wasted, and not purely because of personalities. "It seems there is a fundamental difference in style between a somewhat mission-oriented group and a university group, and it will always be hard to stay completely happy under that situation."[1313] What was needed was a combination of talents, and an institutional arrangement that minimized the problems. Harvard and the Smithsonian now had to search for both more or less simultaneously, looking for a candidate who best fit, and could lead, a new relationship.

Chapter 16

Whipple Steps Down

Goldberg's departure required immediate action. From March through the end of 1970, aside from the finger-pointing there was constructive debate raising all options, from the full merging of the two observatories under one director to their complete separation. Even in its earliest deliberations, the Greenstein's committee wondered whether, somehow, the two institutions needed to be combined into a single entity, an Astrophysics Center.[1314] There was also persistent talk of Whipple's retirement, a move that Goldberg himself would later strongly urge, saying in November 1971 that a merger "is a small step in the right direction," but to be successful Whipple must step down—he must not be continued beyond the normal Harvard retirement age.[1315] Indeed, Whipple's general opposition to many of the suggestions emerging from these discussions only heightened thoughts of his retirement. In a long letter reporting on a fact-finding trip to Cambridge, where he found that Whipple had not budged from his demand for autonomy, David Challinor advised Ripley to ask Whipple "what his plans are when he reaches 65." Whipple had only two more years as a Harvard professor, and because under the agreement the Smithsonian Astrophysical Observatory (SAO) director had to be a professor at Harvard, "This is something we should be working on now so we have a clear understanding of what we would like to do in two years hence."[1316]

Goldberg's position had to be filled immediately. Alex Dalgarno soon emerged as the leading choice to take the chair of the Department of Astronomy as well as become the acting director of Harvard College Observatory (HCO). His name appeared early in the Greenstein's committee deliberations and, in his last report to the Visiting Committee, Goldberg expressed relief that "a man of Professor Dalgarno's scientific eminence and wisdom is willing to accept this responsibility."[1317] Goldberg also suggested that Harvard and the Smithsonian adopt a new management template, like that of the Joint Institute for Laboratory Astrophysics at the University of Colorado, which had close connections with the National Bureau of Standards. But he also looked for a way to decrease the isolation of the 60 Garden Street complex from the rest of the university: "that the Harvard College Observatory might someday evolve into a center for astrophysics which would serve all University departments in which astrophysical research was being carried on. I believe it is not too soon to consider this suggestion very seriously."[1318]

Other management templates soon emerged, such as the one at Greenstein's institution: the California Institute of Technology and the Carnegie Institution of Washington had operated the Mount Wilson-Palomar Observatories since the late 1940s, funded respectively by the Carnegie and the Rockefeller foundations. These ideas were not generated at a single point in time, or by a single person. However, it will facilitate our narrative to focus on the selection of George Field, first to take the directorship of the Harvard College Observatory, then the combined directorship. How that happened is the key issue to understanding the transformation.

Enter George Field

Kuiper first suggested Field at the January meeting of Greenstein's ad hoc committee. Purcell's search committee soon added others. Purcell was also asked, but declined outright. By late July 1971, Field had emerged as the leading candidate and Purcell pressured him to take the job, reporting that Field was asking "hard & good questions" but was "apprehensive about doing lots of administration." Field also wondered about the possibility that Goldberg might return.[1319]

Field also knew Whipple's ambitions; he had been a core member of the Greenstein Decadal Survey, chairing the theoretical astrophysics subcommittee. He was also a member of its subcommittee for radio astronomy starting in July 1969 and well through 1970, just as the NEROC initiative was being debated. Field was also familiar with Harvard's history, in that he had been awarded a postdoctoral position there after his 1955 Princeton PhD under Spitzer in which he modeled radio outbursts from the Sun.[1320] His three-year appointment as a junior fellow had given him the opportunity to collaborate with Purcell, studying hyperfine structure to better understand physical processes responsible for the 21-centimeter radiation. He then had accepted a faculty appointment at Princeton and later moved to Berkeley in 1965 after a sabbatical at Caltech in 1964.[1321] Overall, Field had been exploring the physical processes underlying radio and infrared phenomena, applying this framework to studies of the interstellar medium, star formation, and planetary atmospheres. He regarded astronomy as a "multi-wavelength business" and worked to broaden the department at Berkeley to this end.[1322]

In summer 1971, Field and his wife were in New Hampshire to be near family. He got a call from Purcell, who had a retreat nearby in the White Mountains. "Hey, George! Why don't you come over for a visit?" Field recalled, and as the two sat on Purcell's porch, Purcell put it to him once again: "Look, are you interested in becoming the director of the HCO?" Field recalled not wanting the job at first, even before he was made aware of the management problems. He and his wife enjoyed Berkeley, and after he learned why Goldberg left, he was none too anxious to be his replacement status quo ante.[1323] But Purcell persisted, negotiating with Field through the rest of 1971 in parallel with deliberations over the nature and powers of the Joint Advisory Committee. By late November, Field had been formally offered the position as Harvard director. Greenstein was delighted, knowing that he would be accepted by both institutions as Harvard director. "Harvard and Smithsonian will share a brighter and easier future," he wrote Ripley optimistically.[1324] At the same time, Goldberg wrote Dunlop, happy about Field, but disappointed that the newly approved "Memorandum of Agreement [MOA] Creating a Joint Advisory Committee to Assist the Directors of the Harvard College Observatory and the Smithsonian Astrophysical

Observatory" meant that that there would still be two directors and that "there is no immediate hope of solving the fundamental problem in the Harvard-Smithsonian relationship, namely, the continuance of Fred Whipple as Director of the SAO beyond the usual retirement age."[1325]

Ripley leaked news of Field's appointment to the regents in January 1972 and in April, Dalgarno reported to the Visiting Committee officially that Field would be joining Harvard as a professor of astronomy in July, in residence locally in New Hampshire through the summer and fall. He would become HCO director in July 1973. He also announced that the Smithsonian had funded a new "Chair in Optical Astronomy at Harvard" and that the Smithsonian now had signed an agreement with the University of Arizona for the further development of Mount Hopkins that centered on "a new concept in large telescopes"—the MMT. Dalgarno said nothing of the fate of NEROC, nor the possibility that the two institutions might combine, though he mentioned in passing that there were continuing negotiations to create the joint advisory council to both observatories.[1326] Field's appointment was publically announced to staff in early 1972.[1327]

With Field working in the wings, Challinor and Dunlop explored options for the relationship between the two observatories. They debated the options that had been raised, like the Joint Institute for Laboratory Astrophysics at the University of Colorado (JILA), and the Mount Wilson-Palomar Observatories/Caltech model. The question was, which better fleshed out Goldberg's "Astrophysics Center"? But there was another ingredient to factor in, and this was the most delicate of all: What would happen to Fred Whipple's SAO directorship?

After considerable back-and-forth between Dunlop, Challinor, and Bradley, Ripley visited Harvard to meet with Whipple on the morning of 5 May 1972 and then with Dunlop. In the afternoon he presented Whipple with a formal letter asking that he resign the directorship on 1 July 1973, suggesting as well that Field assume the combined directorship of the Center for Astrophysics (CfA). "He would have thus had a year's indoctrination period at Harvard in order to learn the ropes from all sides." Dunlop assured Ripley that Whipple would retain his professorship until his seventieth birthday, and Ripley also offered Whipple an appointment as Senior Scientist starting on 1 July 1973 through the equivalent period of his Harvard appointment.[1328] Ripley made every effort to assure Whipple that he was very much needed. He asked him to serve on the Joint Advisory Council and to continue working toward the MMT "as well as the whole spectrum of Harvard/SAO projects for the future."[1329] Ripley and Dunlop also created a new deputy director position to be filled on a rotating basis, depending on the affiliation of the new director. Ripley hoped this would help to convince Whipple that the balance of power would not be completely in Harvard's favor. He also hoped that stepping down would not dampen Whipple's enthusiasm for the future health and welfare of the institution, repeating an encouragement he gave that morning, urging Whipple to "'go for broke' and develop new dimensions in astrophysical research."[1330]

Whipple certainly knew this meeting was coming, but possibly did not appreciate its full scope. Bradley and Challinor had tried to prepare the way, and Dunlop had already met with Whipple the previous March to discuss stepping down and eventual retirement. Yet, as his wife Babette recalled, her husband was prepared for Field as Goldberg's replacement, but not his own: "That was a big blow to Fred."[1331]

These were delicate matters, to be sure, converging just at the time that NEROC had been finally closed out when Ripley endorsed the Very Large Array (VLA).[1332] The Orbiting Astronomical Observatory (OAO) program was not going well, and there was constant pressure from NASA and elsewhere to centralize satellite tracking. At least the MMT was working its way through the budget process and was a bright spot. Accordingly, Dunlop would not discuss the next director at that time, or if there was to be a single director, with anyone outside an extremely small circle, especially not with anyone with strong partisan views such as Goldberg. He saw the extension of Whipple's tenure and the directorships as two separate items.[1333] The decisions were made between March and May 1972, and were kept very quiet as all the parties, including Whipple, deliberated over details. When Goldberg offered names to Dunlop to succeed Whipple, Dunlop would only accept them informally over the phone.[1334]

After their meeting, Whipple wrote an uncharacteristically long and frank letter to Ripley, one so effective it held up the public announcement for months. Whipple readily accepted retirement and a senior research associate position, openly and warmly expressing his gratitude for the "extraordinary support" given him by "Jim" Bradley. But there the warmth stopped. Whipple detailed the pros and cons for uniting under a single director, concluding that "my considered opinion is that, on balance, both institutions will lose by amalgamating under a single director." Harvard would end up controlling the SAO; there was no doubt in his mind. This was especially ironic, Whipple felt, because the SAO was on the ascendency, whereas Harvard was deteriorating: "Thus, I see the SAO as a young giant now constrained by being forced to walk in step with a handsome, but older and weaker companion." Whipple knew he was in the minority, which only made him more strident. However, he assured Ripley that, whatever the decision, "I will try to further the individual and mutual welfare of the two observatories to the best of my ability."[1335]

Whipple's "pros" included removing incompatibilities between two directors, simpler and smoother relations, increased attractiveness of the SAO to top scientists, and increased chances for teaching. Whipple's "cons" were far more extensive and designed to worry a man like Ripley. There would inevitably be a weakening of support from the Office of Management and Budget (OMB) and Congress; both would be confused and concerned by a private/public hybrid rather than by "a unique United States astrophysical observatory." Moreover, Congress considered Harvard the richest among universities and shied away from supporting it. And there might be new restrictions on affiliations with other organizations. But there was more; reaching deep, Whipple's institutional identity was at stake: "Much of the current strength of SAO scientific research lies in the broadness of innovative programs, which range into areas well beyond the boundaries of conventional astronomy." Those areas were Earth dynamics via satellites, where Whipple claimed SAO was a "world authority." There was also meteoritics, which he asserted had only recently become "respectable" among astronomers for the lessons they were teaching about the solar system, again, through contributions of the SAO. Both of these areas had grown to become the largest sources of income from institutional overhead. Whipple made clear that this income had been supporting mainstream astronomical activities. Without this support, there would have to be a new strategy for maintaining what was now an expected part of the landscape. And worst of all, SAO would be subjected to the "'diffused' decision making as practiced at Harvard Observatory."[1,336]

Whipple distrusted, and felt alienated from, Harvard astronomy. He truly believed that it would continue to deteriorate, especially if it continued its dependency on the SAO. Then the SAO would be forced to amalgamate "into some other department of the government, as is suggested perennially in Washington." He also expressed the fear (no doubt something he already appreciated from direct experience with OMB) that if the SAO were to grow as a result of the merger, the closer association would create a larger entity in the minds of Washington bureaucrats and feed the impression that "SAO has outgrown the appropriateness of attachment to the Smithsonian." The best guard against this was to have two distinct and autonomous directors. They would garner twice the support and also ensure diversity. Sure, there would be tensions between them. But with a single director, Whipple mused, think of the tensions between competing staff members! Whipple believed that independence of action was all important. The SAO was at that point independent and free to associate with outside institutions, and was then working with several dozen throughout the world, operating some 16 observing stations in 14 different countries. It enjoyed "international rapport" by which he meant independence from "pressures by special interest groups. Such cannot be said of any university today."[1337]

Bradley, Challinor, and Ripley took Whipple's arguments to heart. In the midst of continuing difficulties with NSF over its support for basic rather than applied research in Federal agencies, they learned that, out of the blue, H. Guyford Stever, the new National Science Foundation (NSF) director and former MIT aeronautical engineer, Carnegie-Mellon president, and future presidential advisor, had just announced that he was to visit Harvard and the observatory in March, and wanted to stop in to see his old friend, Fred Whipple, to "talk about astrophysics." Ripley worried what Whipple might say. Whipple, however, remained loyal, performing flawlessly, following Challinor's plea to keep the matter low key and just "brief Stever about the SI in general and in particular its role in astrophysics and other sciences," which included the MMT, NEROC, and the VLA. But Challinor also implored Whipple to learn exactly what "the change in policy" at the NSF was going to be, and if there was one brewing. Specifically, just how "operational does the NSF intend to become?" And how would this apply to individual Smithsonian researchers and to institutions SAO partnered with, such as the University of Arizona?[1338] Soon after the meeting, Bradley happily reported back to Ripley that Stever, Whipple felt, "believes there will be some additional support for basic research." At least there was confirmation that individual Smithsonian proposals would be honored, and there was sympathy for the Smithsonian's efforts with other institutions to build new large optical and radio facilities.[1339]

Whipple's reputation and connections and his evident loyalty to the cause could not be taken lightly. The strength of his reaction to the idea of a single director made Bradley far from sure they were embarking on the right track, so he prepared yet another set of "preliminary thoughts" on the implications if Field were appointed to both posts. Appointments, support for students, joint versus individual initiatives, OMB issues, NSF issues, congressional confusion—all factored as unknowns. Bradley wanted to think that success was still possible: In all ways, both observatories as well as the department of astronomy had to retain sufficient identity to secure and hold respective sources of funding. In whatever they chose to do, success "depends upon the solidarity and clarity of our presentations."[1340]

Ripley played his part. In May 1972, a few days after receiving Whipple's response and ever mindful that Stever had sought out Whipple specifically, Ripley lunched with Stever and asked him for assurance that a closer association between the two observatories would not "prejudice the University's ability to seek valid support from whatever Government granting agency it may wish to apply." Ripley had already approached NASA and found no "prejudice." He hoped he could have the same assurance from NSF so that he could then assure his Harvard counterparts that there will be no "discrimination into entering into such relations."[1341] Ripley had less success with Stever, who regarded astronomy as an especially sensitive area, particularly determining the proper balance between supporting specific research projects by individuals and groups, with supporting infrastructure: the cost of building new facilities and maintaining national observatories. Stever, Ripley reported to Whipple in his response to Whipple's first letter after their meeting, only saw the "contest between support of 'national laboratories' . . . and support for universities" becoming "increasingly competitive, and increasingly subject to political pressures . . . ," and so it was all the more important to "think seriously about the future of SAO."[1342] Indeed, Whipple's letter had convinced Ripley that they were not close to making the transition public. On the eve of the next Visiting Committee, on 12 May, Ripley remained unsettled enough to advise Dean Dunlop to avoid the issue of a single director at the meeting. He included a copy of Whipple's letter, "which contains much food for thought."[1343]

Whipple continued to pour it on, repeatedly calling Challinor and Bradley to plead that it was "a gross mistake to have *one* Director for both observatories." Whipple sent extensive "pro" and "con" listings to Caryl Haskins and James Webb, both regents, and continued to pound Challinor. "I think Fred feels that the Harvard attitude is that Harvard really owns the SAO," Challinor advised Bradley, "and that this feeling of Fred's was exacerbated when HCO started complaining when SAO gave a scholarship to a Brandeis student." Whipple would bristle at any hint of oversight and constantly sought ways to get around any obstacle, as Challinor quipped to Bradley: "Remind me to tell you of Fred's alternative plan for funding MMT. I don't want to put it in writing yet."[1344]

Whither the Relationship?

Communications like these between Bradley, Ripley, and Challinor about Whipple's views continued over the next few months and were the foundation for the real business at hand, which was to decide on a new management structure. The only consensus that emerged over the summer was that deep changes were necessary. For Ripley and Bradley (who had just retired but who had been brought back as consultant), the challenge was to win over Whipple somehow. Which of the two options might be more acceptable to him?

They continued to look at the JILA model, assisted now by George Field. JILA, which had been created and supported by the National Bureau of Standards (NBS), consisted of a faculty composed of university and NBS staff. The wartime radio-propagation group at the NBS had moved to Boulder in the mid-1950s and by the end of the decade, two leading physicists at the NBS in Washington, Lewis Branscomb and former Harvard astrophysicist R. N. Thomas, the same Thomas who had campaigned for the "solar associates," explored ways to combine the talents of the University of Colorado with the legacy of solar research in Boulder. This legacy was the result of Walter Orr Roberts's efforts to associate the High Altitude Observatory with the university after his association with Menzel and Harvard had deteriorated. Thomas moved the "solar associates" concept in the late 1950s to Colorado as a jointly run institute via "a new mechanism for governance" crafted by the Department of Commerce's Office of the General Council.[1345] There would be a single director, and each of its four major elements included full appointments and joint appointments from a variety of local astrophysics-related groups at the NBS together with faculty from departments at the University of Colorado, namely, laboratory astrophysics, theoretical astrophysics, and atomic physics, as well as aerospace engineering. This was not unlike what Goldberg had envisioned for an Astrophysics Center that would strengthen the Harvard College Observatory to be competitive with the SAO.[1346]

By August 1972, Field had examined a copy of the MOA establishing JILA and summarized it for Challinor. JILA acted as a "training facility" as well as a place for laboratory astrophysics that would foster better communication between astronomers and physicists, "particularly to keep the N.B.S. Laboratories for Astrophysical and Plasma Research responsive to the needs of astrophysics."[1347] Field also heard from Louis Branscomb, who described JILA as a joint government/university entity linking public institutions at the state and federal levels. Leadership roles, including a single director, were collectively determined, rotating from one to the other institution.[1348]

Also in August, Bradley interviewed Caryl Haskins, former president of the Carnegie Institution of Washington, now its trustee and an SI regent, and "found him to be enthusiastically in favor of the integrated operation."[1349] The California Institute of Technology and the Carnegie Institution had agreed in 1948 to combine all astronomy activities into a single operation, the Mount Wilson and Palomar Observatories, under a single director.[1350] Through the 1950s, the arrangement had included the management of two autonomous groups of astronomers: faculty such as Jesse Greenstein at Caltech and astronomers at Mount Wilson who were housed in Pasadena a few miles from campus. The union had continued to operate smoothly through the tenure of the first director, the physicist Ira S. Bowen, from 1946 to 1964. But several years after astronomer Horace Babcock succeeded him, from the Mount Wilson side, the Caltech "physicists began agitating for a divorce." The administrative changes in January 1970 that had created the new Hale Observatories improved management, with an associate director representing each of the units. Overall, the institution operated under "a single director, rotating between the two observatories," who had "overall responsibility for the administration of the observatories."[1351] There was a single salary scale, and all appointments were consensual between the director and associate director. Finally, all Caltech faculty teach, and though Carnegie astronomers were not required to teach, they could do so. Most critically, the combined body would be subject to periodic review by an active advisory board.

Bradley felt the successes of the JILA and Hale models gave them the ammunition they needed. Above all, there would be retention of identity, clearly delineated funds and programs, equal compensation for staff members, and frequent communication between the parent organizations. As Bradley articulated acceptable management models, Challinor focused on personalities and proclivities. Tellingly, he reported from an earlier meeting with Field that he actually regarded the SAO to be stronger than the HCO, and so "would really like to be the Director of SAO rather than HCO if he had his choice." Field evidently was also looking for a way to combine the two directorships, Challinor reported, "in a way that would not be a threat to Harvard." Both Challinor and Bradley worked on Whipple, getting him to agree that of the two options, what they called the "Hale Observatories" option, was the best. And because he now knew too that Caryl Haskins endorsed the "integrated operation," Bradley called for a showdown with Whipple.[1352]

Ripley, Bradley, and Challinor visited Cambridge on August 23. Bradley took great care setting up the meeting, making sure that Whipple felt he was in control and that everything was clear and open. He suggested that the meeting be held in Whipple's office. He also asked whether Whipple wanted to invite Field and Dalgarno to the meeting, and confirmed with Whipple that the agenda would include discussion of which model made the most sense, and that Field would be asked to assume the joint directorship. At long last, the deed was done. All that was left was to work out the details. Accordingly, within two weeks of the meeting, Field drafted a preliminary proposal listing himself as director of both observatories, with an assistant director who would be designated Director of the Smithsonian Astrophysical Observatory, following the pattern of the Mount Wilson-Palomar agreement.[1353]

Word started leaking out in late October 1972. So Challinor once again traveled to Cambridge to speak with Whipple and to New Hampshire to negotiate with Field. Knowing that Dunlop was about to address the Harvard Corporation and that the Smithsonian regents were soon to meet, on 30 October Challinor felt it was time to announce Field as combined SAO/HCO director and the formation of the CfA. Ripley penned into the margin of Challinor's note, "Yes, of course."[1354]

Harlow Shapley died the next day, and on the day following that, the Smithsonian issued a press release announcing Whipple's retirement, to be issued after 4:00 PM, after Whipple had assembled his staff to give them the news.

Building the CfA

In his first draft proposal, dating from early November 1972, Field, in line with Dalgarno and others, envisioned that the combined body would be split into divisions. Each division would enjoy considerable autonomy, with access to a single centralized administration that coordinated and apportioned administrative and technical services. There would still be the traditional Department of Astronomy and it would place stronger emphasis on teaching, whereas the SAO would concentrate on "national service."[1355] Field identified seven divisions, but did not name any associate directors. His job over the next few months was to identify them, which included both promotions from within, and some significant new hires.

Field's divisions worried Lundquist. How would SAO remain visible enough so that it could remain viable for competitive funding as well as for congressional appropriations?[1356] Lundquist agreed that some of the divisions would be led by the HCO and some by the SAO, especially the geophysical ones, but within that structure, especially when federal funds were being sought, he saw problems. The discipline groups would cut across organizational structures. The present radio group under Lilley, for instance, included both SI and Harvard staff. But if run by HCO formally now, SAO "management would have largely a support function and might not formulate an independent scientific program."[1357]

Field knew that problems like these existed, and invited Lundquist to help him solve them. Lundquist agreed, but declined his offer to head a geophysics division as SAO's "part of the NASA-sponsored Earth and Ocean Physics Program" as well as the continuing satellite-tracking program. Lundquist found the role "unattractive" because his colleague George Weiffenbach had been leading this effort for two years and especially in the past year had worked very hard to define and establish the area. Lundquist also admitted the role would not be to his "scientific taste." It would be too restrictive unless the division were expanded to include all of the space sciences conducted at the SAO. This would include present and proposed SAO activities such as "the hydrogen maser flights, orbiting observatories, planetary spacecraft experiments, perhaps our balloon astronomy program, and shuttle experiments toward the end of the decade."[1358]

Lundquist well knew that such a division would be a difficult sell for Harvard, for the SAO, and indeed for any academic institution. "Space science" was not a disciplinary field itself, despite the fact that the term was coined by the National Academy of Sciences in creating its Space Science Board. Rather, it was a capability that supported and enabled disciplinary research. It could make parts of that disciplinary research possible, it could greatly aid other parts, but it did not circumscribe the discipline itself. The SAO was building expertise in high-energy astrophysics and infrared astronomy. Although defined by specific technologies, and dependent on access to the high atmosphere and space, disciplinary ties in terms of the questions that stimulated problems in these specialties remained strong. Lundquist knew this: "Unless these uncertainties are resolved, I fear that such ambiguities in lines of authority might ultimately lead to troubles." Further, he knew, there was probably not a single disciplinary area at the SAO that could not, somehow or another, "be encompassed naturally in one of your more basic scientific disciplines."[1359]

But Lundquist also sensed that matters most dear to him, centering on the geodesy work, were not favored by the new director nor by the majority of the Harvard staff. Lundquist had quietly expressed these fears to James Bradley in the fall of 1972, and Bradley confirmed them after a talk with Field. As he reported confidentially to Ripley in September,

"Field is not enthusiastic about the geodesy component," even though, as Bradley well knew, it continued to support much of the planetary research and associated personnel at SAO.[1360]

There were also fiscal uncertainties in Field's organizational structure. Lundquist had asked John Gregory to estimate what fraction of the SAO budget would fall in the space science category and what fraction in the more conventional areas of astronomy and astrophysics. A "large majority" would fall into the space science bag, all of it from NASA. The OMB had already criticized the SAO for asking for too much space science funding; it was almost twice what it allowed for conventional areas.[1361] Of course, as Lundquist well knew, "This split depends somewhat on the definitions of space science versus conventional astrophysics." Lundquist also made "philosophical" points about Field's organizational structure: "organization by discipline under a discipline leader." This was very much like the present Harvard academic structure, "each tub on its own bottom," which would lead to problems based on the past behavior of senior Harvard professors. Harvard, Lundquist sharply observed, echoing Whipple, "has never been able to assemble a majority of these discipline leaders to back any major scientific undertaking beyond what each discipline group could accomplish alone."[1362]

The best example of this, for Lundquist, was the MMT, which required consensus over a broad range of disciplinary specialties, though it was not necessarily optimized for any one of them. This had been possible only by strong leadership at the top. And it was only possible because it was identified as a Smithsonian project in annual appropriations hearings. Thus it was critical, Lundquist advised, that the SAO maintain "enough scientific identity" to preserve its image and faith within the Smithsonian itself, with the OMB, and, most of all, with Congress. To make this happen, the SAO had to be designated the lead organization in selected activities with Harvard in a supporting role. It had be able to assert "that in specific areas it aspires to national and international leadership." This, Lundquist passionately added, had been a successful strategy in the past and had to continue in the future: It was the "correct role . . . to undertake ambitious programs of a scope beyond the capabilities of a typical university department."[1363]

Lundquist served Field faithfully in the identification and procurement of new staff. He went after Alastair G. W. Cameron, a broadly talented astrophysicist who had contributed to many fields and wanted to lead a group that would concentrate in planetary studies, in a sense Whipple's intellectual successor.[1364]

Field had been active in x-ray astronomy and predicted it was likely to be the most exciting part of astronomy in a few years as descendants of Uhuru, the first dedicated, all--sky x-ray mapper, started flying.[1365] Leo Goldberg knew that Riccardo Giacconi, Uhuru's principal scientist, was interested in closer ties with Harvard, as his x-ray group at nearby American Science and Engineering had many connections with Harvard staff. Giacconi was, in Field's words, the "leading x-ray astronomer in the world." He would be more than acceptable to both Harvard and the Smithsonian. Giacconi and three of his core staff would make the SAO a very strong, if not the strongest, x-ray astronomy center, anywhere. Field estimated that he would eventually require at least 10 additional slots.[1366]

With both Field and Whipple away in early 1973, Lundquist was in charge and distributed a broad statement describing Giacconi's anticipated role at the SAO, rationalizing how his work would enhance SAO activity in high-energy astrophysics, building on Trevor Weeke's Large Optical Reflector (LOR) installation at Mount Hopkins and Giovanni Fazio's gamma-ray balloon work. The SAO, Lundquist reminded everyone, had long been interested in x-ray work ever since Whipple had tried to add the field but was denied the chance by NASA program managers "because of the demands of Project Celescope." As Lundquist observed, "It has long been a regret of the SAO directorate that SAO has no activities in x-ray astronomy to complement its gamma ray activities. . . . [T]he x-ray region is the only gap in the SAO-Harvard capability."[1367]

Throughout this time, Whipple and his wife were at their retreat in the British Virgin Islands, called the "Porthole." They were hardly alone, entertaining visitors including George Field. In late February, Lundquist reported to Field, still away in New Hampshire, that "Fred has returned full of vim and vigor after his rest in the Virgin Islands" and that everyone looked forward to Field's return "under similar circumstances." He appended status reports on various SAO initiatives that made it clear that the "unstructured" growth that the OMB had worried about in August 1970 and Goldberg complained about in January 1971 was still very much alive (Figure 56). Field had promised that it would be curbed in his November 1972 draft "Proposal," but at least now it seemed to have stronger disciplinary rationales. Chief among his new proposals were a series of multi-million dollar programs:

GROWTH OF HARVARD COLLEGE OBSERVATORY AND

SMITHSONIAN ASTROPHYSICAL OBSERVATORY

	Personnel in Cambridge		General Budget *		Contracts and Grants *	
	HCO	SAO	HCO	SAO	HCO	SAO
1956	78	12	227	46	48	85
1960	90	214	216	250	735	3,393
1965	164	325	821	921	2,844	6,100
1966	193	375	870	1,100	3,818	10,200

* (in thousands of dollars)

Figure 56 Comparative growth of SAO and HCO, prepared for a periodic review by the HCO Visiting Committee. From the Papers of Jesse L. Greenstein. *Courtesy of the Archives, California Institute of Technology.*

"High Energy Gamma Ray Astronomy—gamma ray balloon flights leading to Shuttle based missions"; "3-meter balloon borne telescope for far IR astronomy"; "Study of Earth Physics by Laser Ranging to artificial earth satellites"; "High Precision Antenna for millimeter wave observations"; "A small but heavy satellite for measuring the Earth's Gravity Field"; and "X-Ray astronomy. . . . rocket flights and eventual follow on to Uhuru."[1368]

Lundquist created these status reports to prepare senior staff for the many reviews they expected were not far off. The HCO Visiting Committee met in its regular cycle in April, and started digging, but now heard mainly positive news.[1369] By May the new Harvard president, Derek Bok, readily approved Giacconi's and Cameron's appointments and Greenstein endorsed promotions for internal candidates for the remaining division chairs. Dalgarno had asked him for support to promote Robert Noyes to full professor, and also to suggest names of promising women and minority candidates. There certainly were good candidates, such as Andrea Dupree, but no one, they thought, seemed ready for a leadership role.[1370]

Also by May, even though there were remaining issues, especially the problem of assessing overhead requirements for funding proposals that consisted of both Harvard and Smithsonian staff, the Harvard Corporation approved the establishment of the CfA.[1371] This led to press releases and the inevitable media play. The *Boston Globe* in particular described the new entity as "Harvard's Smithsonian Astrophysical Observatory." Jim Cornell, totally in the spirit of the moment, gleefully sent a copy of the news piece to a friend in the Smithsonian administration with the inscription: "the take-over is accomplished. Wear your Harvard sweater and *Veritas* tie next week or we won't let you in the door. Best, John Harvard."[1372]

More to the point, but possibly not as newsworthy, the press release finally identified the divisions: Cameron would head Planetary Sciences; Dalgarno would lead Theoretical Astrophysics; Giacconi would direct the High-Energy Astrophysics division; Lilley, the Radio Astronomy division; Noyes would direct Solar and Stellar Physics; W. H. Parkinson would head Atomic and Molecular Physics; and Weiffenbach, as Lundquist suggested, would take Geoastronomy. The only division left for appointment was Optical and Infrared Astronomy.[1373]

When David Challinor asked Field to join him for an inspection tour of Mount Hopkins in February 1973, as if to indoctrinate the new director to Whipple's legacy, Field agreed, knowing it was an important gesture. But he also felt strongly that Whipple should be invited too. Field had just returned to campus, and felt deeply about his visit with Fred and Babbette Whipple in the Virgin Islands. Above all, Field expressed warm appreciation for the man and his accomplishments, but was mostly "impressed by his gallantry in what must be a rather difficult situation"[1374] (Figure 57).

Field wanted to do something for Lundquist, because he had done a heroic job consolidating resources and keeping lines of communication as open as possible. No one had better comprehension of how SAO worked as a government laboratory, and no one at SAO was more adept with that mode of operation.[1375] But Lundquist soon had an offer he couldn't refuse: directing NASA's Huntsville-based Space Science Laboratory at the Marshall Space Flight Center.[1376] So he left the SAO a few weeks before Whipple stepped down as director in

Figure 57 George Field, David Challinor, and Fred Whipple at the Mount Hopkins Baker-Nunn site. February 1973. *From the Fred Lawrence Whipple Observatory Event History, Harvard-Smithsonian Center for Astrophysics (CfA).*

July 1973. From Huntsville, Lundquist wrote warmly to Challinor, reflecting on 11 years at SAO, especially their time together hiking around Mount Hopkins and then when Challinor had made kind comments at his retirement dinner. Lundquist hoped for a "bright future" for the CfA under Field and assured Challinor that in his new capacity at NASA/Marshall he looked forward to many collaborations. But most of all, Lundquist concluded, "I know that Fred Whipple can be rightly proud of his substantial accomplishments as director."[1377]

Figure 58 Fred Whipple liked to punctuate his talks with a meteoritic stone. *Smithsonian Astrophysical Observatory, Publications and Information. Smithsonian Institution Archives Acc. 16-263-B&8Whipple.*

Chapter 17

Whipple's Legacy

Our spiritual godfather, founder of the modern SAO, and by extension the CFA. 1955, the SAO, nearly moribund, all solar stations closed down, staff reduced to a few astro archivists and some pseudo botanists in a Quonset hut behind the Castle. FLW accepts directorship.

—*James Cornell, circa 1990s*[1378]

We have benefited greatly from working in "the house that Fred built."

—*George Field, Myron Lecar, Irwin Shapiro, 2001*[1379]

In 1964, when Leonard Carmichael retired as secretary of the Smithsonian, news accounts of his administration were not kind. The *Washington Post* described Ripley's ascension as the "renaissance" of an institution that "had gone to sleep under a 'timid' and 'safe' man." Even *Science* implied that the Smithsonian under Ripley's predecessors was invisible; at least that part that engaged in fundamental research had declined "as a scientific establishment," wherein the status of research "was deplorable." One indicator of its status, D. S. Greenberg believed, was the fact that the SAO received only $1.2 million in support from the Smithsonian whereas it was awarded "about $4 million from NASA."[1380]

Fred Whipple wanted Carmichael to know that he did not share Greenberg's views. The article "disturbs [me]," Whipple wrote, especially because no mention had been made of the "predominant role you played in rescuing the Astrophysical Observatory from the dismal depths to which it had descended." Whipple was annoyed, but not enough to go public.[1381] Carmichael thanked Whipple for his support, adding, "*I do appreciate it very much!*" (emphasis in original). He acknowledged that he had the good fortune of being secretary when there had been a "tremendous increase" in federal funding, and then added, oddly, as an informal postscript: "I define *you* as: 'the greatest physical scientist of this era at the Smithsonian.'"[1382]

Some 34 years after the *Harvard Crimson* first took notice of the Smithsonian's arrival at Harvard in 1955, the *Crimson* returned to examine how the institutional arrangement had fared over the years after combining into a single Center for Astrophysics. Less concerned with historical details than the overall import of the association, the *Crimson* reporter claimed that the "unusual collaboration" was the first government/university arrangement "to take advantage of pooled resources and helped stake the U.S. to an early lead in international astrophysics."[1383] Quoting John Huchra, a leading Smithsonian astronomer who was one of few who held a dual appointment as professor of astronomy at Harvard, the *Crimson* offered aphorisms reminiscent of Whipple's in 1955, that "the amalgam of the institution is greater than the sum of its parts." But even though the merger created a satisfying "symbiosis" of strengths, the *Crimson* made the point that the Smithsonian had, over the years, paid most of the bills. In return, the *Crimson* pointed out, again quoting Huchra, Harvard offered an "academic atmosphere [that] is very good. . . . One of the draws for me to the Smithsonian is the ability to teach."[1384]

The *Crimson* also reported that the merger did not resolve all the initial administrative complexities. Twelve Harvard full professors "are jointly tenured by Harvard and the Smithsonian but are paid by the Smithsonian." And of the Center for Astrophysics' 140 tenured faculty overall, the Smithsonian paid for 119 of them, but "paid salaries of its staff have lagged behind Harvard's over the past 15 years." "There's always a friction," Huchra reported, bristling at "the perceived 'second-class' citizenship of the Smithsonian staff." There also seems to be resentment that Harvard "prefers to rely for now on the Smithsonian facilities" and had not made good on plans for new Harvard facilities in the Southern Hemisphere. Huchra hoped that the pay scales would soon equalize, but observed that the coming of the Smithsonian in 1955 and the merger in 1973 meant that the "HCO has gotten smaller and the Smithsonian has gotten bigger. . . . If [the Smithsonian] weren't here, there'd only be about eight people."[1385]

The mid-1960s were peak growth years, as growth tables for funding and staffing clearly showed.[1386] By the end of the decade, support for science and its institutions had changed profoundly, and Whipple could no longer hope to maintain his endlessly expansive posture. He knew it, but still, when Ripley finally advised him to step down as director, assuring him of a continuing place of honor at the Smithsonian, but making it clear that the institutional autonomy he treasured was no longer possible, the reality must have stung. He had agreed to back down on NEROC only when he knew he lacked Ripley's strong support, and only when his rift with Goldberg had brought upheaval to both institutions. Even so, from impressions of his widow well after the fact, and from his own testimony at the time, his act of stepping down was not as painful as the merging of the two institutions under one director.[1387]

Whipple never separated himself and his projects from the CfA and remained a senior staff member through his emeritus years right up to his death in 2004. It is the life of the institutional arrangement that he, Menzel, and Carmichael built that effectively ends with the merger of the Harvard and Smithsonian observatories into a single functioning unit. It was, in that regard, the end of an era, because, although it continued as one of the largest astronomical institutions in the world, most of the core programs that built it have since ended. Whipple's legacy remains, however, an infrastructure and a capability that expanded what it meant to be an astronomical observatory (Figures 59 and 60).

Figure 59 Whipple at the May 1982 rededication of the Mount Hopkins station as the Fred Lawrence Whipple Observatory. *Smithsonian Institution Archives, Image # 2002-32272.*

Figure 60 Fred, with Babette, posing at a plaque rededicating the observatory in his name. *Smithsonian Institution Archives, Image # SIA2017-077253.*

Through the rest of his life, Whipple maintained an active and vigorous agenda of publishing, lecturing, and providing testimony to government and other deliberative bodies. His focus remained on comets, meteoroids, and dust particles in the solar system and their interaction with the Earth and Sun. He also engaged broader topics, frequently commenting on the state of astronomy, the way to maintain a vigorous space research program, and the future of science education. Over half of his some 150 scientific publications appeared after he stepped down from the SAO directorship in June 1973.

For most of his career, Whipple consistently pushed for new technologies and innovative ways to apply old technologies. In oral histories, he cited his early advocacy of radio astronomy and his promotion of instrumenting V-2 rockets for all sorts of scientific problems including x-ray studies of the Sun as early examples.[1388] And most definitely his push for vastly improved optical meteor-tracking and satellite-tracking systems and his recognition of their value for geodedical problems of all kinds stimulated new instrumentation as well as computing and reduction techniques. And even though the scientific return from the decade-long Celescope program was meager, by his own admission, it is also clear that here too he pushed the state of the art, promoting the development and use of radical and still experimental electronic-imaging technologies. Finally, his advocacy of the multi-mirror telescope concept stands as a milestone in the exploration of alternative ways to improve optical instrumentation. In many ways, Whipple acted as a catalyst, directing attention of astronomers to new technologies associated with rocketry and satellite flight. Certainly his early advocacy for these radical means of doing astronomy in the late 1940s and early 1950s marks him as a passionate visionary, second only to his remarkable ability at building up the institutions he thought were necessary to pursue, in McGeorge Bundy's words, "his dear dream."

Whither the "Dear Dream"?

I assumed from the beginning of this historical work that Whipple's "dear dream" was space travel, as Bundy suggested. He certainly shared and campaigned for the grand vision fostered by von Braun, Willy Ley, and others in the 1950s, and did all he could to get involved. But Whipple's "dear dream" may also have been building something like the Smithsonian Astrophysical Observatory. As we have shown here, using Whipple's own testimony and his actions, his dear dream was not just space, but was deeper and more personal. It was to be a "success," as he remarked early on in an oral history, describing it as his parents' expectation. And that success came partly with his scientific attainments, but the watershed was clearly his experience in World War II, learning that he could build teams, compete for contracts, and work at a level far beyond his individual capacity, leveraging his own talents. "Consequently I got into administrative things to push projects that I thought ought to be done."[1389] These "administrative things" were at the root of Whipple's dear dream. Building an infrastructure, a capability, an entity, to get things done he wanted to get done. He sensed early on that the future of astronomy was going to be very different: "The big change in my lifetime is from the individual scientist, who occasionally has an assistant or two and has graduate students around, to the team operations, with the big things. It's big science, is the change."[1390]

It is too limiting to state simply that Whipple's "dream" was the institution he built. When asked about his talents as an administrator, he scoffed, minimizing his abilities while stating clearly his one guiding principle: "The simple thing is that when you delegate responsibility, at the same time you delegate authority. As simple as that. That's all there is to it." He believed, moreover, that he was lazy when it came to administration, and that was the key to his success. Not acknowledging his initial resistance to the Smithsonian administration's demands that he acquire administrative assistants and managers in 1959, Whipple rationalized that his success was also because he was "just so lazy on administration that I believe in getting the professionals [to] do it, I would never turn a hand in that area if I could find somebody else who knew what he was doing to do the job for me. And not many administrators would have worked that way." [1391]

Whipple, of course, knew that the key was to bring in the right people, but he also knew, as did the British literary critic F. R. Leavis in his debates in the 1960s with C. P. Snow, that the "talented individual" was not enough; traditional institutional structures also had to change to "sustain the capacity of creative thought."[1392] At some level, he could control who he brought in. But he could not control those administrating above him. When he threatened to resign in early 1959, he was dealing with administrators at the Smithsonian who he felt were destructive and anything but enabling. All that changed when Carmichael asked James Bradley to look after their investments in Cambridge. As Whipple recalled:

> There's the guy who really put the finishing touches on the final growth of the Smithsonian Observatory. I mean, he just knows the Hill like nobody. He was the guy who put the right touches on the right people.[1393]

Whipple credited Bradley for pushing through the MMT, for instance. And as we have seen through this work, indeed, Bradley often was the key individual making things happen, right up to convincing Whipple to step down.

Whipple's Many Worlds

Whipple parlayed his scientific interests in the many tiny worlds strewn throughout the solar system—comets and meteors—into institutional constructions that, in many ways, constituted a whole new set of earthly worlds within which he deftly moved as administrator and entrepreneur. As he told McCrosky in 1958, his first love remained meteors, and everything else was a vehicle to study them, from the high atmosphere, from hyperballistics and reentry problems, to the retrieval of remnants from falls, both natural and artificial: "The way you get money to study meteors is to study the atmosphere."[1394] He retained this view throughout the rest of his career, which more than exemplifies historians' view that "in the last half of the twentieth century, any sharp distinction between military and civilian science has become almost impossible to make."[1395]

Meteors were indeed Whipple's raison d'être. As a featured speaker at the bicentennial celebration of the birth of James Smithson in September 1965, lavishly orchestrated by Dillon Ripley, including academic processions across the National Mall, Whipple stood before his audience brandishing a meteoritic stone "to illustrate progress in our knowledge of the physical universe."[1396] Far more than a prop, the stone represented how, in "the flow of time," knowledge had led to human progress, not only in the sciences, but also in culture writ large, from child rearing and the sustainability of the human race, to health, physical comfort, and to the shift away from religion to science as a means of salvation and self-realization.

Whipple entered professional astronomy when it was changing profoundly. It led to his own transformation. Whipple had long studied the tiny worlds of space, the meteors and the comets, and always returned to them as he tried and failed to conquer larger worlds in mainstream astrophysics. Transformed by the experience of wartime research, however, he found that there were even larger worlds available to him, worlds in fact in need of his skills. He thus entered the world of military-based science in Washington. He was the only elite astronomer to engage space research, combining his interests in meteoric bodies as denizens of deep space and as probes of Earth's uppermost atmosphere, with his dreams of space travel, intent on building a new institutional world that would lead the way. In doing so, as a Space Age entrepreneur, he created a new space for research—a new physical space, actually a series of spaces, operating in new economic spaces, and even new political spaces. All these spaces had new as well as traditional intellectual dimensions, but they tested and stretched the borders of what constituted academic astronomical practice within traditional institutional boundaries.

In company with people like Lloyd Berkner, Whipple was one of "an astonishingly small number of visionary figures" who seized on opportunities to align and profit from the multiple sources of funding offered by the postwar world.[1397] Even though he failed to establish an academic base for space science infrastructure on a national scale, he more than managed to garner support from many civilian and military sources to provide tracking facilities and other expertise in the space sciences, which fueled the spectacular growth, and visibility (Figure 61) of his institution.

Among the goals of this work has been to better appreciate how NASA, as a Cold War institution, influenced the development and growth of the space sciences. The SAO story has shown how NASA priorities vastly accelerated the scope and complexity of early scientific satellite systems while confining the role of the academic community, and also how it encouraged experimentation with new technologies. In addition, we have also illuminated the changing character of the national infrastructure for astronomy fostered mainly by the National Science Foundation's creation of national observatories, reflecting pioneer efforts in physics and anticipating to a degree what NASA would eventually do. Most poignantly, we have seen how Whipple's ambitions for NEROC, aided and abetted by Dillon Ripley's ambition to re-establish a national role for the Smithsonian, brought into question who or what was in control of setting priorities for both new radio and optical telescopes in the eyes of the NSF, the OMB, and Congress.

Figure 61 Whipple surrounded by family holding Life magazine covers recording the SAO's early space exploits. *Courtesy Sandra and Laura Whipple.*

Today, similar efforts involving multiples of observatories and national institutions organized into consortia on both national and international levels have become the norm for the 21st century. And the CfA has certainly been part of this process. Under successive directors, the CfA has been actively involved in the scientific operation of two of NASA's "Great Observatories," the Chandra x-ray observatory and the Spitzer infrared observatory, and has continued as a major site for missions in solar physics. It engages in major projects and programs both ground based and space based. Combining support from direct federal appropriations, as well as support from NASA, the Air Force, the NSF, and more traditional support from private and corporate philanthropy, the CfA has also established major observing facilities in the southwestern United States, in Hawaii, and at the South Pole.

The creation of the CfA did eventually stabilize the relationship between Harvard and the Smithsonian, although not all administrative problems were solved or have been solved to this day. George Field immediately faced the thorny problem of establishing equitable and realistic overhead rates for both competitive federal funding and for private patronage strategies, and also experienced renewed scrutiny from the General Accounting Office and the House Appropriations Committee. Calls for placing Mount Hopkins administratively under Kitt Peak, or even giving over the SAO to the NSF did not disappear altogether in the 1970s. Even so, Field carried the CfA to funding levels even greater than Whipple's.

What is fact is that the institution Whipple created and guided has continued to grow and prosper. The CfA continues today as one of the great astronomical institutions of this world. Its legacy should be appreciated not only in terms of the new knowledge about the physical universe it has helped reveal, but also in terms of the scope of the human universe that can be called an astronomical institution.

NOTES

See Bibliography, Archival Source Abbreviations (p. 361), for explanation of abbreviations in Notes.

Introduction

1. Stuart Leslie (1993), 1.

2. Jacket blurb for Westwick (2003).

3. "Whipple Speaks on Sputnik," *Newsletter of Harvard Foundation for Advanced Study and Research*, 31 December 1957. Folder "Harvard University 1943–1961." Box 2. RU 7431. SIA.

4. Whipple, "Proposal for the Initiation of an Optical Tracking and Scientific Analysis Program for the U.S. Earth Satellite Program." 29 November 1955. Folder "Earth Satellite Project." Box 13. RU50. SIA.

5. Blakely to Whipple, 2 April 1958. Folder "C, general." Box 1. RU7431. Ser. 1. SIA. See http://en.wikipedia.org/wiki/Cabot,_Cabot_%26_Forbes (accessed 21 July 20s12).

6. Needell (2000); Wang (2008); Newell (1980); DeVorkin (1992).

7. Forman (1987), 150.

8. Leslie (1993); Slayton (2012), 258–259; Dennis (1991); Doel (2003); Westwick (2003).

9. On the Golden Triangle, see Leslie (1993), 2.

10. Hershberg (1993), 396–397.

11. On the reticence of astronomers to follow the pattern, see DeVorkin (2000b), "Who Speaks." On the pattern, see Forman (1987). Overall, American astronomy was organized around what John Lankford has described as "a self-conscious power elite: observatory directors." Lankford and Slavings (1997), 182.

12. DeVorkin (1992), 159.

13. On Yerkes-McDonald, see Osterbrock (1997), and Evans and Mulholland (1986). And on the Mount Wilson-Caltech-Palomar arrangement, see Osterbrock (1992); Florence (1994), ch. 29. See McCray (2004), chs. 1, 2, for an overview of the state of telescope building in the 1950s; for new arrangements in solar physics, see Hufbauer (1991), ch. 4, as well as Bassi (2015).

14. McCray (2000, 2004); Edmondson (1997).

15. See Dupree (1963), 456, for his distinction between a central organization and a predominant organization. Consider the venerable Kaiser Wilhelm Institutes in Germany, transformed as the

Max Planck Institutes after World War II, or the National Physical Laboratory in Britain or the civil service structure that supported major observatories in France. See, for instance, Meadows (2009).

16. See Multhauf and Good (1987); Dupree (1959).

17. Bassi (2015).

Chapter 1 – The Many Worlds of Fred Whipple: From Meteors and Comets to Artificial Satellites

18. Whipple OHI (1977), 5–8. He read Andrew Lang's *Blue Fairy Book* and *Yellow Fairy Book* and then Henry Rider Haggard's *She*, which stimulated his science fiction interests. He loved Edgar Allen Poe and H. G. Wells. In his oral history, he seamlessly segued from his reading to a "latent memory of a traumatic early childhood experience," constructed from repeated hospital visits for a range of maladies during which he was examined by hooded and masked nurses and doctors. These, he realized only late in life, fostered memories of being abducted by aliens. See also Fred L. Whipple, "UFO or Early Childhood Trauma?" *Boston Globe*, 30 June 1992, n.p., clipping in SIA Acc. 04-183 Box 1 "Historical." Also in this folder are articles on the "abductee" phenomenon and its possible causes, including everything from New Age beliefs to sleep paralysis, and notes taken on "dreams." See, for instance, "Alien Abduction Claims Are Examined," *Harvard University Gazette*, 20 February 2003, 1, 8.

19. "I was very conscious of it, because it was 40 miles from Omaha to Council Bluffs, and a gun that could fire that distance was fantastic. That was really exciting." Whipple OHI (1977), 7–8.

20. As he recalled in a speech given in 2000. See "My Conversion to Atheism." Acc. 10-017. Fred Whipple Papers. SIA.

21. Whipple OHI (1977), 16.

22. Whipple OHI (1977), 121; Whipple, "My Conversion to Atheism," 4. She was also on the honors list in late 1928. See "University Honors List Announced," *LA Times*, 14 October 1928, 10. See also Babette Whipple OHI (29 October 2007), AIP, 26–27.

23. Herget (1978), 127–147.

24. Bower and Whipple (1930). Whipple OHI (1977), 22. Earle was born in Alameda, just south of Oakland, on 14 January 1930. "Earle Raymond Whipple," California Birth Index (https://familysearch.org/ark:/61903/1:1:V2RH-PQN, accessed 27 November 2014).

25. Whipple (1932).

26. Herget (1978).

27. Whipple OHI (1977), 32–36, on 35. Menzel autobiography (1974), 231–233.

28. Jones and Boyd (1971); Bailey (1931). Comparative perspective from Lankford (1997), notes 102–107, and Gingerich (1988), 23–30. Whipple opted for the "big puddle." Whipple OHI (1977), 34–35.

29. Jones and Boyd (1971), 458, note 61, provides illuminating background on how Pickering resisted taking on instruction responsibilities in the 1880s.

30. Cannon (1922).

31. "The Harvard College Astronomical Laboratory," in "General Notes," *PASP* 40 (1928): 147–148.

32. On Cecilia Payne, see Haramundanis (1996); Payne-Gaposchkin et al. (2001); Boyd (2014).

33. Menzel autobiography (1974), 231–233.

34. Visher (1947), 116–117.

35. Boyden Observatory, in "General Notes," *Popular Astronomy* 42 (1934): 115–116. Harlow Shapley (1928), 297–298.

36. Hershberg (1993), 72–73.

37. Hershberg (1993), 106–110; Keller and Keller (2001), 66–67.

38. Shapley to Whipple, 15 June 1931. Folder "S." Box 2. Whipple Papers. HUG 4876. HUA.

39. Whipple's meteoritics have been examined by Doel (1996), 68–72, in context of atmospheric research, and by McCray (2008), 44–55, as an expression of entrepreneurial engineering.

40. Owen Gingerich recalls Whipple's accounts of these battles. Private communication.

41. Earle Raymond lived with the Cornell family until he moved to Boston to live with his father in 1940. Life impressions at this time gained from Earle Raymond Whipple to the author, 10 October and 21 November 2006; Babette Whipple OHI (2007), 28–30, AIP, and Laura Whipple OHI (2007), 19, AIP.

42. Whipple (1938), on 502, 544. See also Hoppe (1937), 233–240.

43. Whipple (1972) in Berendzen (1972). Whipple's goal was to improve Lindemann and Dobson's work. Whipple (1939), 419–425. In 1922, Öpik had wanted to determine the physical processes involved in the passage of a meteoroid through Earth's atmosphere. He was followed by Lindemann and Dobson to create the first comprehensive density profile from 30 to 150 kilometers. Lindemann and Dobson (1923), 411–437, on 435.

44. Earle Raymond Whipple to author, 10 October 2006. Author's files. Earle had also stayed with Whipple's parents.

45. Bart Bok to Whipple, 12 August 1942; Shapley to Whipple, 18 August 1942. Folder "S." Box 2. HUG 4876. HUA.

46. See Kidwell (1992).

47. Whipple OHI (1977), 61.

48. Whipple to Spitzer, 21 December 1942. Folder "S." Box 2. HUG 4876. HUA. Earle prospered at Hillside, graduating from eighth grade in June 1943 with a good citizenship award. He was class speaker at his graduation, where he introduced his father as the commencement speaker. Earle then returned to California to continue his studies in Ojai, California. Hillside School Graduation Program, 18 June 1943; "Whipple, Earle Raymond," file. SIA Acc. no. 10-017 "Earle" folder. Earle Raymond Whipple to author, 10 October 2006. Author's files.

49. Jones (1978), 290–291.

50. See McArthur (1990), 71–72; Baxter (1968), 93; Stewart (1948). On Whipple's work at the RRL, see "Records of the Radio Research Laboratory, 1942–1946." Box 20, "Whipple" UAV 728.95.1. See also G-500 and subseries UAV 728.95.4, "Records Concerning Demobilization and Laboratory History, 1942–1946 (inclusive)." In a barrage of meetings in London, Whipple and MIT electron tube specialist D. B. Langmuir met with officers at the Grosvenor Square RDM Signal Office to compare American and British chaff designs. "Window of definite value to 8th A.F.," Whipple recorded, but the question was, which system would be chosen? 21 January 1944 entry. See also entries between 11 and 17 February. Box 1, "Misc," containing "F. L. Whipple Diary." Harvard University Archives, Pusey Library.

51. The significance of his contact with General LeMay is not known, but at war's end LeMay became deputy chief of Air Staff for Research & Development at the Pentagon and helped to create think tanks like Project RAND at Douglas to explore missiles as an extension of airpower. See Collins (2003).

52. Whipple OHI (1977), 65; see 61–66 for a general discussion of his experiences. Whipple's experience was exactly opposite Harold Urey's, who was, according to Matthew Shindell, "traumatized" by his managerial duties during World War II. See Shindell (2014), 107–139, on 113.

53. Whipple (1943), 246–264.

54. Sterne to Whipple, 20 October 1941. Folder "S." Box 2. Whipple Papers. HUG 4876. HUA.

55. Whipple OHI (1977), 61.

56. Whipple OHI (1977), 70. Whipple later recalled: "After nearly three decades of reading, talking, and dreaming science fiction . . ." it was the German V-2 missile that made him realize there was a "high probability that I could actually see artificial satellites, and, I hoped, manned space travel in my lifetime." Whipple, "Recollections of Pre-Sputnik Days," 127–134, in Ordway, Liebermann, and Bova (1992). See also Whipple (1972), 967.

57. Whipple "Memorandum," 30 April 1945. Box 2. Folder "Harvard University 1943–1961." RU 7431. SIA.

58. Doel (1996), 69.

59. Ralph Bacon (Frankford Arsenal in Philadelphia) to Whipple, 10 February 1944. Box 2. Folder "B, 1940–1950." Whipple Papers. HUG 4876. Reference courtesy P. McCray.

60. Smith (1989), 30–31; Collins (2002).

61. Whipple concluded that a strike from a body about 1,000 times fainter than can be seen with the eye could puncture a metal skin 1⁄16 inch in thickness, and there was a 1-in-40 (2.5%) chance this would happen in an exposure to space of only 120 hours. He cautioned that his statistics might be off by a factor of 100. Fred Whipple, "Possible Hazards to Satellite Vehicle from Meteorites," app. III in Clauser (1946), 1072–1079, on 1079. General report accessed 8 April 2012 at http://www.rand.org/pubs/special_memoranda/2006/SM11827part1.pdf. Whipple's supplement is reprinted in Whipple (1972).

62. DeVorkin (1992), 94–96.

63. C. S. Piggot to Whipple, 10 September 1947. Box 10. "RDB" folder. Whipple Papers. HUG 4876.808. HUA. Whipple initially declined after considerable discussion with Johnson, claiming lack of time, but eventually agreed.

64. Putnam to V. M. Slipher, 6 February 1946. Lowell Observatory Papers, microfilm edition. Buck to Whipple, 25 July 1949. Box 2. Folder "Harvard University 1943–1961." RU 7431. SIA. Whipple, "Application for Federal Employment," 1 July 1949. Whipple Papers. HUG 4876.810; copy in Whipple Supplement SIA Acc. 10-017, SIA.

65. DeVorkin (1992), 96, refs. 161–163.

66. Jacchia stayed at Harvard collaborating with Whipple as his expert in numerical techniques with computers at MIT and Harvard. He had been at the University of Bologna since the early 1930s, but was expelled in 1939 with the enactment of Italian Racial Laws. See Kidwell (1992), 290. See also http://www.issawiki.org/wiki/index.php?title=Luigi_G._Jacchia_(1911-1996),_Astrophysicist,_United_States (accessed March 2014). See also http://library.duke.edu/rubenstein/scriptorium/mazzoni/exhibit/wwii/C929.html (accessed March 2014). On Jacchia coming to the United States, see also Russell to Shapley, 27 February 1939. PUL/HNR.

67. Baker OHI (1980), 129. Baker had more job offers than anyone at Harvard, but remained undecided to the point that it put his family "in a nervous state as a result of [his] own doubts, cross-purposes, and lack of undisturbed relaxation." (Ibid) In 1948 he moved to the San Francisco area to find a better posting but soon returned to Cambridge to resume a loose connection with Harvard and the optical industry in the Boston area. See, for instance, Shapley to Baker, 7 November 1945; 6 April 1946; Baker, for the record, 25 March 1946; Baker to Shapley, 25 February 1946; 16 December 1946. Shapley Papers. "HCO Optical/Baker." UAV 630.22. HUA.

68. Menzel et al. (1937), 87–98; Menzel (1936), 164. Baker (1939), 255. Baker OHI (1980), 54–69, 98–99.

69. Baker OHI (1980), 98–99.

70. Kidwell (1992), ref. 42.

71. Baker OHI (1980), 105–107.

72. Baker OHI (1980), 107. J. G. Baker. U.S. Patent 2458132. "Schmidt image former." 4 January 1949.

73. Muir-Harmony, Abrahams, and DeVorkin (2012). The glass was the key issue: when commercial optical firms balked, the National Bureau of Standards made the blanks for the optics, yielding a 12.25 inch effective aperture, a field of view of 55 degrees, and a focal length of 8 inches, operating at an astounding focal ratio of f/0.85. In 1951, Perkin-Elmer Corporation agreed to produce them for Harvard. Two cameras were placed at Sacramento Peak and one at Organ Pass, some 50 miles apart, and they were operated in unison to photograph meteors. The remaining two cameras were operated by Peter Millman, who was then at the Dominion Observatory in Canada. Fred Whipple (1951), 144–145 [abstract].

74. McCrosky, (1955a), 170, McCrosky (1955b). See also McCrosky OHI (2005).

75. DeVorkin (1992), ch. 15.

76. Whipple (1950b), 83 [abstract]; (1950a), 375–394. Whipple's comet model is discussed more fully in DeVorkin (1992), 291–293, and DeVorkin, "Pluto, the Problem Planet," in Launius (2013).

77. Quotes from Babette Whipple OHI (2007), 18–20.

78. Whipple never really opened up about his life to Babette; she realized this only after his death, going through his correspondence. He did seek her help with his dreams, or demons, as he called them, specifically a red eye. "He would wake up in the middle of the night seeing this red eye, which was frightening." Whipple worried whether the shock treatments he had received at McLean Hospital in Belmont, Massachusetts, in late 1942 and 1943 had damaged his brain. He never told her that these were a continuation of his treatment in Milwaukee. Babette Whipple OHI (2007), 30–33.

79. Whipple to Dr. and Mrs. William Hutchings, 16 October 1946. Box 4. Folder "H, 1940–1950." Whipple Papers. HUG 4876. See also Whipple to W. Baade, 19 June 1947. Box 2. Folder "B, 1940–1950," 1 of 4. Whipple Papers. HUG 4876. HUA. Courtesy P. McCray.

80. Whipple to Sturla Einarsson, 23 February 1949. Box 3. Folder "E, 1940–1950." Whipple Papers. HUG 4876. HUA. Courtesy P. McCray.

81. Whipple to Greenstein, 11 March 1949. Box 3. Folder "G, 1940–1950." Whipple Papers. HUG 4876. HUA. Courtesy P. McCray.

82. He was none too happy if someone inadvertently linked him with the Meteoritical Society, founded by his old UCLA teacher Frederick Leonard. In May 1949, Harrison Brown of Chicago, with whom Whipple had been discussing meteor irradiation mechanisms, lamented the "amazing

hold that unadulterated scientific amateurism" has on the group, suggesting that if Whipple were to take on the leadership of the society, possibly he could professionalize it. Whipple agreed that the group was "quarrelsome" with lots of "bickering" and "heckling" but wished it would disband and be absorbed by the AAS or even the IAU. Brown to Whipple, 2 May 1949; Whipple to Brown, 10 May 1949. Folder "B, 1940–1950," 4 of 4. Box 3. Whipple Papers. HUG 4876. Courtesy P. McCray.

83. Whipple to Robertson, 6 October 1949. Whipple Papers. HUG 4876. HUA.

84. Baker predicted a 40× increase in capture rate. In fact, they proved to be 25× the best pre-war surveys on predicted shower nights, and over 1000× the original Harvard patrol system. See Millman and McKinley (1963), 685, table 2.

85. Agencies included the Naval Bureau of Ordnance, Naval Office of Ordnance Research, the Air Force Cambridge Research Center (AFCRC), and the U.S. Office of Ordnance Research. See, for instance: Whipple to M. L. Wasser, ARDC, Rome Development Center, Rome, NY, 24 January 1955; Whipple to Capt. Frank S. Attinger USAF—Admin Officer, SAB, office of the chief of staff, USAF, 18 January 1955; Pratt (Harvard Office for Research Contracts) to Menzel, 11 January 1955, clipped to RFP from USAF Rome Air Development Center, dated 5 January 1955, asking for proposals for "Study of Meteors at Altitudes of 100 to 1000 Miles." Whipple Papers. HUG 4876. HUA.

86. Donald E. Taylor to Whipple, 25 March 1947. Whipple Papers. HUG 4876. HUA.

87. Launius, "Outstanding Analysis of the Romance of Aviation," http://www.amazon.com/The-Winged-Gospel-Americas-Aviation/dp/0801869625. See also works by Green and Lomask (1970), Neufeld (2000), and citations therein.

88. Whipple (1992), 128; Whipple, "The Heavens Open" in Ryan et al. (1952), 132–143, on 137.

89. Ryan (1952), 140–143.

90. Whipple, "Recollections of Pre-Sputnik Days," in Ordway and Liebermann, eds. (1992), 127–134, on 128–129.

91. Whipple review of Clarke (1951). *Saturday Review*, 5 July 1952, 13–14.

92. Whipple and Wernher von Braun (1952), 36–47.

93. Whipple (April 1953), 151.

94. Whipple, typescript, "Why Conquer Space?" (1953). Box 19. RU 7431/SIA. Published in *Astronautics* 1, no. 1 (1954): 7.

95. Singer to Whipple, 17 June 1954. Folder "S." Box 14. Whipple Papers. HUG 4876. HUA.

96. Neufeld (2007), 234.

97. Whipple to Singer, 7 October 1954; 28 October 1954. Folder "S." Box 14. Whipple Papers. HUG 4876. HUA.

98. Whipple manuscripts, 22 September 1954: "First Steps in Using Artificial Satellites," folder "Scientific Advisory Board of the USAF." Box 23. RU 7431. SIA. On Mouse, see Green and Lomask (1970), 17–22; and "Report on the U.S. Program for the International Geophysical Year," *IGY General Report*, no. 21 (November 1965). World Data Center A., National Academy of Sciences (U.S.), National Research Council (U.S.), 421–422.

99. "First Steps," 47. On the matter of precision, see MacKenzie (1993).

100. Fred L. Whipple, "On the Security Classification of Upper Atmospheric Techniques and Data." Manuscript draft, Folder "Whipple Security Classification." Box 16. Acc. 04-183. SIA.

101. On IGY aspects, see Needell (2000), ch. 12.

102. See Needell (2000), Neufeld (2007), McDougall (1985), Wang (2008).

103. Needell (2000), 336.

Chapter 2 – Harvard Astronomy Post–World War II

104. See Peggy Kidwell (1992), refs. 72–74. DeVorkin (2006b), 12.

105. Bart Bok memorandum, "Some Notes About Finances and Personnel of the Harvard College Observatory," 27 February 1948. Box 2 Folder "HCO Council 1946–48" RU 7431 Ser. 1/SIA.

106. Bart Bok Memorandum.

107. Harvard ranked third in the production of collegiate degrees in 1943, behind Michigan and California. It was woefully outdistanced in PhD production of starred scientists (Harvard effectively had none) who had received degrees between 1926 and 1943. Visher (1947), tables 4-16 and 5-8. Payne-Gaposchkin was, in fact, starred in 1926, but was listed under Radcliffe. Visher (1947), 116–117. DeVorkin (2006b), 12.

108. Hershberg (1993), 618. James Hershberg, Conant's biographer, has argued that Conant's actions were prompted by political expediency arising from his primary motivation to protect Harvard, rather than by adherence to moral or philosophical considerations relating to Communist sympathies Shapley was alleged to harbor. Shapley, however, was among only a few Harvard faculty with leftist leanings who, owing to their high visibility, as Jonathan Beckwith has observed, survived the 1930s. See Beckwith, "The Science of Racism," in Trumpbour (1989), 242. Also see Trumpbour (1989), 383–389, on student reaction to Marxist leftists on campus.

109. DeVorkin (1992, 2000b).

110. Baker OHI (1980).

111. He was then 67 years old, which was consistent with other Harvard faculty facing the mandatory retirement age. Desmond to Conant. 21 June 1952. Box 128. Oppenheimer Papers. LC.

112. Shapley to Buck, n.d. Dean's Papers (Buck) UAIII 5.55.26. HUA. This normalization was consistent with Conant's goal of instilling a meritocratic dynamic in faculty hiring and retention. Keller and Keller (2001), 66.

113. Harvard College Observatory Council (HCOC) to Buck, 5 February 1952; Shapley to Buck, 14 February 1952. Dean's Papers (Buck) UAIII 5.55.26. HUA.

114. On their contact, see Geiger (1993), 35–37; and Westwick (2003), 30–31. Oppenheimer also took Conant's chemistry courses in the 1920s. See Cassidy (2005), 76.

115. Oppenheimer to Buck, 7 April 1952. Box 128. Oppenheimer Papers. LC.

116. Conant to Oppenheimer, 17 April 1952. Dean's Papers (Buck) UAIII 5.55.26. HUA.

117. Buck to Oppenheimer, 11 March 1952. Box 128. Oppenheimer Papers. LC.

118. Strömgren, written notes to Oppenheimer, n.d., ca. 18 June 1952. Box 128. Oppenheimer Papers. LC.

119. Strömgren, 18 June 1952.

120. "Observatory Committee Report," 18 June 18 1952. Perkins Room, Massachusetts Hall, Harvard University. Box 128. Oppenheimer Papers. LC.

121. Whipple to Buck, 13 February 1952; Whipple to Buck, 6 August 1952. Whipple Papers. HUG 4876. HUA. Although Whipple had been highly successful securing external funding for his meteor studies, the Super-Schmidts were not yet on line and Shapley had just made him chair of the department, which increased pressure.

122. "Observatory Committee Report," 18 June 1952; Shane to Oppenheimer, 3 November 1952. Box 128. Oppenheimer Papers. LC.

123. Which Shapley complained to Henry Norris Russell were "completely deceptive, erroneous, misleading, and out of context!" Shapley to Russell, 3 March 1953. Reel 38. Russell microfilm. PUL/HNR.

124. Conant to Oppenheimer, 28 July 1952. Oppenheimer Papers. Box 128. LC.

125. Paraphrased and quoted statements from Strömgren, n.d. "Observatory Committee Report," 18 June 1952; Draft report "Astronomy at Harvard," 28 October 1952; 11 November 1952; Oppenheimer et al. to "The Corporation," "Draft Report on Astronomy at Harvard," and Conant to "Dear Opie," 7 January 1953. Box 128. Oppenheimer Papers. LC.

126. Conant to "Dear Opie," 7 January 1953. Box 128. Oppenheimer Papers. LC.

127. Conant to "Dear Opie."

128. "I am sorry to bring this matter to you," Shapley closed, "but I have a life-time invested in the fame, happiness and effectiveness of the Harvard Observatory and I feel obliged to do something to protect my investment." Shapley eagerly reported anything he thought would be embarrassing to Menzel. Shapley to Conant, 22 July 1952; Shapley to Paul Buck, 16 July 1952; Worth to Shapley, 9 July 1952. Dean's Papers (Buck) UAIII 5.55.26. HUA. Hoffleit OHI (1979), 61. AIP.

129. Menzel (HCOC) to Observatory Personnel, 24 October 1952; and Menzel (HCOC) to Observatory Personnel: "University has asked HCO to abandon Building B as soon as possible." 24 October 1952. "Donald Menzel 1952–1958." Box 22.5. Hoffleit Papers. MC 529. Schlesinger Library, Radcliffe College.

130. Menzel to Dorrit Hoffleit, 3 November 1952; Menzel to "All Observatory Personnel," 23 January 1953. Hoffleit had already performed an assessment and felt strongly that everything had to be preserved. Hoffleit (2002), 61. Still, Menzel directed Payne-Gaposchkin to organize a major "weeding" and consolidation of the collection to make it fit on two floors of Building A instead of three floors, relieving the third floor for offices. By 24 November, she announced the detailed plan and said it would work. For "weeding and consolidation," see "C.P.G. Reorganization of the Harvard Plate Collection," 24 November 1952. Hoffleit Papers. Box 22.5, "Donald Menzel 1952–1958," MC 529, Schlesinger Library, Radcliffe College.

131. On the concept applied to the night sky, see Schaffer (1998), 189. On the connection to natural development over time, carried into the photographic era, see Nasim (2013), 229–230.

132. Menzel to Observatory Personnel, HCO, 2 February 1953; "Students and junior staff of HCO" (26 signatories) to Dean Buck, 3 February 1953. Box 128. Oppenheimer Papers. LC. On the divestiture of Boyden, see Levy (1994), 43.

133. Menzel to Buck, 8 August 1952. Dean's Papers (Buck) UAIII 5.55.26. HUA.

134. Paul Buck to "Dear Don." 11 March 1953. Dean's Papers (Buck) UAIII 5.55.26. HUA. On the consortium, see Elbers (2015), ch. 7, "Cooperation with Belgium." See also A. H. Jarret, "Boyden Observatory: A Concise History," *Acta Academica* 12 (1979). http://natagri.ufs.ac.za/dl/userfiles/Documents/00001/1087_eng.pdf (accessed 25 May 2012). On the eviction of the AAVSO, see Williams and Saladyga (2011).

135. Buck to Oppenheimer, 16 February 1953; Lawrence Terry to Oppenheimer, 19 February 1953. Box 128. Oppenheimer Papers. LC. Menzel autobiography (1974), 530.

136. Bundy to Oppenheimer, 25 September 1953. Box 128. Oppenheimer Papers. LC.

137. These included Strömgren, Jan Oort, Otto Struve, Martin Schwarzschild, Leo Goldberg and Jesse Greenstein. See DeVorkin "Serving the Muse" in Philip (2006), 11–20. Menzel to Buck, 24 April 1953. Dean's Papers (Buck) UAIII 5.55.26. HUA; Strömgren to Buck, 10 July 1953; Bundy

to Struve, 11 December 1953; Struve to Bundy, 21 December 1953; Bundy to Oppenheimer, 14 December 1953. Box 128. Oppenheimer Papers. LC.

138. Oppenheimer to Buck, 15 April 1953. Folder "Ad Hoc Committee." Box 128. Oppenheimer Papers. LC.

139. Payne-Gaposchkin to Buck, 13 April 1953. Folder "Harvard College Observatory 1952–1953." Dean's Papers UAIII 5.55.26. HUA; Luyten to Hoffleit, 2 June 1984, Hoffleit Papers 22.19 MC 529, Schlesinger Library, Radcliffe College.

140. Menzel autobiography (1974), 530.

141. Menzel, "Interim Report on Observatory Problems," 13 February 1953. Dean's Papers. UAIII 5.55.26. HUA.

142. Keller and Keller (2001), ch. 8, quote from 175.

Chapter 3 – The Creation of the SAO

143. Clymer (1955).

144. Clymer (1955).

145. Clymer (1955).

146. A useful overview is Hacker (2000), 309–313.

147. Doel (1996), 198.

148. One exception found in the Astrophysics Data System, http://www.adsabs.harvard.edu/, was the Department of Theoretical Astronomy and Astronomical Geophysics at the Jagellonian University in Cracow, created in 1960. Searching and allowing for synonyms reveals that astronomy and geophysics were first associated somehow in the late 1940s, though rarely, with a small jump in the 1960s, a spike in the mid-1980s, and a significant jump after 2000.

149. For a general history of the APO, see Jones (1965), 13–14, 70. See also Portolano (2000), 480–503, who clarifies Adams's motives for supporting astronomy.

150. Jones (1965), 81.

151. Langley (1888). See also "Langley's Researches on Solar Heat" in "Notes," *The Observatory* (1885): 309–310. There were at least several definitions of the term in the 1880s. See Butler (1993).

152. Butler (1993), 141. Langley to Baird, 27 January 1886 [*sic*; actual year 1887], as cited in Jones (1965), 102, ref. 17, 319.

153. Hellman (1978), 117–121.

154. Langley to W. C. Endicott, Secy. of War, 18 December 1888, in Jones (1965), 119.

155. These included A. A. Michelson, Edward Morley (in a separate venture unrelated to optics), Wolcott Gibbs, and astronomer Edward Holden. Jones (1965), 128.

156. Abbot (1947), 1. This section and chapter partially derive from DeVorkin (1990, 1998).

157. On the balloonsonde work, see DeVorkin (1990).

158. H. H. Clayton had been the Chief Forecaster of Argentina and a colleague of A. Lawrence Rotch, of the Blue Hill Meteorological Observatory outside Boston. Clayton had excitedly written Abbot in 1912 about correlations he had found with Abbot's changes in the solar constant. Abbot and Clayton remained close allies. True (1929), 70.

159. Abbot to Marvin, 27 April 1925. Abbot Papers. SIA.

160. The patronage of individuals and institutions is detailed in Jones (1965), ch. 9.

161. On Langley's campaign for interchangeability, see Langley (1891), 197.

162. For instance, he was elected the first president of the new International Astronomical Union's Commission 10 on Solar Radiation at its inaugural meeting in Rome in 1922, and received endorsements from other commissions. See *Transactions of the International Astronomical Union 1* (London: Imperial College Bookstall, 1922), 28–33.

163. Members included Merriam, Abbot, Isaiah Bowman, and various CIW and Weather Bureau people. The Cycles Conferences encompassed the disciplines of climatology, paleontology, botany, geology, physiology, economics, archaeology, and astronomy. They were held between 1922 and 1933.

164. M. M. Paranjpe and D. Brunt, "The Variations of the Solar Constant and Their Relation to Weather," *Quarterly Journal of the Royal Meteorological Society* 64, no. 276 (1938): 459–476. Paranjpe was an assistant lecturer in physics at the Royal Institution of Science.

165. Sterne (1939), 563.

166. Sterne et al. (1940).

167. Aldrich to W. S. Adams, ca. 30 January 1946, Abbot Papers, SI.

168. Upon his succession Wetmore darkly noted that the Smithsonian had been stagnant for many decades. He dedicated himself to changing that situation. Hellman (1978), 211.

169. Alexander Wetmore, "Report of the Secretary of the Smithsonian Institution for the Year Ended June 30, 1946," in *Annual Report of the Smithsonian Institution, 1946* (Washington, DC: Smithsonian Institution, 1946), 10.

170. Wetmore, "Report," 10–11.

171. Alexander Wetmore, "Report of the Secretary of the Smithsonian Institution for the Year Ended June 30, 1950," *Smithsonian Institution* (1950), 1.

172. "An Act to Provide for the Reorganization of Government Agencies, and for Other Purposes," 20 June 1949, 81st Cong., 1st Sess., 1949, 63 Stat. 203, ch. 226. This act gave the president the powers to periodically review and reorganize the federal government. Agencies could be reorganized, abolished, or consolidated. Deliberations would take the next few years and be transmitted to Congress by 1 April 1953.

173. Keddy to Wetmore, 19 December 1949, citing a report in the file dated 19 February, "Office of General Services." Box 1. "Keddy" files. RU 175. SIA.

174. Keddy to Wetmore, 4 November 1946. Box 1 "Keddy" files. RU 175. SIA.

175. See the folders "APO: Proceedings of the Annual Meeting of the Board of Regents" and "Board of Regents Meeting Jan. 12, 1951." Boxes 15 and 148. RU 50. "Office of the Secretary 1949–1964." SIA. One aspect of the deliberations is presented in Jones (1965), 286–287, who cites the debate as recorded in the *Congressional Record.*

176. Draft of "Survey of Research Interests of the Staff of the Smithsonian Institution," 1 November 1954. Circulated on 27 September 1954 from the Sub Committee on New Work Programs, mainly on how to use the Barro Colorado facility, cooperating with military interests. RU 50. "Office of the Secretary 1949–1964," and RU 175. Box 2 "Research Policy." SIA.

177. "Subcommittee on New Work Programs," and "Program Subcommittee," ca. 9 July 1954, n.d. Box 2. RU 175. SIA. A summary of meetings 24 May through 9 July after a year-long exploratory discussion on the future of research at SI.

178. "Statement by L. B. Aldrich concerning the work of the Astrophysical Observatory," 29 October 1953. "SAO General Corr. A–K 1950–1962." Box 15. RU 50. SIA.

179. D. H. Dunkle to Philip Drucker, 8 July 1953. "Analysis of SI Research Programs 1952–1953" and "Research, New Work Programs." Rough notes, n.d. Box 2 "Research Policy." RU 175. SIA.

180. Emphasis in original. Lomask (1976), 98–101, on 100. https://babel.hathitrust.org/cgi/pt?id=umn.31951d00291901g;view=1up;seq=112.

181. Abbot to Carmichael, 14 August 1953. "Keddy" files. Box 1. RU 175. SIA.

182. Abbot to Carmichael, 29 October 1953. "Statement by L. B. Aldrich concerning the work of the Astrophysical Observatory." "SAO General Corr. A–K 1950–1962." Box 15. RU 50. SIA.

183. Folder "A.P.O. Future." Box 15 "Office of the Secretary 1949–1964." RU 50. SIA.

184. Stanley S. Ballard to Carmichael, 3 November 1953. Folder "A.P.O. Future." Box 15 "Office of the Secretary 1949–1964." RU 50. SIA.

185. Keddy notes 9 November 1953, appended to Aldrich's appeal, "Statement by L. B. Aldrich concerning the work of the Astrophysical Observatory" 29 October 1953. Carmichael memorandum to staff, 7 December 1953. Folder "A.P.O. Future." Box 15. RU 50. "Office of the Secretary 1949–1964." SIA.

186. Roberts, in residence at the HAO, had obtained his PhD from Harvard in 1943 with a study of the dynamics of solar spicule activity and their relationship to subsequent activity in the solar corona, a valuable hint in predicting coronal activity and hence ionospheric disturbances on Earth. See Bassi (2015).

187. Quoted in Carmichael to Roberts, 25 August 1954. RU 50. Box 16 [Carmichael] "APO Correspondence M–R (1951–1964)." "Report on Trip to Boulder, Colorado and the High Altitude Observatory at Climax, Colorado, August 11–13, 1954." Folder "APO Future." Box 15. RU 50. SIA. Doel (1990), 140.

188. Bundy to Ward Darley, President U. Colorado, 29 March 1954; Thomas Knowles to Ward Darley, 4 May 1954. "University of Colorado President's Office Correspondence" files. Archives, University of Colorado at Boulder. Courtesy Joseph P. Bassi. Notes on Carmichael's conference with Shapley, 6 October 1954. Folder "A.P.O. Future." Box 15 "Office of the Secretary 1949–1964." RU 50. SIA. Menzel and Roberts did not share the same views regarding classified research and both had experienced considerable disruption from hurtful and damaging investigations questioning their loyalty to the United States. They also were suspect by factions within the astronomical community debating the form and character of a national optical observatory. See Bassi (2015), 91–97.

189. The discussion here parallels Doel (1990), 137–153, and Bassi (2015).

190. "Committee on New Work Programs" to Scientific Personnel, 27 September 1954. Box 1. RU 175. "Graf/Keddy" files. "Survey of Staff Research Interests" and "Notes on Conference." Folder "A.P.O. Future." Box 15 "Office of the Secretary 1949–1964." RU 50. SIA.

191. Bowen to Bush, 7 October 1954; Leonard Carmichael to Robert R. McMath, 18 October 1954; J. E. Graf to Carmichael, 13 October 1954. Folder "A.P.O. Future." Box 15 "Office of the Secretary 1949–1964." RU 50. SIA. Carmichael wrote only after J. E. Graf initiated a security investigation finding no evidence of subversive connections nor that he was unfit for employment at the Smithsonian. He was found, moreover, to have "something the Institution could use."

192. Robert R. McMath to Carmichael, 12 November 1954. Folder "A.P.O. Future." Box 15 "Office of the Secretary 1949–1964." RU 50. SIA.

193. Edmondson (1997), 29. Quoted from a "memorandum of conversation" between Carmichael and R. R. McMath.

194. Notes on phone call from Menzel; notes from conference with Spitzer, 4 November 1954. Folder "A.P.O. Future." Box 15 "Office of the Secretary 1949–1964." RU 50. SIA. George Field, a Spitzer student at Princeton, was about to complete his thesis on applying plasma physics to stellar atmospheres.

195. Carmichael to Bowen, 21 December 1954. Folder "A.P.O. Future." Box 15 "Office of the Secretary 1949–1964." RU 50. SIA.

196. Buck to Menzel, 14 April 1953. Dean's Papers UAIII 5.55.26. HUA; "Notes on conference between LC and Dr. Richard Nelson Thomas of Harvard," 15 December 1954. Folder "A.P.O. Future." Box 15 "Office of the Secretary 1949–1964." RU 50. SIA. R. N. Thomas to Whipple, 29 December 1954. FLW SIA. Box 2. Folder "Harvard University 1943–1961." Box 2. RU 7431 SIA. Doel (1990), 137–153, has discussed the solar associates, including a telling diagram by Thomas.

197. Bowen to Bush, 7 October 1954. Folder "A.P.O. Future." Box 15 "Office of the Secretary 1949–1964." RU 50. SIA.

198. "Notes on conference between LC and Bowen," 7 December 1954. Folder "A.P.O. Future." Box 15 "Office of the Secretary 1949–1964." RU 50. SIA.

199. Goldberg to Carmichael, 20 December 1954. Folder "A.P.O. Future." Box 15 "Office of the Secretary 1949–1964." RU 50. SIA.

200. Hunsaker to Carmichael, n.d. clipped to 17 December 1954, Carmichael to Hunsaker. Folder "A.P.O. Future." Box 15 "Office of the Secretary 1949–1964." RU 50. SIA.

201. Leonard Carmichael, "Telephone Conversation with Dr. Donald H. Menzel," 20 December 1954. Folder "Whipple 1954–1955." Box 24. RU 50. SIA.

202. Doel (1990), fig. 2. "Notes on conference between LC and Dr. Richard Nelson Thomas of Harvard," 15 December 1954. Folder "A.P.O. Future." Box 15 "Office of the Secretary 1949–1964." RU 50. SIA. R. N. Thomas to Whipple, 29 December 1954. Folder "Harvard University 1943–1961." Box 2. RU 7431. SIA.

203. Roberts to Carmichael, 5 January 1955. Folder "APO General Correspondence." Box 15 "Office of the Secretary 1949–1964." RU 50. SIA. Graf to Carmichael, via Keddy, clipped to Roberts to Carmichael, 5 January 1955. "APO Correspondence M–R (1951–1964)." Box 16. RU 50. SIA.

204. Carmichael response to Roberts's 5 January letter, 27 January 1955. "APO Correspondence M–R (1951–1964)," Box 16 "Office of the Secretary 1949–1964." RU 50. SIA.

205. Roberts to Carmichael, 5 February 1955. "APO Correspondence M–R (1951–1964)." Box 16 "Office of the Secretary 1949–1964." RU 50. SIA.

206. Graf to Carmichael, 16 February 1955. "APO Correspondence M–R (1951–1964)." Box 16 "Office of the Secretary 1949–1964." RU 50. SIA.

207. Menzel to Dean McGeorge Bundy, 19 January; 24 January 1955. Dean's Papers, "Correspondence 1954–1955 Harvard College Observatory," Dean of the FAS. UAIII 5.55.26. HUA.

208. Menzel to Bundy 24 January 1955. Menzel had also been spearheading an effort by East Coast institutions to establish a radio astronomy consortium. The cross-field nature of this research effort naturally promoted multi-institutional collaborations. See Munns (2003), 101, and Munns (2013), 157; Needell (1987), 261–288.

209. Bundy to "Dear Nate" 26 January 1955. Dean's Papers, "Correspondence 1954–1955 Harvard College Observatory," Dean of the FAS. UAIII 5.55.26. HUA.

210. Bundy to "Dear Nate" 26 January 1955. Dean's Papers, "Correspondence 1954–1955 Harvard College Observatory," Dean of the FAS. UAIII 5.55.26. HUA.

211. Whipple OHI (1977), 81.

212. Whipple to Burbidge, 27 March 1964. Folder "B." Box 7. RU 7431. SIA. Despite Whipple's contentions, Harvard was certainly not unfriendly to contracting. During the period 1940 to 1954 they amounted to some 16 percent of Harvard's operations. See Keller and Keller (2001), 179.

213. On Orbiter, see Neufeld (2000), 231–257, and (2007), 304–305.

214. He was acceptable as a research associate. Bundy to Menzel, 3 December 1956. Harvard College Observatory Council (HCOC) 55/56. RU 7431-2. SIA.

215. E. O. Hulburt, director of research at the Naval Research Laboratory, had been trying to bring Whipple to the NRL since 1950. Whipple's indirect methods of utilizing meteor trails for upper air studies were then proving valuable as calibrations of in situ measurements from rockets. At first, Hulburt wanted him to head their Optics Division. Then in June 1955, anticipating retirement, Hulburt asked Whipple to be his successor. Whipple refused, fearing the "politically befuddled atmosphere of Washington." Quotes from Whipple to Hulburt, 16 September 1955. See also Hulburt to Whipple, 10 January 1950; 3 June 1955; 13 July 1955; 28 July 1955; Whipple to Hulburt, 16 January 1950; 26 July 1955. "NRL Correspondence." RU 7431. SIA.

216. Keddy, "for the Record," 11 March 1955. "APO Future." Box 15. RU 50. SIA.

217. Aldrich to Carmichael, 25 October 1954; Carmichael to Aldrich, 2 November 1954; Aldrich to Carmichael, 14 January 1955. "APO General Corr." Box 15. RU 50. SIA.

218. Graf to Carmichael, n.d. "APO General Corr." Box 15. RU 50. SIA. "APO General Corr."

219. Carmichael to Whipple, n.d. "Never Sent." "APO—Harvard College." Box 16. RU 50. SIA.

220. Menzel to Bundy, 16 March 1955. "APO Future." Box 15. RU 50. SIA.

221. Carmichael to Whipple, 9 May 1955; Whipple to Carmichael, 5 May 1955; "Memorandum, Telephone Conversation with Dr. Whipple," 22 May 1955; Carmichael to Menzel, 21 June 1955. Folder "A.P.O. Future." Box 15 "Office of the Secretary 1949–1964." RU 50. SIA.

222. Carmichael to Whipple, 9 May 1955.

223. Carmichael to Franklin Floate, 11 June 1956. APO Harvard Observatory Building 1955–1963. Box 17. RU 50. SIA.

224. "Minutes—HCOC." 11 March 1955. "HCOC 55/56." Box 2. RU 7431. SIA.

225. Whipple to Keddy, undated, clipped to Keddy to Whipple, 18 March 1955, containing a draft of Aldrich to "Staff of the Division of Astrophysical Research, Astrophysical Observatory," and comment dated 25 March 1955. Folder "A.P.O. Future." Box 15 "Office of the Secretary 1949–1964." RU 50. SIA.

226. Sterne to Menzel, 18 June 1955. Folder "A.P.O. Future." Box 15 "Office of the Secretary 1949–1964." RU 50. SIA.

227. Menzel to Carmichael, 28 June 1955. Folder "A.P.O. Future." Box 15 "Office of the Secretary 1949–1964." RU 50. SIA. On Sterne's background, see "Sterne, Theodore." Box 5. RU 7431. Ser. 1. SIA. See also Bergin (2000), 70.

228. Menzel to Bundy, 8 December 1955. HCOC 55/56. Box 2. RU 7431. SIA.

229. Max Krook, "Interstellar Matter and the Solar Constant," in Shapley, ed. (1953).

230. Menzel to Bundy, 8 December 1955. RU 7431. Box 2. HCOC 55/56. Whipple Papers.

231. On Project Lincoln, see Slayton (2012), 255–262. There are numerous parallels in this case study, with the notable exception that SAO was not considered, then or now, a military laboratory explicitly focused on defense research.

232. Menzel to Bundy, 8 December 1955. HCOC 55/56. Box 2. RU 7431. SIA. Hynek is popularly remembered as an advocate for the scientific study of unidentified flying objects. Balanced insights into his efforts can be gleaned from Jacobs (1975) and Dick (1996).

233. Menzel to Bundy, 8 December 1955. Other names Menzel mentioned whom they might be looking at included Guido Műnch of Caltech and Allan Sandage of the Mount Wilson and Palomar observatories.

234. Whitney OHI (2005), 22. Hoffleit (2000).

235. Whitney OHI (2005), 22, 27–28.

236. Keller and Keller (2001), 337.

237. Menzel to Bundy, 8 December 1955. HCOC 55/56. Box 2. RU 7431. SIA.

238. Sterne to Carmichael, 31 January 1956, clipped to Carmichael to Whipple, 8 February 1956. "Carmichael" folder. Box 1. RU 7431. SIA.

239. "HCOC Minutes" folder. HCOC 55/56. Box 2. RU 7431. SIA.

240. Whipple to Bundy, 15 February 1956. "SAO/CfA Historical Information 2 of 2." Box 1. Acc. 04-183. SIA. On the issue of retaining staff, see Keller and Keller (2001), 67–68.

241. Whipple to Bundy, 15 February 1956.

242. Whipple to Bundy, 15 February 1956.

243. Pusey to Bundy, 20 March 1956. HCOC 55/56. Box 2. RU 7431. SIA.

244. Pusey to Bundy, 20 March 1956.

245. The fact that Bart Bok, Harvard's strongest advocate for radio astronomy, had loudly announced that he planned to leave Harvard did not help one bit. Replacing him became a "matter of urgent concern." Pusey to Bundy, 20 March 1956. HCOC 55/56. Box 2. RU 7431. SIA. Bok had options such as becoming "Director of Research" of the new "National Radio Observatory" when it was funded. There was also an offer to be director of the Commonwealth Observatory in Australia. Bok even felt he was at odds with Whipple and Menzel over what type of radio facility was needed. Bok wanted to optimize for 21-cm studies of galactic structure, whereas, as he recalled, Whipple urged meteor-echo work to study the physics of the upper atmosphere. Bok to Carmichael, 4 February 1955; Carmichael to Bok, 17 February 1955. "APO General Correspondence A–K (1950–1962)." Box 15. RU 50. SIA. On Bok's claims, see Bok OHI (May 1978), 38–40, 99, 103–105. AIP. Levy (1993).

246. Menzel and Whipple to Bundy, 6 April 1956. HCOC 55/56. Box 2. RU 7431. SIA.

247. Menzel and Whipple to Bundy, 6 April 1956, 3. One matter Menzel raised illuminates the excruciatingly slow liberalization of the entrenched elitism on college campuses, especially Harvard. Cecilia Payne-Gaposchkin was designated Phillips Astronomer in 1938, but was still not of professorial rank. Struve chided Harvard: "She enjoys a world-wide reputation and clearly deserves this recognition from Harvard." Although Pusey initially turned her case back to Menzel and the HCOC, they fought back with a strong endorsement in April, which ultimately made her the first tenured woman at Harvard to become a full professor. DeVorkin (2000a), 204–205.

248. Menzel and Whipple to Bundy, second letter, 6 April 1956. HCOC 55/56. Box 2. RU 7431. SIA.

249. To make sure this suggestion was dead on arrival, he had his office draft a new policy limiting the nature of any academic relationship for a full-time employee of the Smithsonian Institution. Carmichael to Whipple, 17 April 1956. "Carmichael" folder. Box 1. RU 7431. SIA.

250. Munns (2003), 109.

251. Menzel to Bundy, 20 June 1956. HCOC 55/56. Box 2. RU 7431. SIA. Menon headed the radio group till late 1958. On T. K. Menon's life, see T. K. Menon, "The Discovery of the First HI Shell in the Galaxy," in *Seeing through the Dust*, ed. A. R. Taylor, T. L. Landecker, and A.G. Willis, ASP Conference Series, vol. 276 (San Francisco: Astronomical Society of the Pacific, 2002), 23–26.

252. Struve to Menzel, 9 July 1956; Bowen to Menzel, 18 July 1956. Support for Gold also came from Arthur Kantrowitz to Menzel, 31 July 1956; D. R. Bates to Menzel, 31 July 1956; Harold Spencer Jones to Menzel, 1 August 1956. HCOC 55/56. Box 2. RU 7431. SIA.

253. Menzel to Bundy, 9 November 1956. HCOC 55/56. Box 2. RU 7431. SIA.

254. Vestine to Pusey 17 October 1956; "HCOC Minutes," 8 November 1956, 2. HCOC 55/56. Box 2. RU 7431. SIA.

255. Menzel to Bundy, 16 November 1956; Menzel to Reynolds, 26 October 1956. HCOC 55/56. Box 2. RU 7431. SIA.

256. "HCOC Minutes," 1 March 1956, 1. HCOC 55/56. Box 2. RU 7431. SIA.

257. Velma A. Adams to Harvard College Observatory Council, 18 June 1956. HCOC 55/56. Box 2. RU 7431. SIA. Adams was replaced by Helen Federer as administrative assistant and Rose Murray as executive secretary, who would oversee the growing secretarial and clerical staff, with administrative oversight for the maintenance staff and the "maids." Managing contracts and grants alone required full-time attention but she went on to list dozens of other duties and presumed responsibilities. Memorandum, Velma to "Dr. Whipple," 3 May 1956. HCOC 55/56. Box 2. RU 7431. SIA.

258. "HCOC Minutes," 1 March 1956, 1. HCOC 55/56. Box 2. RU 7431. SIA.

259. Menzel and Whipple both became deeply involved in the deliberations over a national observatory, taking major roles in various aspects of the negotiation leading to its definition and management. See Edmondson (1997), 60–67, and England (1982), 290–292. On Menzel's role in the establishment of the National Radio Astronomy Observatory, see England (1982), 284–287.

260. England (1982), 280–286.

261. Menzel to Putnam, 13 November 1956. HCOC 55/56. Box 2. RU 7431. SIA. On Yerkes-McDonald, see Osterbrock (1997) and Evans and Mulholland (1986). And on the Mount Wilson-Caltech-Palomar arrangement, see Osterbrock (1992); Florence (1994), ch. 29. See McCray (2004a), chs. 1, 2, for an overview of the state of telescope building in the 1950s.

262. Menzel to Putnam, 13 November 1956.

263. Menzel to Putnam, 13 November 1956.

264. Menzel to Putnam, 13 November 1956.

265. This would include Armagh in "Northern Ireland," Dunsink in "Southern Ireland," and observatories in Sweden, Germany (Hamburg), and Belgium. They would manage an old 60-inch telescope as well as its newer and highly innovative Armagh-Dunsink-Harvard (ADH) wide-field Baker-Schmidt reflector. On the ADH, see Lindsay (1953), 140–146. This was the largest expression of James Baker's modification of the original Schmidt design.

266. Menzel to Carmichael, 28 June 1955. Folder "A.P.O. Future." Box 15 "Office of the Secretary 1949–1964." RU 50. SIA.

267. Mother to Dorrit, 20 June 1955. Hoffleit Papers. Schlesinger Library. Radcliffe.

268. On the decision for Vanguard instead of Orbiter, see Green and Lomask (1970), ch. 3.

269. Whipple to Keddy, 26 February 1957. "Organization Charts." Box 19. RU 50. SIA.

270. McCray (2008), ch. 4.

271. Muir-Harmony (2012); Hayes (1967).

272. Burgers (1957).

273. Whipple to Carmichael, 25 May 1956. "Carmichael" folder. Box 1. RU 7431. SIA.

274. It was later designated the third such world symposium on cosmic aerodynamics; the first two, held under the auspices of the International Astronomical Union, occurred in Paris in 1949 and Cambridge, England, in 1953.

275. Fred Whipple, "The Committee on Needs of Astronomy," 3 April 1955. Folder "N." Box 15. HUG 4876. HUA.

276. Whipple, "The Committee," 4.

277. DeVorkin (2000b); Whipple (1956).

278. Quote from https://siarchives.si.edu/history/joseph-henry (accessed 24 March 2017).

279. Whipple (1956), iii–iv.

280. *Toward a Fuller Understanding of the Universe* (Cambridge, MA: Harvard College Observatory, 1956).

Chapter 4 – Whither the Solar Constant: A Referendum on Basic Research

281. Carmichael, draft letter to Whipple, n.d., ca. spring 1955. "Whipple 1954–1955." Box 24. RU 50. SIA.

282. Hufbauer (1991), 154–156.

283. R. N. Thomas to Whipple, 1 March 1955. "SAO/CfA Historical Information 2 of 2." Box 1. Acc. 04-183. SIA.

284. Fred Whipple, "Future Research Plans," 11 November 1955. Copies to "DHM and CHPG." "SAO/CfA Historical Information 2 of 2." Box 1. Acc. 04-183. SIA.

285. Indeed, as John Keddy reported to Leonard Carmichael in the latter part of 1955, SAO and Whipple were "almost wholly absorbed in the Satellite Tracking Program." Keddy to Carmichael, undated. "APO Budget 1955–56." Box 12 "Office of the Secretary 1949–1964." RU 50. SIA.

286. Whipple, "Future Research Plans," 11 November 1955, 3.

287. Whipple to Sterne, 27 September 1955. Folder "Sterne, Theodore." Box 5. RU 7431. Ser. 1. SIA.

288. Struve's remarks, in P. Th. Oosterhoff, ed., *Transactions of the International Astronomical Union IX 1955* (Cambridge: Cambridge University Press, 1957), 14–15.

289. Whipple to Sterne, 27 September 1955. Folder "Sterne, Theodore." Box 5. RU 7431. Ser. 1. SIA.

290. "Board Room SI October 3, 1955." This document was part of a historical file that was deposited in the SI archives well after his primary collection. Evidently, given the nature of these two folders, he made an effort to reserve these documents for a possible history of his involvement in creating the SAO. "SAO/CfA Historical Information 2 of 2." Box 1. Acc. 04-183. SIA.

291. Whipple to Sterne, 13 February 1956. "SAO/CfA Historical Information 2 of 2." Box 1. Acc. 04-183. SIA.

292. Whipple to Sterne, 4.

293. Whipple to Keddy, 26 February 1957. "Organization Charts." Box 19. RU 50. SIA.

294. Newcomb, a legendary name in American mathematical astronomy, had no obvious connection to either the Smithsonian or Harvard, though he trained at the latter under Benjamin Peirce. He was also antithetical to astrophysics. "Astrophysicist Gets 2 Posts," *New York Times*, 20 July 1956, 4.

295. Whipple to Sterne, 21 November 1956. Folder "Sterne, Theodore." Box 5. RU 7431. Ser. 1. SIA.

296. Aldrich to Whipple, 24 June 1955. "Aldrich." Box 1. RU 7431. SIA.

297. Keddy to Carmichael, 28 July 1955. "Phone conversation with Dr. Whipple July 27, 1955 relative to APO matters." "APO Administrative." Box 11. RU 50. SIA.

298. Butler was one of several to write Whipple about the fate of the instrumentation there, particularly a Dobson spectrograph that had been in use since 1928 and was potentially available to anyone who might have a use for it. See *Annals of the Astrophysical Observatory of the Smithsonian Institution*, vol. 5, ch. 1 (1932), 55.

299. Preston Butler to John Graf, 6 January 1957. "APO General Corr." Box 15. RU 50. SIA.

300. On his visit to Table Mountain, see Whipple to Stanley Aldrich, 25 September 1956. "APO General Correspondence A–K." Box 15. RU 50. SIA.

301. Sterne and Dieter (1957), 147.

302. Sterne and Dieter (1959), 9–12.

303. Carmichael to Sterne via Whipple, 16 July 1957. "APO Administration." Box 11. RU 50. SIA.

304. Carmichael to Sterne via Whipple, 2.

305. Carmichael to Sterne via Whipple, 2.

306. See numerous letters between 1954 and the 1960s in "Secretaries—Abbot, Charles Greeley 1945–." Box 192. RU 50. SIA. Abbot also sent a particularly strong endorsement from A. Ångström: Abbot to Whipple, 27 June 1955. "Abbot" folder. Box 1. RU 7431. SIA. Ångström, the retired head of the Swedish Meteorological Service, quipped, "I think scientists are apt to be too skeptical against periods—especially if the periods are not their own."

307. Carmichael to Commander, Air Force Cambridge Research Center, 3 August 1956. "Abbot" folder. Box 1. RU 7431. SIA.

308. Abbot to Carmichael, 27 May 1963. "Secretaries—Abbot, Charles Greeley 1945–." Box 192. RU 50. SIA. Carmichael to Abbot, 3 August 1956. "Abbot" folder. Box 1. RU 7431. SIA.

309. Carmichael to Sterne via Whipple, 16 July 1957. Both Graf's and Keddy's initials at the bottom of the internal copy of this letter indicate that they had a strong hand in drafting it.

310. Keddy to Whipple, 11 July 1955; 5 January 1956; 3 July 1956. "APO Administration." Box 11. RU 50. SIA.

311. Carmichael to Sterne via Whipple, 16 July 1957. "APO Administration." Box 11. RU 50. SIA.

312. "Meeting, Regents Room." 23 August 1957. "August 23, 1957 meeting." Box 19. RU 50. SIA.

313. "Meeting, Regents Room," 4.

314. Sterne to Carmichael through Whipple, 10 September 1957. "Carmichael" folder. Box 1. RU 7431. SIA. "Backwaters" from Whipple to Sterne, 27 September 1955. "Theodore Sterne" folder. Box 5. RU 7431. SIA.

315. Sterne to Carmichael, 10 September 1957.

316. Sterne to Carmichael.

317. Sterne to Carmichael, 5.

318. Sterne to Carmichael, 10.

319. Carmichael to Whipple, 11 October 1957. "Carmichael" folder. Box 1. RU 7 431. SIA.

320. Abbot (1958b), quote on 21.

321. Carmichael to Whipple, 14 October 1957. "Carmichael" folder. Box 1. RU 7431. SIA, copy in RU 50. Box 11. "APO" folder, with copies to Eisenhart and Keddy, noting that Graf and Keddy prepared the draft.

322. Carmichael to Whipple, 14 October 1957. Emphasis in original.

323. Carmichael to Whipple, 14 October 1957.

324. Sterne to Whipple, 31 October 1957. "Carmichael" folder. Box 1. RU 7431. SIA.

325. In 1962, the Smithsonian transferred its U.S. Forest Service lease for Table Mountain to NASA's Jet Propulsion Laboratory. See http://tmf-web.jpl.nasa.gov/index.html and http://tmoa.jpl.nasa.gov/History2.htm (accessed 17 March 2012). For a sense of the descendants, see Klett (1992), 119–120.

326. Carmichael to Whipple, 8 March 1961. "Carmichael" folder. Box 1. RU 7431. SIA.

327. Carmichael to Whipple, 8 March 1961; Whipple to Carmichael, 9 March 1961. "Carmichael" folder. Box 1. RU 7431. SIA.

328. Hufbauer (1991) offers the most comprehensive review and analysis, 165–176, and especially ch. 7, "The Inconstant Sun." See also Hoyt (1979) for a discussion of the failure of the program and its importance for climate change studies, and Newkirk (1983), 429–467, on 446.

329. Foukal et al. (1977), 954–955. But there was some substance in the long record of Smithsonian data. Discrepancies between the Mariner 6 and Mariner 7 data stimulated a retracing that confirmed that there was a modulation in solar irradiance at Earth's surface caused by the presence of active regions on the Sun. The modulation was neither direct nor visible from Earth. Newkirk (1983), 446.

330. Dupree, "Paths to the Sixties," in Arm (1965), 3.

331. In conformance with Peter Galison's view of how experiments end, "as [the] progressively refined articulation of a set of phenomena." Galison (1987), 127.

Chapter 5 – Optical Satellite Tracking: Establishing the Smithsonian's Central Role

332. Cornell OHI (2010), 188–189, reading excerpts from working notes prepared for Irwin Shapiro's remarks in an appreciation of Whipple's contributions.

333. Thomas (1963); Hayes (1967); Green and Lomask (1970); McCray (2008); Muir-Harmony (2009). For a technical treatment: Veis (1963).

334. TPESP minutes, meetings 4 and 5, cited in Green and Lomask (1970), 149.

335. Leland Cunningham to Whipple, 30 July 1955. Folder "C-D-E-F." Box 14. Whipple Papers. HUG 4876. HUA.

336. Needell (2000), 336–340, on 339.

337. Whipple (1992), 132. On the committee and its decisions, see Michael Neufeld (2000), 240; Needell (2000), 343–345; Green and Lomask (1970), 97–98. See also Hayes (1962), 282.

338. Bergaust and Beller (1956), 36; Needell (2000), 344.

339. "Minutes of the Special Meeting LPR Committee Technical Panel on Rocketry USNC for the IGY," reprinted in Logsdon et al (1995), 299–315.

340. Green and Lomask (1970), 146, citing Corliss (1967), chs. 2, 14.

341. Henry B. Allen (Franklin Institute) to Whipple, 12 October 1955. Folder "C-D-E-F." Box 14. Whipple Papers. HUG 4876. HUA.

342. Whipple to Carmichael, 9 November 1955, enclosing a copy of a paper he was to present at the American Astronomical Society in Troy on Nov. 12. Folder "Whipple, Fred L. (APO) 1954–1955." Box 24. RU 50. SIA.

343. Quotes from Whipple to Hulburt, 16 September 1955. See also Hulburt to Whipple, 10 January 1950; 3 June 1955; 13 July 1955; 28 July 1955; Whipple to Hulburt, 16 January 1950; 26 July 1955. Whipple Papers, NRL Correspondence. HUG 4876.808. HUA.

344. Singer OHI (1991), 30–32. Whipple did not meet the same fate as Fred Singer, who, in the wake of the decision for Vanguard, was effectively frozen out of the continued satellite instrumentation deliberations, even though he made entreaties. On his proposals to the NSA, see Singer, via J. Crone, to Director, NSA, 14 September 1955. "Ranger Box 110.7." "Maryland 55-59" folder. Singer Papers, NASM.

345. "Outline of Conference in Los Angeles 16 July 1955 on Proposed Artificial Satellite Program—Project Orbiter," notes by Frank Estabrook, OOR western branch, 16 July 1955. Office of Ordnance Research. Courtesy M. Neufeld. See Neufeld (2000).

346. Tombaugh, in fact, had been no stranger to competition; at White Sands, he faced the same sort of skepticism from radio technicians and traditional tracking specialists, but as he and others well knew, a combination of as many techniques as possible, limited only by budget, was the right course to take. See DeVorkin (1992), 112–113.

347. By the mid-1950s, he and his staff took some 15,000 photographs at varying tracking rates, and believed they were picking up some evidence for near-Earth bodies. Roemer (1971), 643. See also Tombaugh et al. (1959), 110.

348. "Outline of Conference," Estabrook notes.

349. E. P. Martz Jr. to Gerhard Heller, 22; 24 August 1954. Von Braun Papers, Huntsville, courtesy M. Neufeld. See also document 11-7, Wernher von Braun, "A Minimum Satellite Vehicle: Based on components available from missile developments of the Army Ordnance Corps," 15 September 1954. In Logsdon, et al, eds. (1995), 274–281, on 280.

350. Joseph Kaplan to "Dear Dr. . . . ," 2 October 1955, reprinted in Logsdon (2001), 57–58, contained within USNC, "Minutes," First Meeting TPESP Technical Panel on the Earth Satellite Program.

351. As derived from general comments in McCrosky OHI (2005).

352. Rinehart (1999), 146.

353. Rinehart (1999), 119.

354. Whipple to Carmichael, 23 September 1955. "Carmichael" folder. Box 1. RU 7431. SIA.

355. Whipple to Rinehart, 11 July 1955. Folder "R." Box 16. Whipple Papers. HUG 4876. HUA.

356. Whipple to Frederick J. Maxwell, 7 December 1955. Folder "M—Meteor Project." Box 15. Whipple Papers. HUG 4876. HUA.

357. Generally regarded as terminal ballistics, which included the study of high-speed penetration of gaseous and metallic media, the nature of resultant impact cratering, mechanisms of spalling and fracturing, and means to prevent these erosive processes.

358. Memo from Keddy to Whipple and Rinehart, 4 June 1956. "APO Air Force Cooperation File 1956–1959." Box 12. RU 50. SIA.

359. Keddy memo, 4 June 1956.

360. Whipple to Carmichael, 25 July 1956. "Carmichael" folder. Box 1. RU 7431. SIA.

361. E. P. Henderson to Keddy. "Meteorites and Ballistics," 20 March 1958. "APO General Correspondence M–R 1951–1964." Box 16. RU 50. SIA.

362. Henderson to Rinehart, 25 April 1958, with comment by Cooper dated 28 April, clipped to Rinehart to Henderson, 14 May 1958. "Henderson, E. P." Box 13. RU 188. SIA.

363. Rinehart (1999), 148.

364. http://www.nmt.edu/new-giving/endowments-and-funds/3337-dr-john-s-rinehart-memorial-endowment.

365. Henderson felt considerable pressure to build his collection and maintain the authority of the National Museum and was annoyed with Whipple for being critical of H. H. Nininger. Roy S. Clarke Jr., Howard Plotkin, and Timothy J. McCoy, "Meteorites and the Smithsonian Institution," in McCall et al., eds. (2006), 250–251. On contentions between factions in meteoritics, see Plotkin (1997), 32–38.

366. Clarke et al., in McCall et al., eds. (2006), 250–251.

367. Rinehart (1999), 150. See also Marvin (1992).

368. Whipple and Spitzer, "Tentative Optical Observing Program for Earth Satellites, undated report (November 1955)." Folder "Earth Satellite Program, 1955." Box 13. RU 50. "Office of the Secretary 1949–1964." SIA.

369. "Tentative Optical Observing Program."

370. "Tentative." See McCray (2008), 67–68.

371. "Tentative," 5–6.

372. "Tentative," 7.

373. "Tentative," 6–9. Whipple and Spitzer also briefly considered, but rejected, an old suggestion by Ernst Öpik to use a fixed camera with periodic oscillatory motion.

374. "Tentative," 9–10.

375. "Proposal for the Initiation of an Optical Tracking and Scientific Analysis Program for the U.S. Earth Satellite Program," 26 November 1955. Folder "Earth Satellite Program, 1955." Box 13. RU 50. "Office of the Secretary 1949–1964." SIA.

376. Whipple to Carmichael, with cover Keddy to Carmichael, 26 November 1955. Folder "Earth Satellite Program, 1955." Box 13. RU 50. "Office of the Secretary 1949–1964." SIA.

377. *Annual Report of the Board of Regents of the Smithsonian Institution*, Publication 4272 (Washington, DC: U.S. Government Printing Office, 1957), 69.

378. Letter from Hugh Odishaw to USNC Executive Committee, 30 November 1955. "6.11 Earth Satellite Program, Project 30.3." "Smithsonian Inst: Initial Dev. of Optical Tracking System, 1955–1959, IGY/NAS." Cited in Muir-Harmony (2009), 19, ref. 31.

379. Whipple, "Proposal for the Initiation of an Optical Tracking and Scientific Analysis Program for the U.S. Earth Satellite Program," 29 November 1955. Two short descriptions of these functions after they were later assumed by the SAO are Gingerich (1968) and Marsden (1988).

380. "Earth Satellite Program, 6.11 Project 30.3: Smithsonian Inst: Initial Dev of Optical Tracking System, 1955–1959." Odishaw to Executive Committee, 30 November 1955, "Proposal for the Initiation of an Optical Tracking and Scientific Analysis Program for the U.S. Earth Satellite Program." Quotation from appended letter, Kaplan to Waterman requesting grant for the STP, 5 December 1955. NAS Records courtesy Muir-Harmony.

381. Green and Lomask (1970), 146, citing Corliss et al. (1967), 14.

382. Charles A. Whitney and George Veis, "A Flashing Satellite for Geodetic Studies," *SAO Special Report* no. 19 (1958): 9–20. Acknowledges discussions with Davis, H. E. Edgerton, Jacchia, McCrosky, and Whipple.

383. Davis OHI (2005), 26–28. NASM/AIP.

384. On Stanton, see Hon. Ralph M. Hall, "Tribute to Inventor Austin Stanton," *Congressional Record* 141, no. 40 (3 March 1995). Extensions of Remarks, p. E508. http://www.gpo.gov.

385. Whipple, "Proposal for the Initiation of an Optical Tracking Network," 29 November 1955, 3, 6, attached to Carmichael to Odishaw, 29 November 1955.

386. Alan T. Waterman to Leonard Carmichael, 14 December 1955. "Earth Satellite Program." Box 13. RU 50. SIA.

387. Whipple to Hynek, telegram, 11:45 PM, 30 December 1955. "Hynek, J. Allen." Box 2. RU 7431. Ser. 1. SIA.

388. Keddy to Whipple, 30 December 1955. "Keddy" folder. Box 3. RU 7431. SIA.

389. Keddy to Whipple, 28 December 1955. "Keddy" folder. Box 3. RU 7431. SIA.

390. Keddy to Whipple, 28 December 1955. Whipple and his staff would soon come to know the pleasures of dealing with standard forms such as SF 52, SF 57, the SI form 4 ("white") for purchases from federal appropriations, or the "yellow" form 8 for other purchases, and the importance of always using the proper billing terms for his many projects, whether they be "APO-NSF Grant" for the $49,000 to plan the satellite-tracking facilities, or the designation "APO S&E" for actual federally appropriated activities.

391. Rinehart to Keddy, 3 January 1956; Keddy to Whipple, 5 January 1956. "APO General Correspondence M–R (1951–1968)." Box 16. RU 50. SIA.

392. Whipple to Hynek, 16 December 1955. Folder "Hynek, J. Allen." Box 2. RU 7431. Ser. 1. SIA.

393. Hynek (1951).

394. Whipple to Hynek, 30 December 1955. Folder "Hynek, J. Allen." Box 2. RU 7431. Ser. 1. SIA.

395. Whipple, undated notes, "STP" folder. Acc. 04-183. Box 4. SIA. Whipple to Baum, 13 February 1956. "Baum, William A., 1956–59." Box 1. RU 7431. SIA.

396. Keddy to Whipple, 16 December 1955; 23 December 1955. "Keddy" folder. Box 3. RU 7431. SIA.

397. Carroll (1951), 122–123; McCrosky (1955). On McCrosky's relationship with Whipple: McCrosky OHI (2005), 84.

398. Undated telegram clipped to Whipple to Otto Struve, 28 February 1956. Folder "Cunningham, Leland." Box 1. RU 7431. Ser. 1. SIA.

399. Whipple to von Braun, 9 January 1956. Folder "Von Braun, Wernher." Box 5. RU 7431. Ser. 1. SIA.

400. Von Braun to Whipple, 4 February 1956. Folder "Von Braun, Wernher." Box 5. RU 7431. Ser. 1. SIA.

401. Vanguard Staff (1956), 46–47.

402. Whipple to Keddy, 7 February 1956. Folder "Earth Satellite Program, 1955." Box 13. RU 50. "Office of the Secretary 1949–1964." SIA.

403. Whipple to Keddy, 7 February 1956.

404. Whipple to Alan Waterman, 8 February 1956. Folder "National Science Foundation." Box 4. RU 7431. Ser. 1. SIA.

405. Whipple to Hynek, 26 March 1956. Folder "Hynek, J. Allen." Box 2. RU 7431. Ser. 1. SIA.

406. Keddy to Whipple, 9 March 1956; 14 March 1956. Keddy, "Note for the Record." Folder "Earth Satellite Program, 1956." Box 13. RU 50. "Office of the Secretary 1949–1964." SIA.

407. Whipple to Graf, 30 March 1956. Folder "Graf." Box 2. RU 7431. SIA. Whipple's diffidence toward Keddy is evident in his oral histories: Whipple OHI (1976), 11, and OHI (1977), 112.

408. Whipple to Graf, 30 March 1956. Folder "Graf." Box 2. RU 7431. SIA.

409. Whipple, "Proposal." ca. April 1956. "STP" folder. Box 4. Acc. 04-183. SIA.

410. "Proposal."

411. Green and Lomask (1970), 80–81, 89–90.

412. Green and Lomask (1970), 90.

413. Whipple, "The Satellite Program," 12 November 1955. Text of a talk given at meeting of AAS in Troy, New York. Folder "Whipple, Fred, 1954–55." Box 24. RU 50. "Office of the Secretary 1949–1964." SIA. See also Whipple and Spitzer, "Tentative Optical Observing Program for Earth Satellites."

414. Needell (2000), 336. See also Muir-Harmony (2009), 33–35.

415. "Winner: SPOT—Smithsonian Precision Optical Tracking," "Norris, Kathryn" folder. Box 4. RU 7431. Ser. 1. SIA.

Chapter 6 – Satellite Tracking: Building and Operating the Program

416. Both the CIA and the Office of Defense Mobilization evidently intended to fund additional Vanguards beyond the IGY period. See Needell (2000), 345.

417. Baker OHI (1980), 152–153.

418. Baker OHI (1980), 153.

419. Hayes (1962), 287.

420. Hayes (1962), 293–296.

421. Hayes (1962), 22.

422. Carmichael to Whipple, 8 August 1957, re: "Immediate Crash program in Satellite Tracking Program Until First Camera is Approved." Folder "Earth Satellite Program, 1957." Box 13. RU 50. "Office of the Secretary 1949–1964." Eisenhart memorandum (29 August 1957) for the file: placed in "Extension of Statement for the Record made August 21 1957 relative to proposed increase in contracts with Nunn, Boller and Chivens, Incorporated, and with Perkin-Corporation." "STP" folder. Box 25. RU 522. SIA. Green and Lomask (1970) point out, without explicit documentation (ch. 9), that R. W. Porter had also visited Perkin-Elmer during this time, found that they had grossly underestimated the magnitude of the job, and so directed the Smithsonian to renegotiate the contract to allow Perkin-Elmer to at least break even. The Academy through NSF provided the funding.

423. "Minutes of administrative meeting based upon remarks by Whipple, Carmichael, Keddy, Eisenhart, 23 August 1957." Folder "August 23, 1957 Meeting." Box 19. RU 50. "Office of the Secretary 1949–1964." SIA.

424. "Minutes of administrative meeting."

425. In a comprehensive technical review of optical satellite tracking, Veis (1963), 284–285, identifies only the Baker-Nunn in the three-axis category.

426. Strömgren (October 1956), 354–355. Clyde Chivens and Harry Boller, both Caltech graduates, built specialty instruments for many uses, from motion picture editing systems to deep well diagnostic instruments. Chivens was well known to Mount Wilson astronomers and to the physicist Ira Bowen, who engaged him in several Office of Scientific Research and Development contracts. Bowen OHI (1969), sess. I. AIP. On Boller and Chivens, see http://bollerandchivens.com/?p=42 (accessed 10 July 2012).

427. Eisenhart to Keddy, 28 June 1956. Folder "Earth Satellite Program, 1956." Box 13. RU 50. "Office of the Secretary,1949–1964." SIA.

428. Eisenhart to Keddy, 28 June 1956. Hynek to Eisenhart, August 1956. Rough-draft statement by Hynek on "the history of our relations with Joseph Nunn and Associates, Mechanical Engineers, of Los Angeles, California," 3, 4–5, 7. Folder "Eisenhart 1956–1957." Box 1. RU 7431. SIA.

429. In contrast to Hynek's own testimony to Eisenhart, Hayes asserted that Whipple and Hynek "had wide contacts throughout the rather specialized field of optical instrumentation" and were convinced they had made the best choice. Hayes (1962), 298–299. Reference deleted in Hayes (1967), 25.

430. Muir-Harmony thesis (2009), 33–34. "December 3, 1956 minutes of the Tracking and Computation Working Group," Folder "4th meeting of Tracking and Computation," series 4.11, IGY/NAS. At this December meeting, Whipple also had to defend his design for the Moonwatch network of visual observers. The Naval Research Laboratory (NRL), long a critic of this plan, along with skeptical National Academy officials, had argued that professionals had to be involved, and the NRL even proposed its own visual tracking network. As Patrick McCray has shown, Whipple successfully resisted these claims, and the panel turned down the NRL proposal. See McCray (2006, 2008).

431. Muir-Harmony (2009), 35–36, 38–41.

432. Muir-Harmony (2009), 49. On the delays, see 60–62.

433. Hayes (1968), 25. Hayes, however, did not emphasize the tensions that existed through the summer of 1957 between the Smithsonian, National Academy, industry, and SAO participants over the delays.

434. Hayes (1962), 320–321. Berkner's announcement and his recollection of it are from Needell (2000), 346–347.

435. McCray (2008), 142; Green and Lomask (1970), ch. 9. See also "Minutes of the 12th Meeting of the TPESP," 3 October 1957. NAS/IGY series 4.10."12th Meeting." Courtesy Muir-Harmony.

436. Needell (2000), 346–349; McCray (2008), 141; Newell (1980), 55–57.

437. K. Auchincloss and A. Friendly Jr., "Russians Launch Artificial Satellite: Scientists Here Direct Spotting," *Harvard Crimson,* 5 October 1957. http://www.thecrimson.com/article/1957/10/5/russians-launch-artificial-satellite-prussias-announcement/ (accessed 15 July 2012). Carmichael more than once reminded Whipple that public pronouncements had to be passed by him. See Carmichael to Whipple, 9 August 1955. Folder "Earth Satellite Program." Box 13. RU 50. SIA. Whipple to Carmichael, 16 October 1956. Folder "Carmichael, Leonard." Box 1. RU 7431. Ser. 1. Box 1. SIA.

438. Hayes (1967), 50–51.

439. A detailed accounting is available in Hayes (1967), ch. 2. See also McCray (2008), ch. 5.

440. "Minutes of the Fourteenth Meeting," reprinted in Logsdon, ed. (2001), 85.

441. "Minutes," 86.

442. Hynek to Chief of Ordnance, 2 December 1957. Folder "Small Missile Telecameras." Box 130. RU 188. SIA.

443. See Muir-Harmony (2009), 63, on jurisdictional issues at the Iran site.

444. Porter to Whipple, 22 November 1957. "General Correspondence M–R." Box 16. RU 50. SIA. Whipple did hire an assistant director for management in 1959, but he was vaguely positioned as "administrative officer" with poorly defined all-purpose duties.

445. E. Stuart Ferguson to Whipple, 11 December 1957. Resignation as executive officer of STP. Ferguson became an engineer for the Harvard Radio Meteor Project, but soon resigned. See Whipple, "Termination," 7 December 1960. Folder "F, general." Box 2. RU 7431. Ser. 1. SIA.

446. Carmichael to Whipple, 16 December 1957. A few days later, in a less stern mood, Carmichael sent Whipple an "English analysis of our problems," sharing the fact that "you and I are both tarred with the brush of research administration." The tongue-in-cheek analysis "A Fable," created by E. S. Hiscocks for the *National Physical Laboratory News,* ended with a moral: "The camel, even when mangy, bears the burden of many asses." Carmichael to Whipple, 23 December 1957, with attachments. Folder "Carmichael." Box 1. RU 7431. SIA.

447. "Minutes of the Fifteenth Meeting, Technical Panel for the Earth Satellite Program January 7, 1958," 2. Folder " F-TPESP" (1 of 4 folders). Box 17. RU 188. SIA.

448. Memo regarding the SAO Public Information Unit and the SI Editorial and Publications Division, 9 January 1948. Folder "APO General Correspondence S–W 1954–1964." Box 16. RU 50. "Office of the Secretary 1949–1964." SIA.

449. Carmichael to John Willard, 6 December 1957. "APO General Correspondence S–W 1954–1964." Box 16. RU 50. "Office of the Secretary 1949–1964." SIA.

450. Carmichael to Willard, 21 January 1958. Folder "APO General Correspondence." Box 15. RU 50. "Office of the Secretary 1949–1964." SIA.

451. Eisenhower telegram, 1 February 1958, congratulating SAO. Folder "Earth Satellite Program, 1958." Box 14. RU 50. "Office of the Secretary 1949–1964." SIA.

452. Paul Hodge to Whipple, 10 February 1958. Folder "H, general." Box 2. RU 7431. Ser. 1. SIA. On Whipple's opinion of Keddy and Odishaw, see Whipple OHI (1976), 2, 11, 25, and Whipple OHI (1977).

453. Whipple to McCrosky, 3 March 1958. Folder "McCrosky, Richard." Box 3. RU 7431. Ser. 1. SIA.

454. Whipple to McCrosky, 13 January 1958. Folder "McCrosky, Richard." Box 3. RU 7431. Ser. 1. SIA.

455. Henize to Morgan Thomas. Folder "Iran Outgoing Correspondence 1958–59." Box 21. STP/SIA. See Muir-Harmony (2009), 62–64, on the difficult initial relations in Iran and India.

456. Baker to Richard S. Perkin, 5 April 1958. Folder "Baker James G." Box 1. RU 7431. Ser. 1. SIA.

457. Whipple to Baker, 9 May 1958. Folder "Baker James G." Box 1. RU 7431. Ser. 1. SIA. On the Air Force system, see Wilkes and Gleditsch (1978), 205–225.

458. On the details, see Hayes (1967), 64–69. And on the variations, see Muir-Harmony (2009). See also various status reports in *SI Astrophysical Observatory Quarterly Progress Report No. 5 for the Establishment of Optical Tracking Stations for the IGY Earth Satellite Program*. Project Director: Whipple. Prepared by: Henize and David Self. NAS/IGY records 6.11 "Earth Satellite Program:" Project 30.6. STP/SIA.

459. Hayes (1967), 40–42; McCray (2008).

460. John J. Barry, VP Shawmut Bank, to Whipple, 9 December 1957. Folder "B General." Box 1. RU 7431. SIA.

461. Sterne (1958), 1245.

462. Jacchia (1959), 335.

463. Jacchia (1958).

464. What eventually appeared in the 16 August issue carried his past affiliations at the end of the article. Carmichael to Whipple, 7 March; 18 March 1958. Folder "Whipple, Fred L. (APO) 1958." Box 24. RU 50. SIA. Whipple to Carmichael, 26 March 1958. Folder "C." RU 7431. SIA. Whipple (1958a), 32–33, reprinted in Whipple (1972a).

465. Maynard and Co., "An Administrative Analysis of the Optical Satellite Tracking Program," 11 April 1958. "Earth Satellite Program, 1958." Box 14. RU 50. SIA. Hereinafter "Willard Report."

466. "Willard Report," 1, 4.

467. "Willard Report," 4–5.

468. "Willard Report," 5.

469. "Willard Report," pt. 2, 2.

470. "Willard Report," pt. 2, 3–4.

471. "Willard Report," 10 and chart D. See also Organization Chart, "Smithsonian Astrophysical Observatory," 25 March 1957, with annotation dated 3 April 1957. Folder "APO Organization." Box 19. RU 50. SIA.

472. Rinehart (1999), 152, recalled he privately felt it was something of a "hodgepodge . . . a fact that often made me uneasy." See also McCrosky OHI (2005), 64–65; Avrett OHI (2005), 28–30; Whitney OHI (2005), 30. Whitney describes Whipple's management style as a network of "invisible threads in a well-built suit" crediting Whipple with the metaphor, 36.

473. "1/3 of 112 people × average of 20 minutes per day × 5 days × 52 weeks = 3235 hours." "Willard Report," pt. 2, 22.

474. "Willard Report," pt. 2, 13.

475. "Willard Report," 14.

476. "Willard Report," 26, emphasis in original.

477. See "NSF" folder. Box 4. RU 7431. SIA.

478. "Statement by the President released to press," 29 July 1958. Folder "National Aeronautics and Space Admin. 1958." Box 258. RU 50. "Office of the Secretary 1949–1964." SIA.

479. Roland (1985). Carmichael to T. Keith Glennan, 20 August 1958. Folder "National Aeronautics and Space Admin. 1958." Box 258. RU 50. "Office of the Secretary 1949–1964." SIA.

480. Hugh Odishaw to Carmichael, 30 December 1958. Folder "Earth Satellite Program, 1959." Box 14. RU 50. "Office of the Secretary 1949–1964." SIA.

481. Hayes (1967), 124–126; McCray (2008), 226–228.

482. Hayes (1967), 110–111.

483. FLW to A. L. Loomis, 9 April 1958. Folder "L, general." Box 3. RU 7431. Ser. 1. SIA.

484. James Bradley to Fred Whipple, 26 January 1959. "STP-NSF Correspondence, 102." Box 1. RU 522. SIA. Announcing Bradley's retirement in *Smithsonian Year 1972*, Ripley added: "His ability to handle the various threads of administration and to support the Secretary's Office in all its aspects has been outstanding and one of my personal pleasures is the fact that he will continue to work for the Institution on a part-time basis in the future." S. Dillon Ripley, "Statement of the Secretary," in *Annual Report of the Smithsonian Institution for the Year Ended 30 June 1972* (Washington, DC: Smithsonian Institution Press, 1972), 14.

485. J. E. Luton (NSF) to Keddy, 9 February 1959. "STP-NSF Correspondence, 102." Box 1. RU 522. SIA.

486. D. L. O'Hara to James C. Bradley, 1 March 1959. "STP-NSF Correspondence, 102." Box 1. RU 522. SIA. Most critically, the Baker-Nunn stations needed to add staff partly to handle the ever-increasing numbers of satellites to track. Keddy to Odishaw, 12 August 1958. Folder "Operation of Optical Tracking Stations: 1956–1963." NAS IGY records 6.1 Earth Satellite Program: Project 30.8. SIA.

487. "Helter Skelter," *The Washington Post and Times Herald*, 3 March 1959, A14. "Critical Statements in the Press," James Bradley to Carmichael. Folder "Earth Satellite Program, 1959." Box 14. RU 50. "Office of the Secretary 1949–1964." SIA.

488. On 12 March, Hynek responded to the question whether "the rivalry among American agencies engaged in space programs delayed the tracking of the Discoverer satellite." SIA. Hynek called it all a "misunderstanding," in that he meant only "as far as the capabilities of the system were concerned, we certainly could have tracked it." *Missile Development and Space Sciences: Hearings before Committee on Science and Astronautics*, 86 Cong., 1st Sess. (1959) (statement of J. Allen Hynek), 480–482. George B. Findley, "Project Space Track. A Method of Predicting Satellite Decay," April 1961, Proj. 1772, Task 17721 ESD TH61 38. Folder "Earth Satellite Program, 1959." Box 14. RU 50. "Office of the Secretary 1949–1964." SIA.

489. Whipple OHI (1976), 17–19.

490. Bradley OHI No. 5 (1975), 119. RU 9515.005. SIA.

491. Fred L Whipple to Carmichael. 9 April 1959. "Admin. 1958-88." Box 1. Acc. 04-183. SIA. Both the original and carbon were found only in Whipple's file.

492. McCrosky OHI (2005), 103. The Mitre connection is from McGrath OHI, 42–43. On Tillinghast, see "Carlton W. Tillinghast, Jr., 1933–1969," *SAO News* 9, no. 9 (September 1969): 1.

493. Porter to Whipple, 22 November 1957. Folder "General Correspondence M–R." Box 16. RU 50. SIA.

494. Whipple OHI (1976).

495. J. Alan Hynek to STP Staff, 3 October 1958. "First Anniversary of Sputnik," "Historical Sputnik." Box 4. Acc. 04-183. SIA. McCray (2008) has explored the ramifications of this letter in some depth, especially Hynek's recognition of Moonwatch contributions.

496. James Bradley, Memorandum," 3 February 1959. "Bradley, James C." Box 1; also folder "Earth Satellite Program, 1959." Box 14. RU 50. "Office of the Secretary 1949–1964." SIA.

497. "Smithsonian Institution Astrophysical Observatory Civilian Personnel," 8 December 1958. "NASA APO Reports." Box 258. RU 50. SIA.

498. Whipple to Bundy, 10 October 1958. Documents: "APO Harvard Observatory Building 1955–1963" and "Space Required by the Smithsonian Astrophysical Observatory." Box 17. RU 50. SIA.

499. Whipple to Bundy, 10 October 1958, 1–2.

500. McCray (2008), 53, 187.

501. Galison, "The Many Faces of Big Science," in Galison and Hevly, eds. (1992), 2.

502. Menzel to Bundy, 10 November 1958. Notes the understaffing of HCO and how the SAO "has come to our rescue" with regard to staff. "HCO Council, 1957–59." Box 2. RU 7431. Ser. 1. SIA.

503. T. Keith Glennan, "Statement," 27 January 1960, 11. "National Aeronautics and Space Administration Press Releases." Box 259. RU 50. SIA.

Chapter 7 – Organizing for Space Research

504. Portions of this chapter are based on DeVorkin (2005), 229–250, and DeVorkin, "The Space Age and Disciplinary Change in Astronomy," in Dick (2010).

505. Davis OHI (2005), www.aip.org/history/ohilist/32148.html (accessed 27 December 2011).

506. "Space Research Meeting," 4 February 1958. Davis Papers, author's files, source: R. Davis.

507. Davis, "Space Research Meeting." See also DeVorkin (2005), 234–237.

508. Davis, "Space Research Meeting."

509. Undated memo from Cecilia Payne-Gaposchkin (1958 undated). "HCO Council, 1957–59." Box 2. RU 7431. Ser. 1. SIA.

510. Davis, "Space Research Meeting," table II.

511. LeCompte (1962); Bushnell (1962). Variants in spelling of Cat Eye are in Gebel and Wylie, (1958), 84.

512. Gebel, at Wright-Patterson since 1948 and searching for solutions to low-light-level problems, by the late 1950s claimed light-level gains on the order of 1 billion working with RCA and Westinghouse contractors and using Image Orthicon imaging components. LeCompte (1962), 10–12.

513. Through 1957, Gebel tested his device using a small telescope at a local observatory at Wittenberg University, Springfield, Ohio, and was confident of success. Gebel and Wylie (1958), 84–85.

514. Simons to Whipple et al., 29 November; 13 December 1957. Folder "Report on Operation Manhigh II." "Correspondence, 1948–1974." Box 1 (2.1.08). Hynek Papers. Northwestern University.

515. R. Aikens et al. (1960), 339. Hynek continued to develop applications for the image orthicon after he moved to Dearborn Observatory.

516. Robert J. Davis, "Ultraviolet Stellar Magnitudes," in Van Allen, ed. (1956), 157–165.

517. Robert J. Davis, SF57 (issued and revised March 1961), "Application for Federal Employment (in the Scientist-as-Astronaut program)." R. J. Davis files, NASM. Biographical information is largely from this source as well as from R. J. Davis OHI (2005).

518. Robert J. Davis, SF57. Continuation Sheet for Form 57.

519. Davis OHI (2005), 22. Davis was attracted to radio astronomy by Bok and Purcell. His thesis was essentially a mapping project to supplement others initiated by his mentors and other Harvard graduate students. Robert James Davis, "A 21-Centimeter Survey" (1959).

520. Davis OHI (2005), 40.

521. Whipple to distribution, 10 June 1958, attaching Whipple, "Notes on the Development and Operation of an Astronomical Telescope in a Satellite Orbit." 5 June 1958. Folder 9. Box 15. Martin Schwarzschild Papers, Princeton University.

522. L. V. Berkner, 3 July 1958, night letter to: USNC/IGY/ESO. "Space Science Board Requests for Support: Proposals 1958," folder 1.3, SSB Papers, National Academy of Science Archives (SSB/NAS). See also Berkner to recipients, 3 July 1958. SSB files, Berkner Papers. LC. On the formation and first years of the Space Science Board, see Atkins (1966); Needell (2000), 356–367.

523. Historians have provided a general framework and insight into how scientists responded to Lloyd Berkner's appeal and the continuing efforts of the President's Science Advisory Committee deliberations since late 1957. See Smith (1989), 37–44; Wang (2008) and Needell (2000). Code's response and his creation of the Wisconsin Experiment Package are analyzed in Marché and Walsh (2006), 185–199.

524. Berkner to Menzel, 2 September 1958. Folder 1.3. Berkner Papers, SSB/NAS.

525. Keller to Whipple, 26 June 1958; Whipple to Keller, 1 July 1958. "Rocket and Satellite Research Panel, 1957–1960." Box 4. RU 7431. SIA.

526. Whipple and Davis, "Contract Proposal: Feasibility Studies of the Development and Operation of an Astronomical Telescope in a Satellite Orbit," attached to Whipple to Keller, 1 July 1958.

527. Whipple and Davis, "Contract Proposal."

528. Whipple, notes "Developments on Space Telescope Project after Discussion with Dr. Pickering," 30 July 1958; 1 August 1958. "G.F.S. files." Box 8. RU 260-8-08/58. SIA. Here Pickering was alluding to "Project Red Socks," which he proposed, with Caltech president Lee DuBridge, barely weeks after Sputnik 1 flew, to send an Explorer-class payload (later called Pioneer) to the Moon, claiming that it was "'imperative' for the nation to 'regain its stature in the eyes of the world.'" Cited in Hall (1977), 5

529. Whipple notes.

530. Whipple notes.

531. Excerpted from various documents, including "Tentative Work Schedule for Feasibility Study" and "Outline of Work," ca. August 1958. "G.F.S. files," 1 August 1958. Box 8. RU 260-8-08/58. SIA.

532. On the properties of this tube, see R. C. Jones (1959), 462–466.

533. Whitney OHI (2005), 18–19.

534. McDougall (1985), 141–176, for a general discussion; Koppes (1982), ch. 7, for JPL; and Westwick (2003), 234–235, for the early option of placing the space program within the Atomic Energy Commission.

535. Whipple to L. Carmichael, 29 March 1958. "Carmichael" folder. Box 1. RU 7341 SIA.

536. Needell (2000), 347.

537. Statement of Dr. Fred L. Whipple, Director, Smithsonian Astrophysical Observatory, Cambridge, Mass. In *Legislative History on H.R. 12575 "The Space Act of 1958,"* NASA Law Library, vol. 3, 367–389, on 368, fully quoted in DeVorkin (2005). Whipple's scenario was similar to one of several options envisioned by presidential science advisor James R. Killian Jr., in his "Memorandum on Organizational Alternatives for Space Research and Development" (30 December 1957), reprinted in David S. F. Portree, ed., *NASA's Origins and the Dawn of the Space Age*, NASA Monograph #10 in Aerospace History (September 1998), 27–33, see 32.

538. "Statement of Dr. Fred L. Whipple." In NASA, ed. (1960), on 369–370, 372.

539. Moran, J. Anthony to Carmichael, 10 September 1958; Carmichael to Whipple, 18 September 1958. Carmichael Papers. Box 1. RU 7007. SIA.

540. Whipple to Carmichael, 26 September 1958. "Carmichael" folder. Box 1. RU 7341. SIA.

541. McCray (2008).

542. Whipple to Carmichael, 26 September 1958. "Carmichael" folder. Box 1. RU 7341. SIA.

543. For the full listing, see DeVorkin (2005).

544. Whipple to Carmichael, 26 September 1958.

545. Roland (1985); McDougall (1985), Neufeld (2000, 2005), and others have described how NASA acquired the NACA and its centers as well as the Vanguard program and chunks of the ABMA and associated centers.

546. Logsdon (2001), 1.

547. Berkner to Menzel, 2 September 1958. Berkner files, courtesy Allan Needell.

548. Berkner to Goldberg, 29 August 1958. Berkner files, with attached list "Proposals for Study by Space Science Board," 1 August 1958. ADM:ORG:NAS:SSB, "Requests for Support: Proposals Summary 1958."

549. Keddy to Odishaw, 12 August 1958. NAS IGY records 6.1 Earth Satellite Program: Project 30.8: Smithsonian Inst.: Operation of Optical Tracking Stations: 1956–1963.

550. Newell to Dryden, 18 February 1960. 6216 Documentation. OAO files. NHO.

551. Newell to Dryden.

552. T. Keith Glennan, *The Birth of NASA: The Diary of T. Keith Glennan*, NASA SP-4105 (Washington, DC: NASA History Office, 1993), 139.

553. Rudney (1971).

554. Rudney (1971), 9. In retrospect, Newell (1980), 356, felt this decision was a mistake.

555. Newell to Dryden. 18 February 1960. 6216 Documentation OAO files. NHO; Schwarzschild OHI (1977), 174–175; Goldberg OHI (1983), 37–38, and Goldberg OHI (1978), no. 2, 119.

556. Newell (1980), 97, 207.

557. Davis to Roths, 3 March 1959. Davis collection, author's files.

558. Davis, "Rough Draft of notes for use at the NASA meeting, 9–10 February 1959," 4. This reflects standard astronomical observing practice. Davis files, author's collection.

559. "Celescope" was, in fact, proposed by Menzel, who won a free lunch from Whipple. Whipple to Smithsonian Staff, 17 December 1959, "Code Name for the Orbiting Telescope Project." The name was "intended to convey the concept of a truly celestial telescope," "Interoffice Memoranda." "Celescope Records." Box 8. RU 260. SIA.

560. Code OHI (1982).

561. Smith (1989), 40; and Code et al. to Nancy Roman, 15 June 1960. Document III-6 in Logsdon (2001), 568–570.

562. Rudney (1971).

563. The launch vehicle was originally envisioned to be NASA's never-built Vega; six Vegas were planned, with some 30 sounding rocket flights as tests. Newell to Silverstein, 16 March 1959. Document III-4 in Logsdon (2001), 561–566.

564. Hynek to Kupperian, 24 February 1959. "Old STP." Box 118. RU 188. On shared interests between SAO and the Air Force regarding scintillation, see "Observation Technique, satellites, missiles, etc. 1959–1960 Jeffery." Box 28. RU 188. SIA. Courtesy T. Muir-Harmony.

565. Marché and Walsh (2006); Roman to Whipple, 15 April 1959; Menzel to Carmichael, 30 November 1959. Folder "APO Harvard College." Box 16. RU 50. "Office of the Secretary 1949–1964." SIA. W. R. Markey's final report (1960) on the balloon program is in file "Air Force 19(604)5459," Box 32, RU 188; and in Box 28, "Observation Technique, satellites, missiles, etc. 1959–1960 Jeffery." Courtesy T. Muir-Harmony. NASA also refused to fund the SAO's proposals to test its instruments with inexpensive sounding rocket flights from Wallops. Keddy to Glennan, 17 September 1959. "Whipple." Box 6. RU 260. 6-07/59, SIA.

566. Davis to Whipple, 8 June 1959. "Celescope" files. Box 1. RU 260-1-07/58. SIA.

567. Whipple to Baum, 19 February 1959. "Baum, William A., 1956–59." Box 1. RU 7431. SIA.

568. See, for instance, Tuve et al. (1958), 592, and Baum et al. (1959), 749. Baum OHI (1997), 5. AIP.

569. Whipple and Davis (1960).

570. Davis to Whipple, 10 June 1960. "Celescope" files. Box 1. RU 260-1-07/58. SIA.

571. Lautman to Whipple, 19 February 1960. "Report on the first international space science symposium of COSPAR." "Whipple-Whitney Correspondence 1959–1962." Box 124. RU 188. SIA. Courtesy T. Muir-Harmony.

572. Menzel to Carmichael, 30 November 1959. Folder "APO Harvard College." Box 16. RU 50. "Office of the Secretary 1949–1964." SIA.

573. Raymond Wilcove to Keddy, 3 November 1959, attached to Bradley to Whipple, 5 November 1959, asking for a response to the House Committee on Science and Astronautics. "Bradley, James." Box 1. RU 7431. Ser. 1. SIA.

574. Personal experience and MacKenzie (1993).

Chapter 8 – Permanency and Authority

575. See "Contracts SAO '54–'66." Box 37. RU 188. SIA.

576. "Moved Again!" *SAO News* 1, no. 1 (June 1961): 1.

577. "Great Day for Space!" *SAO News* 1, no. 1 (June 1961): 4.

578. Haramundanis OHI (2005), 37.

579. Haramundanis OHI (2005), 64–75.

580. Whipple to Bundy, 10 October 1958. "Space Required by the Smithsonian Astrophysical Observatory," in "APO Harvard Observatory Building 1955–1963." Box 17. RU 50. SIA.

581. Whipple to Bundy, 2–3.

582. Vaughan Harmon to Eisenhart, 9 November 1958. "APO Harvard Observatory Building, 1955–1963." Box 17. RU 50. SIA. See also John R. Adler, "University to Build at HCO, Smithsonian Will Rent Observatory Project," *Harvard Crimson*, 5 February 1959. http://www.thecrimson.com/article/1959/2/5/university-to-build-at-hco-pcontracts/.

583. Whipple to Bundy, 10 October 1958, 4. "Space Required by the Smithsonian Astrophysical Observatory," in "APO Harvard Observatory Building 1955–1963." Box 17. RU 50. SIA. Carmichael to Bundy, 13 October 1958. "APO Harvard Observatory Building 1955–1963. Box 17. RU 50. SIA. Box 17.

584. Menzel to Carmichael, 30 November 1959. Folder "APO Harvard College." Box 16. RU 50. "Office of the Secretary 1949–1964." SIA.

585. John R. Adler, "University to Build at HCO." http://www.thecrimson.com/article/1959/2/5/university-to-build-at-hco-pcontracts/.

586. Whipple, "Remarks by Fred L. Whipple at the Ground-Breaking Ceremony of the New Building January 11, 1960," recd. 25 January 1960. "APO Harvard Observatory Building, 1955–1963." Box 17. RU 50. SIA.

587. Whipple, "Remarks," 2; Menzel, "Remarks by Donald H. Menzel at the Ground-Breaking Ceremony of the New Building January 11, 1960," recd. 25 January 1960. "APO Harvard Observatory Building, 1955–1963." Box 17. RU 50. SIA.

588. "Great Day for Space!" *SAO News* 1, no. 1 (June 1961): 3.

589. Davis to Nancy Roman, 22 August 1960. "Celescope Project, 1958–61." Box 1. RU 260-1-07/58. SIA.

590. Historians such as Robert Smith, Michael Dennis, and Bruce Hevly have long pondered what it means to engage in Big Science. Smith quotes a 1988 Office of Technology Assessment report that identified these factors: 1. "Large and expensive facilities"; 2. "large, multidisciplinary teams"; 3. "cooperative planning"; 4. "alignment of personal goals of scientists with institutional goals." Robert Smith, "The Biggest Kind of Big Science: Astronomers and the Space Telescope," in Galison and Hevly, eds. (1992), 186–187.

591. Whitney OHI (2005), 37. See also Avrett OHI (2005). On the "Research Reserve," see Menzel to Solberg, 5 May 1949, "Navy 1941–1949." D. H. Menzel Papers. HUG 4567.12. HUA.

592. Y. Kozai OHI (1997). Kozai (1959) developed a sophisticated gravitational model that included both atmospheric drag and varying geopotential. See Brouwer (1959).

593. "Minutes of Administrative Staff Meeting," 1 April 1963. "Admin. Conf. + Meetings." Box 7. RU 188. SIA. Courtesy T. Muir-Harmony.

594. Powers to Ripley, 18 November 1966, "Summary History of Smithsonian Research Bill." Folder "Smithsonian Research Legislation 1965–1966." Box 7. RU 137. SIA. What actually constituted a "mistake," in Powers's words, has not been explored.

595. Draft letter, Ripley to Haworth, 13 November 1967. "NSF General." Box 14. RU 108. SIA.

596. Excerpt of a quotation from James Hobbins, in Irvin Molotsky, "S. Dillon Ripley Dies at 87; Led the Smithsonian Institution during Its Greatest Growth," *New York Times*, 13 March 2001.

597. Draft letter, Ripley to Haworth, 13 November 1967.

598. Draft letter, Ripley to Haworth. Securing direct appropriations for research funding remained tricky. See Powers to Bradley, 21 August 1964; Ripley to Haworth, 26 February 1965. "NSF 1972." Box 14. RU 108. SIA.

599. Draft letter, Ripley to Haworth.

600. Gary Bisson (Asst. Gen. Counsel) to Bradley, 25 August 1964, through Powers to Whipple, "For information you requested" and "Statutory Authority for programs of the Smithsonian Astrophysical Observatory." Cites: "(63 Stat 623; 20 U.S.C. 53a), sec. 2." "Smithsonian Research Legislation 1965–1966." Box 7. RU 137. SIA.

601. Gary Bisson, 25 August 1964, citing Hon. Ken Regan (House) to John E. Graf, Acting Secy., SI, 4 April 1949. Several years later, Powers interpreted the 1949 wording more narrowly, that it "provides appropriations, but not basic authority, and by now sounds almost like a limitation of such authority." Powers to Ripley, 18 November 1966, 2. "Smithsonian Research Legislation 1965–1966." Box 7. RU 137. SIA.

602. Tillinghast to James Bradley via T.W. Taylor, 1 September 1964. Bradley to Powers, "General Authorizing Bill," 2 September 1964. Folder "NSF 1963–1972." Box 47. RU 253. SIA.

603. Whipple to staff, 14 September 1965. Folder "Smithsonian Bicentennial 1965." Box 8. RU 7431. SIA.

604. Galler to Bradley, 20 December 1965. "Smithsonian Research Legislation 1965–1966." Box 7. RU 137. SIA. SIE was the Science Information Exchange, an international service the Smithsonian had long fostered and fought to retain, with funding from the NSF. See Ripley to DuBridge, 20 March 1969; and Mansfield to Ripley, 24 March 1969. Folder "Mansfield Letter, codify SI research authority." Box 8. RU 137. SIA.

605. Coffey to Tillinghast, "SAO Legislative Proposal," 5 January 1966; 12 January 1966. Folder "Smithsonian Research Legislation 1965–1966." Box 7. RU 137. SIA.

606. Phil Ritterbush to Bradley and Galler, 21 January 1966. "Comments on draft legislation in the sciences." Folder "Smithsonian Research Legislation 1965–1966." Box 7. RU 137. SIA.

607. Ritterbush to Galler, 22 January 1966. "Draft Legislation in the Sciences." Folder "Smithsonian Research Legislation 1965–1966." Box 7. RU 137. SIA.

608. Ritterbush to Galler, 22 January 1966. The council was created to "advise the Secretary on broad programmatic activities and to offer guidance on Smithsonian contributions to research and public understanding in science, history and art." Board of Regents, "Minutes of Meeting, January 27, 1966." RU 1. SIA.

609. Powers to Bradley, 9 March 1966. "Proposed Research Legislation." "Smithsonian Research Legislation 1965–1966." Box 7. RU 137. SIA.

610. Ripley to BoB (W. H. Rommel), 6 May 1966; Ripley to Charles Schultze (Director, BoB), 6 May 1966. "Smithsonian Research Legislation 1965–1966." Box 7. RU 137. SIA.

611. Grasty Crews II to Frank Bow, 16 May 1966. "Smithsonian Research Legislation 1965–1966." Box 7. RU 137. SIA.

612. Bradley to Powers, Ritterbush, Galler, 13 June 1966. Folder "Research Bill." Box 7. RU 137. SIA.

613. W. H. Rommel (Asst. Director for Legislative Reference) to Ripley (attn. Bradley), 26 July 1966. "Smithsonian Research Legislation 1965–1966." Box 7. RU 137. SIA.

614. Rommel to Ripley, 26 July 1966.

615. Cornell OHI (2010), 177–179. The Smithsonian's original building is known as "The Castle," and the term colloquially refers to the Institution's central administration.

616. A. E. Lilley log entries for October 1965. "Lilley Log." Acc. 83-121. SIA. Hereinafter AELL/SIA.

617. "The Role of Smithsonian Astrophysical Observatory in the Federal Structure/Astronomical Community/Cambridge Complex/Great Society," November 1965 draft. Box 7. RU 137. SIA.

618. "The Role of Smithsonian Astrophysical Observatory," 2.

619. "The Role of Smithsonian Astrophysical Observatory," 3–4.

620. "The Role of Smithsonian Astrophysical Observatory," 6. On Adams's campaign, see M. Portolano (2000).

621. "The Role of Smithsonian Astrophysical Observatory," 7. 1965 was, in fact, the bicentennial of James Smithson's birth, not the Institution's.

622. "The Role of Smithsonian Astrophysical Observatory," 7.

623. "The Role of Smithsonian Astrophysical Observatory," 8.

624. Supplemental Justification, 9. Folder "Goals and Programs, 1967–1976." Box 7. RU 137. SIA.

625. Supplemental Justification, 9.

626. Supplemental Justification, 12.

627. Supplemental Justification, 15.

628. Supplemental Justification, 10. Equipment that the SAO possessed at the time included radiation counters, mass spectrographs, high-powered microscopes, and facilities to prepare and polish specimens. These needed constant upgrades, the white paper asserted, if SAO were to enjoy continued leadership in this research field. They looked forward to additional support to obtain expertise and "equipment for etching radiation tracks in solid specimens, an electron probe microanalyzer, and an improved microanalyzer."

Chapter 9 – Breaking Cognitive Barriers from Meteors to Lunar Samples

629. Whipple to Kuiper, February 1971. "Summary of Programs and Achievements of the SAO" presented as "Appendix 2" to 1 March 1971. (Greenstein) Ad Hoc Committee on Harvard-Smithsonian Observatory Relations. Folder "Personal and Confidential—Not for Publication—Recommendations to Harvard." Greenstein Papers 62.2. CIT.

630. Doel (1996), 23, 26–27. On Shapley's interdisciplinarity, see also DeVorkin (1984).

631. DeVorkin (2000b).

632. Cited in DeVorkin (1993), 159.

633. Just as Chandra Mukerji has shown for oceanography. See Mukerji (1989).

634. Doel (1996), 76.

635. Whipple and Hawkins (1959).

636. Whipple and Hawkins (1959), 519. As noted earlier, Whipple avoided amateur groups that collected and studied meteorites, including the American Meteor Society, led by Charles P. Olivier and devoted to observing meteor trails and radiants. See Brown to Whipple, 2 May; 10 May 1949. Folder "B, 1940–1950." Box 3. HUG 4876. HUA. See also Plotkin (1997).

637. Whipple and Hawkins (1959), 560.

638. Whipple and Hawkins (1959), 560.

639. Carmichael to Whipple, 3 May 1960. Folder "APO Whipple, Fred L 1959–1964." Box 24. RU 188. SIA.

640. Whipple OHI #2 (1976).

641. McCrosky and Boeschenstein (1965), 2–3.

642. Plotkin (1997), 32–38.

643. Marvin (2004), A207.

644. Fireman (1960), 1828.

645. Fireman (1960), 1831.

646. Plotkin (1997), 34, quoting Whipple to Bradley, 12 February 1960.

647. Whipple (1959), 267.

648. Gold (1955). Whipple seems to be among the first to relate it to the problem of physically landing on the Moon. Compton (1989), ch. 6, pt. 2, indirectly suggests the question was first posed in the early 1960s. Hall (1977) cites Gold's 1955 paper in app. A, but not in context of the controversy. Prior to Gold, Wesselink (1948) concluded that the surface of the mare must be a shallow layer of fine powder.

649. Whipple (1959). See also J. W. Salisbury, "Geological Problems in the Location of a Lunar Base," in Morrow, ed. (1961), 311–340, on 323.

650. Whipple (1959), 272.

651. Varsavsky (1958), 291–303.

652. Jacchia and Whipple (1961).

653. McCrosky and Boeschenstein (1965), 4.

654. McCrosky and Boeschenstein (1965), 5–9. McCrosky calculated that compared with the atmospheric volume patrolled by a single Super-Schmidt, some 6,000 square kilometers at 80 kilometers of elevation, a Prairie Network pair could cover 100 times this amount. Comparing their performance with that of the Super-Schmidts, McCrosky and his colleagues estimated that the entire network would record at least one bright trail every 12.5 hours, and any one single camera would catch a trail in about 800 hours, covering an area of 750,000 square kilometers spread across the Midwest from South Dakota to Oklahoma and from Illinois to Nebraska.

655. McCrosky and Boeschenstein (1965), 9–18.

656. Cornell OHI (2010). On the TV character Sergeant Bilko, see http://en.wikipedia.org/wiki/The_Phil_Silvers_Show (accessed 12 December 2013). Tougas first appears in annual reports in 1962.

657. Plotkin (1997), 35, describes the tensions in some detail.

658. Carmichael added in a penned note, "I do hope this is a good idea in any case!" Carmichael to Whipple, 3 May 1960. Folder "APO Whipple, Fred L 1959–1964." Box 24. RU 188. SIA.

659. Whipple to Carmichael, 5 May 1960. Folder "APO Whipple, Fred L 1959–1964." Box 24. RU 188 RU 188. SIA.

660. Plotkin (1997), 34–36. On earlier efforts to nationalize the meteorites collected from the Prairie Network, see Marvin (2004), A209–A210.

661. R. E. McCrosky, C.-Y. Shao, and A. Posen, "Prairie Network Fireballs. I—General Information and Orbits," translated abstract, *Meteoritika*, no. 37 (1978): 44–59; Ceplecha and McCrosky (1976).

662. Plotkin (1997), 37.

663. Sears (2012). Biographical information also from Tammy Peters, "Documenting a Geologist's Adventures," http://siarchives.si.edu/blog/documenting-geologists-adventures (accessed 30 December 2013).

664. Marvin (2004), A207.

665. Langway and Marvin (1964), 205. Marvin and Klein (1964).

666. Marvin, et al. (1970), 346–350.

667. King (1969).

668. "Center for Short-lived Phenomena (Cambridge, Massachusetts), Records, 1968–1974." Record Unit 607. SIA. See also *Annual Report of the Smithsonian Institution* (1968), 262–264.

669. A fall in Portugal occurred just a few months before Allende. Citron quickly arranged for a 128 gram meteorite to be flown from Portugal to the United States for rapid analysis and it arrived barely 30 days after the fall and within 12 days of the center being notified. Fireman (1969).

670. Fireman (1969), 275.

671. This and other aspects of Wood's career are taken from Sears (2012), 903–915; quote from 904. See also Wood to Frondel, 17 August 1964; Frondel to Wood, 29 September 1964. Acc. 05-265. Box 1. SIA.

672. As deduced from annual reports of that period in *SI Year* and *Bulletin of the American Astronomical Society (BAAS).*

673. Whipple to Wood, 12 December 1958; Wood to Whipple, 19 May 1965. Folder "Web–Win 60-01: 1960–1964." Box 3. Acc. 05-265. SIA. During these years, Wood also had offers from other key figures, such as Hess at Princeton. Hess to Wood, 14 November 1960. Folder "Hes–Hy." Box 1. Wood acc. 05-265. SIA.

674. Emphasis in original. Wood was the only East Coast scientist to sign the strong letter of protest to James E. Webb, joining Gerald Wasserburg, Bruce Murray, Roger Revelle, and Harrison Brown, as well as his former teacher Gordon MacDonald. Wasserburg et al. to James E. Webb, 24 January 1962. Folder "Web–Win 60-01." Box 3. SIA. *Prospector* has been described in Ulivi and Harland (2004), 39–40, and more extensively in Burnham (1995), 213–228.

675. Hall (1977), 122. Burnham (1995), 221–222, shows how Jet Propulsion Laboratory's cost overruns for Ranger coupled with NASA's lack of clarity in its widely varying proposals to Congress for a robotic sample return mission led to its cancellation by Congress in 1962.

676. Homer Newell to "Dear Colleague," 3 May 1966. Folder "Apollo." Box 3. Acc. 05-265. SIA.

677. Sears (2012), 907.

678. Sears (2012), 908–909.

679. For a historiographic review, see Shindell (2010), 191–230, esp. 194–196.

680. NASA (1969), 220–241.

681. SAO/NASA Astrophysics Data System. Compilation. http://adswww.harvard.edu/.

682. Wood continued his studies of the thermal histories of chondrites and the origin of planets, publishing "Asteroids, Comets and Meteors" in the high-profile 1967 compendium *Odishaw* (1967), 211. See also his detailed but accessible textbook: Wood (1968).

683. *Annual Report of the Smithsonian Institution* (1968), 456.

684. Newell (1980), 214–222.

685. Newell (1980), 294–295; Newell to Wood, 9 March 1967. Folder "Apollo." Box 3. Wood acc. 05-265. SIA.

686. John A. Wood and Ursula Marvin, "Project Information Sheet: Apollo Lunar Sample Analysis," 9 September 1968. Folder "Apollo." Box 3. Wood acc. 05-265. SIA.

687. See, for instance, McCorkell et al. (1967).

688. Review literature on the nature of the lunar surface reveals that in the 1960s there was little consensus over the nature of the mare surface that would be encountered by the Apollo astronauts. There were solid arguments for predicting that the surface would be strewn by ejecta, but these could be volcanic as well as meteoritic, reflecting the major options for the formation of lunar features. See, for instance, Kopal (1965), 737–855, on 822, and Baldwin (1964) 73–94, for a general review of all popular mechanisms.

689. Sears (2012), 909. Wood describes in some detail here how he transported his vials of dust back to Boston, and how he arranged viewings by colleagues, friends, and family. On LRL, mainly the nature of its operations in handling contamination, see Susan Mangus and William Larsen, Lunar Receiving Laboratory Project History, NASA/CR-2004-208938. See also Compton (1989), ch. 4.

690. Sears (2012).

691. Sears (2012), 910.

692. Sears (2012), 910.

693. Sears (2012).

694. Wood et al., "Lunar Anorthosites and a Geophysical Model of the Moon," in Levinson (1970), 965–987. See also Wood (1972), 462–501. See also Ulivi and Harland (2004), 142.

695. Sears (2012), 914. On the Allende connection, see Marvin et al. (1970), 346–350.

Chapter 10 – "Project Celescope": SAO's Orbiting Astronomical Observatory

696. Whipple (1972a), 1035.

697. Carmichael to Whipple, 9 December 1957. Folder "Whipple, Fred, 1957." Box 24. RU 50. "Office of the Secretary 1949–1964." SIA. On Cat Eye, see Bushnell (March 1962) and LeCompte (1962). This was very much in the spirit of the new NASA: pushing to get away from the conservatism of the National Advisory Committee on Aeronautics (NACA) for preferring known technologies. NASA programs would foster "larger, more versatile—and more expensive—tools," a policy implicit in the Space Act of 1958. Newell (1980), 97.

698. Marché and Walsh (2006), 191.

699. Marché and Walsh (2006), 187, 191. There was a long tradition of photoelectric photometry at Wisconsin. See Osterbrock (1976).

700. On the challenges of the contracting process for emerging technologies, see Dennis (1991).

701. Mack (1990); Hall (1977); Bester (1966).

702. McGee and Wilcock, preface, in McGee and Wilcock, eds. (1960), viii. Americans attended this symposium, including astronomers William A. Baum, John Hall, E. W. Dennison, and Kent Ford, meeting engineers from Westinghouse, the Wright Air Development Center (R. K. H. Gebel), and the Electron Tube Division, RCA. McGee and Mandel (1962), 328.

703. Code OHI (1982), 36.

704. McGee to Baum, 14 March 1964. McGee Papers, Imperial College Archives. Courtesy Samantha Thompson.

705. Davis to Whipple. 27 October 1958. "Celescope Project, 1958–61." Box 1. RU 260-1-07/58. SIA. This description closely fits the Corona system. See Day et al., eds. (1998).

706. Davis to Whipple, 27 October 1958. "Celescope Project, 1958–61." Box 1. RU 260-1-07/58. SIA.

707. Day, Dwayne A. "Black Apollo." The Space Review, 29 November 2010. http://www.thespacereview.com/article/1734/1 (accessed 17 December 2010).

708. Nitchie to Davis, 2 May 1960. Davis "Report," 8 January 1960. "Celescope Project, 1958–61." Box 1. RU 260-1-07/58. SIA.

709. R. E. McCrosky to Whipple, 10 June 1960. "Cat Eye" TM-6/10/60 memo. Box 114. RU 188. SIA.

710. "Quarterly Progress Report No. 3." Celescope NASA Grant NsG-51-60—March 15 through June 30 1960 (August 1960). "Progress 1960." Box 10. RU 522. SIA.

711. Nitchie to Davis et al. (Davis, Grossi, Baez, Rustgi, Cobb, Miles, Arazi), 17 August 1960; H. W. Cobb to Davis, 22 August 1960. "Celescope Project, 1958–61." Box 1. RU 260-1-07/58. SIA.

712. On Fisher's early work in spaceborne x-ray technology, see Hirsh (1983), 50.

713. Nitchie to Davis, 28 September 1960. "Washington Meeting of 20 September on Celescope Budget." Box 1. RU 260-8-70/60. SIA.

714. Grossi and Goldstein to Davis, 29 October 1960. "Celescope Project, 1958–61." Box 1. RU 260-8-70/60. SIA.

715. Nitchie to Davis, 20 September 1961; 17 March 1961. "Celescope Project, 1958–61." Box 1. RU 260-8-70/60. SIA.

716. Grossi to Davis et al., 21 July 1960; S. Sydor to Edward Shenton, 17 April 1961. "Celescope Project, 1958–61." Box 8. RU 260-8-70/60. SIA.

717. Arazi to Grossi, 24 January 1961. "Celescope Project, 1958–61." Box 8. RU 260-8-70/60. SIA.

718. J. M. Franklin to Davis, 22 March 1961. "Celescope Project, 1958–61." Box 8. RU 260-8-70/60. SIA.

719. Trip reports, Davis et al. to Wallops, 25 April 1961; H. W. Cobb to Davis, Nitchie, Franklin, 3 May 1961. "Celescope Project, 1958–61." Box 8. RU 260-8-70/60. SIA.

720. Davis to Whipple, 29 June 1961. "Celescope Project, 1958–61." Box 8. RU 260-8-70/60. SIA. Rudney (1971) states that the scale-up was originally suggested by the primary proposers to Berkner's SSB appeal of July 1958, including Code, Spitzer, Whipple, and Goldberg, citing Jenkins, "Rocket Ultraviolet Astronomy," in Chiu, H. Y., et al, eds., *Stellar Astronomy II: Proceedings of the Summer Institute* for Observational Astronomy. New York: Gordon and Breach (1969). Marché and Walsh (2006), 191, note that the scale-up for the Wisconsin package allowed for redundancy.

721. Davis to Whipple, 29 June 1961; Arazi to Davis, 10 July 1961; O. P. Rustgi to Davis, 7 June 1961. "Celescope Project, 1958–61." Box 8. RU 260-8-70/60. SIA.

722. Nitchie and Davis to Whipple and Tillinghast, 18 May 1961. "Celescope Project, 1958–61." Box 8. RU 260-8-70/60. SIA.

723. Arazi to Davis, 10 July 1961. "Celescope Project, 1958–61." Box 8. RU 260-8-70/60. SIA.

724. "Quarterly Progress Report No. 3." Celescope NASA Grant NsG-51-60—March 15 through June 30 1960 (August 1960). "Celescope Project, 1958–61." Box 8. RU 260-8-70/60. SIA.

725. Davis to Whipple, 25 May 1961, reporting on a trip to NASA 23 May 1961. "Celescope Project, 1958–61." Box 8. RU 260-8-70/60. SIA.

726. Code OHI (1982).

727. By 1971, EMR was a division of Weston Instrument, part of the Schlumberger Group. EMR press release 16 February 1962. Progress report on Uvicons, "UV Sensitive TV Camera Tubes," 7 June 1962. Megerian to Davis, 2 February 1962, "Monthly Report—January 1962." "Celescope Project, 1958–61." Box 8. RU 260-8-70/60. SIA. The Schwarzschild optical design, created by Karl Schwarzschild, provided for flat uniform aplanatic fields.

728. "Summary of Highlights—Month of August 1963 Progress report no. 23, Sept 30 1963," 1. "Celescope Project, 1958–61." Box 8. RU 260-8-70/60. SIA.

729. Hurlbut to Lundquist; 19 December 1963; to Tillinghast, 29 October 1963. "Celescope Project, 1958–61." Box 8. RU 260-8-70/60. SIA.

730. Lundquist to Tillinghast, n.d., attached to Hurlbut to Tillinghast, 19 December 1963. On the same day Hurlbut wrote to Tillinghast, he lambasted one of the SAO's management staff, Mark Malec, because the SAO financial records and status reports had been tardy. "These repeated delinquencies of a contractual reporting requirement do not shed a very favorable light on Smithsonian," he told Malec, partly to give Malec the ammunition he needed to get the reports done. Clearly, Hurlbut saw communications problems rampant. "Celescope Project, 1958–61." Box 8. RU 260-8-70/60. SIA.

731. Nancy Roman, "The Orbiting Astronomical Observatory," lecture, ESRO Summer School, 13 August 1964. "OAO." NHO.

732. The overall situation, including delays and cost overruns, are detailed in Rudney (1971), 12–15.

733. Harry Goett, Director, NASA/Goddard to Whipple, 14 July 1964, 2. "OAO 1964–1965." Box 8. RU 7431. SIA.

734. Davis to Stroup, 1 October 1964; "November Monthly Report no. 37" (13 November 1964). "Celescope Project, 1958–61." Box 8. RU 260-8-70/60. SIA.

735. Whipple to Harry Goett, 8 December 1964, 2. "OAO 1964–1965." Box 8. RU 7431. SIA.

736. Goett to Whipple, 13 December 1963. "Orbiting Astronomical Observatory, 63–64." Box 7. RU 7431. SIA.

737. Rudney (1971), 17, referencing Newell to Robert Seamans, 24 December 1964. Created in April 1960, the Space Sciences Steering Committee consisted of space science program managers within the Office of Space Science (and, variously, Applications), and, significantly, included senior consultants and community leaders from the disciplines involved. See Newell (1980), 113–114, 121.

738. Rudney (1971), 17–19; Marché and Walsh (2006).

739. "OAO-A1 Press Kit." "NASA Press Release 66-60" (Sunday, 20 March 1966), 23.

740. T. W. Taylor to James Bradley, 11 January 1965. "Celescope Correspondence." Box 11. RU 522. SIA.

741. "Celescope Balloon Experiment—Backup," n.d.; 13 January 1965; T. W. Taylor "For the Record," 28 January 1965. Celescope 18 Balloon NASA. Box 12. RU 522. SIA.

742. Tillinghast to Megerian, 29 January 1965. "Celescope 18 Balloon NASA." Box 12. RU 522. SIA.

743. Megerian to Tillinghast, 8 February 1965. "Celescope 18 Balloon NASA." Box 12. RU 522. SIA. "Change Proposal No. 2," 26 February 1965. "Celescope 18 Balloon NASA." Box 12. RU 522. SIA. The SAO continued on exploring budgets for balloon flights through December, until NASA declined their March proposal. Edward R. Redding to Director, Smithsonian Institution, 16 December 1965. Box 7. RU 260-7-07/65. SIA.

744. Davis to Burke, 21 May 1965. "Celescope 11 Minutes." Box 11. RU 522. SIA.

745. Tillinghast to W. A. Landymore, 16 June 1965. "Celescope 06 CTC S-1-65." Box 11. RU 522. SIA.

746. Tillinghast to W. A. Landymore.

747. Tillinghast to W. A. Landymore. Inferences about Goett's reactions are taken from a collective profile of his recorded behavior at meetings at Headquarters and direct testimony by Homer Newell, who regarded him as a well-loved, "untiring battler." See Newell (1980), ch. 14, 251.

748. Davis to Burke, 21 May 1965, 3–4. "Celescope Minutes." Box 11. RU 522. SIA. Goddard people assigned to the task included John Townsend, John Lindsay, and George Ludwig, among others.

749. Telegram, Ziemer to Burke, 8 October 1965. "Celescope 7/65." Box 11. RU 522. SIA.

750. Naugle (for Newell) to Whipple, 8 December 1965. "07/65." Box 7. RU 260. SIA.

751. Newell to Goett, 10 May 1965. "Newell 4490 Correspondence" folder, NHO. TOS was eventually transferred to the National Oceanographic and Atmospheric Administration when it formed in 1970.

752. Newell to Goett, 10 May 1965, 2. On the state of the TOS system at the time, see "TIROS Operational Satellite System Fact Sheet" (June 1965), http://docs.lib.noaa.gov/rescue/TIROS/TL798M4T5771965.pdf (accessed 26 December 2011). See also Mack (1990).

753. "Chronology of Administrative and Other Events, 1957–1969," app. E, in Levine (1982). See also Sam Keller OHI (1983), tape 1, side 1, sec. 3. Davis OHI (2005). Davis recalls running into Goett at a local restaurant and learning that his abrupt dismissal was no surprise.

754. On final preparations, details of the launch, and fate of the mission, see Rudney (1971), 19–22, and on the aftermath, 23–30. "OAO Defects Traced," *Flight International*, 2 February 1967,

181–182. Newell, "Approval of the Recovery Plan for the Orbiting Astronomical Observatory (OAO) Program" (17 November 1966), 2. Quoted from Rudney (1971), 33.

755. Rudney (1971), 34.

756. Rudney (1971), 34.

757. Rudney (1971), 36, from *Technology Week*, 3 April 1967, 19.

758. Karth was characterized as "both a strong supporter and, at times, a tough critic" of the space program. Cortright, ed. (1975), ch. 5.4.

759. They had already established a new Astronomy Missions Board (AMB) at the end of 1966 to complement a recently established Lunar and Planetary Missions Board, created after a 1965 summer study at Woods Hole to improve communications with scientists and Congress. See Tatarewicz (1990), 101; Leo Goldberg OHI (1978), 137; Goldberg OHI (1983), 47.

760. Sures to SGO/OAO Program Manager, 31 May 1966, 6. Distribution confined to NASA offices. Folder OAO-2. 6224 NHO.

761. Goetze (1962), 131–136.

762. "Westinghouse Camera Tube Aboard OAO to Map Invisible Starlight," Westinghouse Electric Corporation, Pittsburgh, Press Release TP-610-296N. Lundquist Collection; also in the Whipple Papers, "Celescope, 1962–1968" folder, 2 of 2. HUG 4876. HUA. An excellent popular review of the tube and its operation is Green (1968), 140–145. A series of papers announcing the SEC design was presented by Westinghouse researchers in *Advances in Electronics and Electron Physics* 22 (1966), including one by Davis (1966).

763. Evans (1967), 1128. Green (1967), S14; Beardsley and Hansen (1968), S4–S5. Beardsley (S5) endorsed the detector, saying that it provided increased sensitivity: "However, perhaps the outstanding advantage of the scanning camera system is that the signals can be read out electrically so that remote viewing and signal enhancement are possible."

764. Grumman Aircraft Engineering Corporation. Advanced Systems/Space Sciences Group. "The OAO in Association with a Manned Space Station. NY: Bethpage, 1965." NASA contract NASW-1143. [Space Science Board], introduction, in *Space Research—Directions for the Future*, vol. 1, Publ. No. 1403 (Washington, DC: National Academy of Sciences, 1966), 63.

765. OAO, Introduction, 63. The equipment included a CSM/LEM Lab to house three men for a 45-day maximum mission duration, after which the OAO would be "boosted" into a higher orbit for automatic operation using, presumably, electronic detectors.

766. Haramundanis OHI (2005), 77, 83. Whitney OHI (2005), 36.

767. Stuart H. Clarke to Davis, 25 November 1964. Davis to "Dear Sir," n.d., approximately November 1964. W. C. Kelly to Davis, 1 November 1966. Davis Collection, author's files. See "NASA to Select Scientist-Astronauts for Future Missions," NASA Press Release 64-248 (19 October 1964). Davis Collection. See also Brooks et al. (1979), ch. 7.5; Shayler and Burgess (2007).

768. Davis to "Dear Sir." Davis listed references as Hynek, Lundquist, John J. Burke, and Leo Goldberg. He did not list Whipple. See Davis to Hynek, 7 November 1966. Davis Collection, and Davis OHI (2005), 118.

769. The were some 900 initial applicants. Davis, "Statement of Space Experiments I would like to Perform," sent 14 February 1967. Written in response to W. C. Kelly, Office of Scientific Personnel, telegram, 8 February 1967. Davis Collection, author's files.

770. Among the letters of commiseration Davis received from other candidates, more than one vowed to try again, though some expressed annoyance with the test procedures, and "that GD cardiologist Lovelace." Neville to Bob, 9 August 1967. Davis Collection. On Lovelace, see Weitekamp (2004). The testing and selection process in Houston has been described in some detail. See Shayler and Burgess (2007). See also Brooks et al. (1979), chaps. 7–5, and Compton (1989), ch. 5.

771. Davis OHI (2005), 166.

772. Newell (1980), 356–357.

773. Burke to Stroup, 29 December 1967; Burke to White, 8 October 1968. "Celescope." Box 11. RU 522. SIA.

774. Whipple, *Smithsonian Year 1969*, 190. Each of the selected areas was 100 square degrees. The nominal effective field was some 2 degrees square in W. W. McCurdy, "A Brief Summary of the Smithsonian Astrophysical Observatory's Celescope Program," November 1960, SAO internal memorandum. Davis Collection. However, Wilson and Boksenberg (1969), 453, described the field as 2.8 degrees in diameter.

775. Whipple, *Smithsonian Year 1969*, 192.

776. Gerald M. Truszynski to Bradley, 19 May 1969. "7/68 Celescope." Box 11. RU 522. SIA. Clear and deep cuts began after 1967, and the greatest decline was in the FY 1968–1969 period. "Smithsonian Astrophysical Observatory Financial History, July 1 1965 to Present" (5 April 1977). See Lynds and Goldberg (1975), table XI, 22, "Total Federal Funding for Astronomical Research."

777. Thomas O'Toole, "Satellite's Report on Stars Amazes Astronomers," *Washington Post*, 19 April 1969, 16.

778. O'Toole, "Satellite's Report." O'Toole actually called the Pleiades "dim" to the eye, which indicates a certain unfamiliarity with the sky, or his overly eager effort to impress his readers.

779. Telegram, Davis to Satyendra, Science and Technology Assoc., Inc., 8 April 1970. "Celescope 08 No. 3." Box 11. SIA.

780. Davis et al., "Ultraviolet Television Data from the Orbiting Astronomical Observatory, I. Instrumentation and Analysis Techniques for the Celescope Experiment," in Code, ed. (1972), 1.

781. Robert Ayer to Distribution, "Project Celescope Organization," 15 April 1970. "Celescope 12." Box 12. RU 522. SIA.

782. Davis OHI (2005), 92.

783. Davis (1973).

784. Avrett OHI (2005), 39–46. Kurucz et al. (1974). Robert T. Ayer to Distribution "Project Celescope Organization," 15 April 1970. "Celescope 12." Box 12. RU 522. SIA.

785. Whipple had planned that Whitney would head up the analysis of the Celescope data. But by the time his group got involved, Whitney had taken an extended sabbatical, leaving Eugene Avrett in charge. One of Avrett's graduate students, Robert Kurucz, assisted in the computer work, generating a series of theoretical stellar atmospheres models to complement the observational catalog. Davis OHI (2005), 91. Ayer to Distribution, 15 April 1970. "Celescope 12." Box 12. RU 522. SIA. Smithsonian Astrophysical Observatory, *Star Catalog* (1966).

786. Haramundanis OHI (2005), 78.

787. "Project Celescope Status and Plan for Refining Calculation of Ultraviolet Magnitudes" 17 July 1970. Lundquist Papers. SL. Copy in author's files.

788. Lundquist, "Celescope Presentation, Jan 13, 1971." Lundquist Papers. SL. Copy in author's files.

789. Avrett OHI (2005), 46.

790. Haramundanis OHI (2005), 86.

791. She and her mother did work together, determining the intrinsic colors of the stars in the catalog. Haramundanis and Payne-Gaposchkin (1973).

792. Lundquist OHI (2011), 169. Horak and Lundquist (1954), 42.

793. Lundquist OHI (2011), 344. On Argus, see Van Allen (2004).

794. Lundquist OHI (2011), 317–318.

795. David Latham to Attendees, 3 December 1970. "Minutes of December 3 meeting held to discuss methods for interpreting Celescope data." Lundquist Papers. SL. Copy in author's files.

796. "NASA Briefing on OAO program at NASA HQ," 26 March 1971. Lundquist Papers. SL. Copy in author's files.

797. Lundquist OHI (2011), 320.

798. Lundquist OHI (2011), 321–332, quote on 328. Davis to Distribution, 11 September 1970. "Agenda for meeting 14 September 1970." Lundquist Papers. SL. Copy in author's files.

799. Robert T. Ayer to Distribution, 18 September 1970. "Revision No. 2 to Project Celescope Organization and Standard Operating Procedures." Lundquist Papers. SL. Copy in author's files.

800. Avrett OHI (2005), 51–52.

801. Avrett OHI (2005), 53.

802. See http://nssdc.gsfc.nasa.gov/nmc/experimentDisplay.do?id=OAO-B-01 and Rudney (1971), 60.

803. "ESRO Report Presented to the Eighteenth COSPAR Meeting, Varna, Bulgaria, June 1975," cited in http://nssdc.gsfc.nasa.gov/nmc/spacecraftDisplay.do?id=1972-014A (accessed 22 January 2012).

804. Davis et al., "Ultraviolet Television Data from the Orbiting Astronomical Observatory. I. Instrumentation and Analysis Techniques for the Celescope Experiment," in Code, ed. (1972), 1–22.

805. Lundquist, "Highlights of the Celescope Experiment," in *Performance Evaluation of the Celescope Experiment* (Contract NAS 5-1535) (Cambridge, MA: Smithsonian Institution, Astrophysical Observatory, July 1971).

806. Kupperian to Davis, 22 July 1968; Davis to Kupperian, 11 August 1968; Davis to SAO Scientific Staff, 16 September 1968; C. Dixon Ashworth to Davis, n.d., ca. July 1968. Copies courtesy Charles Lundquist. Lundquist Papers. SL.

807. Rudney (1971), 61–62.

808. Davis et al., in Code, ed. (1972).

809. S. B. Parsons, "Comparison of Celescope Magnitudes with Model Atmosphere Fluxes for A, F, and G Supergiants," 461–462, in Code, ed. (1972).

810. Classic examples abound in history in which the existence of new classes of phenomena were debated in terms of the collective faith in the observations. See Harwit (1981) for a discussion of the astronomical background; Galison (1987), chap. 4, sec. 17; and Dick (2013).

811. NASA Briefing on OAO program at NASA HQ, 26 March 1971. Lundquist Papers. Copy in author's files.

812. Haramundanis OHI (2005), 87. Davis et al. (1973) openly acknowledged the difficulties, but concluded that their analysis and comparisons were "essentially correct" and that the adjustment they indicated were the best that could be made. See also Haramundanis and Payne-Gaposchkin (1973).

813. Peytremann and Davis (1974), conclusions and quote on 224–225. On the extinction bump, see Savage (1975), 103. The bump was real and was interstellar, not circumstellar, even though there were as yet no models for grains that could fully explain the consistent appearance of the peak over the sky. The Wisconsin conclusions were supported by observations taken during the Skylab missions in 1973 and 1974. See Parsons et al. (1976), on 436.

814. Deutschman et al. (1976).

815. Reed (2003).

816. Lundquist OHI (2011); McCrosky OHI (2005).

817. Marché and Walsh (2006), 185.

818. Naugle to Don C. Morton, 28 June 1971. OAO files. NHO. Don Taylor was far more passionate, telling Naugle and copying Rep. Joseph Karth, that NASA's decision led to five "general considerations" in order of emotional impact, ending with a veiled threat: "A callous disregard for the feelings of the scientific community as would be indicated by turning off one of the most outstanding successes of the unmanned program may make the best scientists reluctant to return to space research when science and NASA recovers from its current depression." Donald J. Taylor to John Naugle, 28 June 1971. OAO files. NHO.

819. Smith and Tatarewicz (1985).

820. Space Science Board (1966), 155.

821. Davis OHI (2005), 88.

822. Fred Whipple, "Selection of Astronomy Experiments for Space," 383–384 in *Astronomy from a Space Platform, Proceedings of the Symposium held 27-28 December, 1971 in Philadelphia, PA. Science and Technology Series,* Vol. 28, *A Supplement to Advances in the Astronautical Sciences.* Edited by George W. Morgenthaler and Howard D. Greyber. An American Astronautical Society Publication, 1972

823. Newell (1980), 97.

824. Newell (1980), 356–357.

825. Gingerich to author, 29 September 2015. Gingerich OHI (2005), 61–64.

826. Gingerich OHI (2005), 69.

827. Gingerich OHI (2005), 78.

828. See, for instance, Mihalas (2003) and Underhill (1962), 11. Gingerich to author, 16 September 2015.

829. According to citations in the published literature, in 1957 access to a 704 mainframe was mentioned only by those institutions connected somehow to large, high-energy physics facilities such as Livermore (Berkeley) or Los Alamos, or in one case to a large industry, for example, General Electric. ADS searches, ca. January 2015.

830. Avrett OHI (2005), 36.

831. Kurucz (1970).

832. Gingerich et al. (1971). As of July 2015, the paper had garnered more than 800 citations, averaging between 30 and 70 citations per year in the 1970s and 1980s.

833. Noyes OHI (2005), 49.

834. Gingerich OHI (2005), 78.

Chapter 11 – Building the Infrastructure for Satellite Geodesy and Aeronomy: Work of the Research and Analysis Division

835. Muir-Harmony (2009), 76; McCray (2008), 17.

836. Hayes (1967), ch. 4, and Organization Charts, author's files.

837. Bradley to Marshall Greenspan, 30 November 1960. NASA Executive Asst. Folder "STP no. 134 Current Years Proposals." Box 5. RU 522. SIA.

838. Carmichael to Webb, 4 April 1963. Folder "STP no. 134 Current Years Proposals." Box 5. RU 522. SIA. On ANNA, see Warner (2000), Hacker (2000), and David (2015).

839. Ursula B. Marvin, "Continental Drift," in Lundquist and Friedman, eds. (1966), 32.

840. Lundquist and Friedman, eds. (1966), 69.

841. Lundquist and Friedman, eds. (1966), iii–iv.

842. Lundquist, "Scientific Horizons from Tracking Space Objects," in Lundquist and Friedman, eds. (1966), 1, citing Whipple and Hynek (1958).

843. Lundquist, "Scientific Horizons."

844. See McCray (2008) throughout, especially 94–95, 214–222, 235.

845. See Hayes (1967) as well as two extended reports for the Smithsonian's *Annual Report to the Board of Regents* in 1961 and 1963, which were followed by the Smithsonian monograph in 1967, and a commercial reprint in 1968. See McCray (2008) for Moonwatch and Muir-Harmony (2009, 2012) for the Baker-Nunn network.

846. Dick (2003), 530.

847. Ivan I. Mueller, "Levallois Medal Citation for George Veis," http://www.gfy.ku.dk/~iag/HB2004/part3/32-levalloir-price.pdf (accessed 3 March 2014). The medal is one of several bestowed by the International Association of Geodesy.

848. Warner (2002), 375–378.

849. Yoshihide Kozai OHI (1997), 9. Kozai observed that in the late 1950s, of the few schools in celestial mechanics in the United States, Yale under Dirk Brouwer was well balanced between theory and practice, whereas UCLA was heavily weighted on the practical side. Hagihara, however, himself highly theoretical, had managed to produce a corps of well-trained but highly effective practitioners who could apply orbit theory in creative ways.

850. Kozai was proud of his speed using both hand-cranked and electric desk calculators: "I was very fast—nobody defeated me [*laughs*]." Kozai OHI (1997), 16.

851. Kozai (1959). He quickly found he was in competition with leaders in the field such as Yale's Dirk Brouwer, who was also editor of the *Astronomical Journal*, where Kozai hoped to publish. But once they both had a chance to read and critique the other's papers, Brouwer published them both in a special November 1959 issue. Kozai garnered some two dozen citations within a few years, comparable to Brouwer's.

852. Whipple, "Report on the Astrophysical Observatory," 86. Secretary's Report to the Board of Regents (1958).

853. Cherniack and Gaposchkin (1963).

854. Cherniack and Gaposchkin (1963), 16.

855. Gaposchkin (1964), 1, 60.

856. Duncombe and Duncombe (1967), 582. For a sense of the state of the fundamental reference system, see chapters by Clemence, Scott, and Vasilevskis in Strand, ed. (1963). See also Dick (2003), 530–535.

857. Langley used his position as a bully pulpit. See Langley (1891) and Bartky (2000).

858. Herbert A. Luft, commercial advertisement. *Sky & Telescope* (September 1966): 154. At the time, the classic *Bonner Durchmusterung* cost in the range of $125.

859. Haramundanis OHI (2005), 13–15, 45. FK4 is the *Fourth Fundamental Catalogue*. See A. Kopff, Supplement-Katalog des FK3 (FK3 Supp), *Sonderbruck Astron.-Geodät. Jahrbuch 1954* (1953, supplement published in 1963).

860. J. Meeus (1967), 268.

861. Haramundanis OHI (2005), 51; 61; 69 outlines the details. She also carried computer tapes from the catalog project to Washington to see them converted into properly sized film masters for publication. Then, with the film-based charts in hand, she headed up a staff of about "half a dozen girls" to add objects such as double stars and nebulae that were not on the tapes, and to carefully employ drafting templates to illustrate each chart.

862. Smithsonian Astrophysical Observatory (1969). Box set of 152 charts with a 13-page booklet.

863. Haramundanis OHI (2005), 54.

864. Whipple, "Astrophysical Observatory," *Annual Report to the Board of Regents* (1963), 162.

865. "The Construction and Use of Star Catalogues Conference" (1967), 552.

866. Haramundanis (1967), 588–596. Haramundanis OHI (2005), 58. Eichhorn was in the audience, and Haramundanis's recollection is that she felt obligated to defend the *Catalog* in the Q&A session.

867. Haramundanis (1967), 593.

868. Haramundanis (1967), 593–594.

869. Haramundanis (1967), 594–595.

870. Haramundanis (1967), 596. The group did agree that "practically all geodetic requirements are being filled by that catalogue" now, and so it should be regarded as the basis of future improvements. Haramundanis ended the discussion by giving thoughts for further improvement.

871. Haramundanis, OHI (2005), 65.

872. See *Sky & Telescope* 34 (August 1967), 89–90; Lundquist and Whipple (1973), "Historical Introduction," to Gaposchkin (1973, 11

873. Whipple and Hynek (1957), 37–43.

874. Hayes (1967), 144–152, outlines the steps of the process.

875. Whipple and Hynek (1957); *SI Annual Report for 1961*, 85–86.

876. Whipple, in *Annual Report of the Smithsonian Institution* (1964), 158.

877. Kaula (1959); Jeffreys (1941), 55–66.

878. Izsák (1962), 2. The heuristic value of this family of orbits was later questioned. See Lubowe (1969), 6–10; and Brouwer (1963), on 230.

879. Gaposchkin and Lambeck (1970), 1–3; Mudgway (2001); Butrica (1996).

880. Arnold (1967), 64.

881. Dale L. Johnson, Barry Roberts, and William W. Vaughan, "Reference and Standard Atmosphere Models." http://ntrs.nasa.gov/archive/nasa/casi.ntrs.nasa.gov/20020092087.pdf.

882. DeVorkin (1992), 292–294.

883. Jacchia and Slowey (1962).

884. Dalgarno OHI (2007), 24–25.

885. Dalgarno OHI (2007), 32, 35.

886. Lundquist to Foster Haley, Public Information Office, 4 May 1955. Courtesy Charles Lundquist. Lundquist Papers. SL.

887. The conference had been created to introduce geodesists to the promise of spaceborne efforts. Whitten and Drummond (1959).

888. Silverstein to Goett, 27 December 1960, re: "Satellite Tracking by the Smithsonian Astrophysical Observatory." Clipped to C.W.T. to Whipple, n.d. "NASA Relations" folder (March 1951–1961). RU 7431. SIA.

889. Newell, quoted in Warner (2000), 405, refs. 61, 62, quoting news report of Newell's testimony.

890. Whipple, 30 September 1960, date uncertain. "Comments on the ad hoc panel called by General Donald Yates to study the problem of scientific space probes for NASA and the DOD." "Scientific Advisory Board." Box 23. RU 7431. SIA. See also Warner (2002), 389, 402–403.

891. Warner (2000), 405, refs. 61, 62. Another perspective on DOD's concern for classification was related to keeping the orbits of Corona surveillance satellites hidden. See Sturdevant (2008), on 11; and Cloud (2001). On Project ANNA (acronym for "Army, Navy, NASA, Air Force"), see *"Project ANNA: Geodetic Satellite System": Hearings before the Subcommittee on Space Sciences of the Committee on Science and Astronautics*, 87th Cong. (1962); and HHR-32, *NASA Office of Defense Affairs: The First Five Years*, ch. 18. http://history.nasa.gov/HHR-32/ch18.htm.

892. Whipple to Bradley, 8 November 1960. "Scientific Advisory Board." Box 30. RU 7431. SIA.

893. Jon Grady to Cliff Marsh, 17 April 1961. Folder "Hawaii: Administration, 1961–64." Box 16. RU 263. SIA. See Muir-Harmony (2009).

894. Whipple to Jerome Wiesner, 26 November 1961, asking for his support for flashing light geodetic satellite (ANNA). Folder "Whipple, Fred, 1959–64." Box 24. RU 50. "Office of the Secretary 1949–1964." SIA.

895. Newell (1980), 220 and elsewhere, notes the support of PSAC in resisting classification.

896. "A History of the Committee on Science and Technology," 85th–110th Congresses (1958–2008), 97. http://archives.democrats.science.house.gov/docs/Committee_History_50years.pdf (accessed 20 March 2014).

897. Whipple (1972), 1336. Whipple, "Statement of Dr. Fred L. Whipple . . . ," "Fourth Meeting of Panel on Science and Technology, House Committee on Science and Astronautics, March 2, 1962," in Whipple (1972), 1332–1343, on 1336. Newell (1980) explicitly mentions how the existence of Baker-Nunn stations in places such as Japan and India required that they not be classified so as not to cause international discord.

898. Whipple (1972), 1337–1338.

899. Whipple (1972), 1332, 1337–1339; Henricksen is remembered in Fisher (2005) and is generally regarded as one of the first "to apply space age data to the mapping sciences." https://aas.org/obituaries/soren-w-henriksen-1916-2011.

900. Whipple (1972), 1339.

901. Whipple (1972), 1342.

902. Whipple (1972), 1343.

903. Warner (2000), 406.

904. Newell (1980), 119.

905. Warner (2000), 407. ANNA 1B is identified as a Navy geodetic satellite. No mission results are listed. See http://nssdc.gsfc.nasa.gov/nmc/spacecraftDisplay.do?id=1962-060A (accessed 21 March 2014).

906. Lundquist (1967), 10.

907. Anderson et al. (1965). Beacon Explorer-A, originally planned as an atmospheric Explorer, was augmented to include retroreflectors by Henry H. Plotkin at NASA's Goddard Space Flight Center. See H. H. Plotkin, "The S-66 Laser Satellite Tracking Experiment," NASA Goddard Space Flight Center report X-524-63-59 (13 February 1963), and Snyder (1967).

908. Newell (1981), 119. The board had representatives from NASA, DOD, and the Department of Commerce's Coast and Geodetic Survey. "NASA to Launch Geodetic Satellite, First of Its Type," NASA Release 65-333 (29 October 1965), 4.

909. NASA Release 65-333, 3–5.

910. See Computer Sciences Corporation (1970), Reece and Marsh (1974). See also http://nssdc.gsfc.nasa.gov/nmc/datasetDisplay.do?id=ESGP-00012 and http://archive.org/stream/nasa_techdoc_19740014777/19740014777_djvu.txt (accessed 21 March 2014).

911. John Grady to Cliff Marsh, Hawaii, 6 February 1961. Folder "Hawaii: Administration, 1961–64." Box 16. RU 263. SIA. On the USAF network, see Sturdevant (2008), 6–22; and Wilkes and Gleditsch (1978).

912. Paul Preuss, Major General, USAF to Whipple, 8 November 1963. Folder "STP Cooperation with USAF." Box 5. RU 522. SIA. This and other citations in this section courtesy Teasel Muir-Harmony.

913. George Woron to T. W. Taylor via Tillinghast, 17 December 1963. Folder "STP Cooperation with USAF." Box 5. RU 522. SIA. SPACETRACK was the code name for the USAF developmental system for optical reconnaissance, including Baker-Nunn stations, centered at Hanscomb Field in Massachusetts. Richard S. Eckert, "Focusing on the Skies: The Air Force Baker-Nunn Camera System, an Operational and Diplomatic History," Office of History, Air Force Space Command (February 1991).

914. Whipple to Ripley. 17 April 1964. Folder "NASA-Sat. Tkg. Program 1963–1964." Box 7. RU 7431. SIA.

915. Bradley to Hon Eugene Zuckert, Secretary of the Air Force, 18 August 1965. RG 340 1965 SECAF (Cd S). Entry A1 1-F. Box no. 1090 2109-65 Baker Nunn—declassified NND 974384 4-02-09 NARA—Records of the Office of the Secretary of the Air Force [OSAF]. Courtesy James David.

916. Lundquist to "The Record," 30 April 1965. Folder "STP Cooperation with USAF." Box 5. RU 522. SIA.

917. Allan L. Reed, Rear Adm. (USN) to Whipple, 30 July 1965. Entry UD-05D #60. Box 1. RG 373 (Defense Intelligence Agency). National Archives, Greenbelt, MD (NARA). Courtesy James David. Bradley to Zuckert, 18 August 1965, 2.

918. Bradley to Zuckert, 18 August 1965,

919. Whipple to Webb, 27 September 1965. "NASA 1965–1966." Box 8. RU 7431. SIA. Muir-Harmony (2009), 109, makes this point.

920. Muir-Harmony (2009), 109–113, explores these contentions at the working level for the stations in Japan and India, concluding that any classification would have killed the relationship. Lundquist OHI (2011), 382–385, relates how the issue was settled.

921. Vernon Lawrence, Vice Adm. (USN) to Assistant Secretary of Defense. 2 February 1968. Entry UD-05D #60. "Chron file." Box 4. RG 373 (Defense Intelligence Agency). NARA. Challinor to Ambassador Graham Parsons, 19 July 1968. Folder "NASA-Invitation to Propose Experiments." Box 32. RU 253. SIA. "NASA Security Classification for Contracts," 2 August 1967. Folder "STP Security." Box 5. RU 522. SIA.

922. See Compton and Benson (1983), 83–102, cited in Compton (1989), "Setback and Recovery, 1967," http://www.hq.nasa.gov/pao/History/SP-4214/ch7-6.html (accessed 22 March 2014). See also Lynds (1975), table XI, 22.

923. Gerald Truszynski to Bradley, 19 May 1969. Folder "STP 140-STP." Box 5. RU 522. SIA.

924. Gerald Truszynski to Whipple, 20 May 1965. Folder "STP 140-STP." Box 5. RU 522. SIA.

925. Jack Coffey to Tillinghast, 25 July 1969. Folder "STP 140-STP." Box 5. RU 522. SIA.

926. William F. McKeon to John Jameson, 15 September 1969. "S&E Budget, FY 1971," work file—SAO "Office of Programming and Budget 1957–1973." T90068. Box 12 FY 1971. SIA.

927. "12/ Smithsonian Astrophysical Observatory," January 1969. Work file—SAO "Office of Programming and Budget 1957–1973." T90068. Box 12 FY 1971. SIA.

928. Jameson to T. Ames Wheeler, 9 June 1969. Work file—SAO "Office of Programming and Budget 1957–1973." T90068. Box 12 FY 1971. SIA.

929. Bradley to Galler, 17 September 1969. Work file—SAO "Office of Programming and Budget 1957–1973." T90068. Box 12 FY 1971. SIA. "Geoastronomy" remained exclusively an SAO categorization from 1971 through 2015. A scan of the ADS reveals only 48 instances of its use in full text during that period, all but one or two by the SAO.

Chapter 12 – Ground-Based Aspirations in Southern Arizona

930. Whipple, "Mount Hopkins and the MMT," in Weekes, ed. (1979), 3, 5. L. H. Solomon, "Effect of Bright Sky on Baker-Nunn Operation," 9–12, in Solomon (1967).

931. Radio antennae are typically called "dishes" when they assume a concave, parabolic shape.

932. Van Helden (1984), 134. McCray (2004a), 1–12, nicely depicts the lure of large telescopes.

933. McCray (2004a), 48. On the origins of these decadal surveys, their impact, and moreover the burgeoning matrix of intertwined advisory panels created between the Space Science Board and the National Aeronautics and Space Administration (NASA) in the 1960s, see Smith (1989), Tatarewicz (1990), and DeVorkin (2000b). The Whitford Panel limited its attention to ground-based observatories, reflecting strong biases within the astronomical community at the time.

934. Geoffrey Burbidge to Whipple, 7 February 1964. Folder "B." Box 7. RU 7431. SIA.

935. Burbidge to Whipple, 7 February 1964.

936. Whipple to Burbidge, 27 March 1964. Folder "B." Box 7. RU 7431. SIA.

937. Whipple to Burbidge, 27 March 1964.

938. Menzel autobiography (1974), 556–558, on 558.

939. Whitney OHI (2005), 20–21; Layzer OHI (2007), 51–52.

940. Goldberg OHI (1978), 107. On Pusey's sympathies for a meritocratic culture, see Keller and Keller (2001), 177.

941. Goldberg OHI (1983), 14–17.

942. Goldberg OHI (1983), 11 (by O. Gingerich).

943. Foster F. Comstock to Arthur Trottenberg, 4 January 1966. "Harvard College Observatory 1965–1983." Box 19. HUG 4876. HUA.

944. Menzel to Goldberg and Whipple, 9 October 1967. "Harvard College Observatory 1965–1983." Box 19. HUG 4876. HUA.

945. "Minutes of the Meeting." 21 November 1967. Nine people were on the committee, including Goldberg, Menzel, Whipple, Dalgarno, Layzer, Liller, Lilley, and Whitney. See also "Minutes of the Meeting of the Committee of Senior Professors" (20 January 1968), 16. "Harvard College Observatory 1965–1983." Box 19. HUG 4876. HUA.

946. On the Prairie Network, see Plotkin (1997).

947. Ripley to the Honorable Clinton P. Anderson, 21 January 1966. FLWO Event History. See http://www.sao.arizona.edu/FLWO/flwohis.html.Hereinafter FLWO Event History. Compiled by FLWO staff, esp. Don Hogan. See also Whipple in Weekes, ed. (1979), 3.

948. Whipple OHI (1976), 67.

949. Smith and McCrosky (1954).

950. Latham OHI (2005), 31.

951. Anderson to Ripley, 22 January 1966. FLWO Event History, 4. See also Oehser (1970), 193–194. *Annual Report of the Board of Regents of the Smithsonian Institution* (1959), 51.

952. These included the design and procurement of buildings and out-structures to house scientific equipment, and engineering services for surveying and developing access roads. "Contracts File, 1966–1970." Boxes 1–3 in series 1. RU 262, Smithsonian Astrophysical Observatory, Mt. Hopkins Department, Records. SIA.

953. Memoranda, Tougas to Coffee, 10 May 1966; 30 September 1966. FLWO Event History. "Tribute to Mark Malec by Tom Hoffman." "Malec" folder. Series 7. Box 1. Lundquist Collection. U. of Alabama library.

954. Whipple (1966), "Smithsonian Astrophysical Observatory," 181–213, on 196–197.

955. Hirsh (1983), ch. 3.

956. According to Hirsh (1983), 49–50, these detectors were never operational because of the delicacy of the spark chambers and their susceptibility to vibration and high vacuum environments. Whipple, "Report on the Astrophysical Observatory," 147–173, in Smithsonian Institution (1963).

957. Defelice et al. (1963), 673–674. Goldberg describes the circumstances of his appointment in Goldberg OHI (1983), 2–3, with O. Gingerich.

958. Greisen (1971) credits Morrison (1958), 858. See also P. Morrison (1961), "The Origin of Cosmic Rays," in Flügge, ed. (1961). Pollack and Fazio (1965) cite the 1961 paper.

959. Pollack and Fazio (1965), 730–744.

960. Fazio et al. (1968a), L83–L89.

961. Watson (2011), 13–19.

962. Letters in "Fazio" folder. Box 2. Acc. 85-215. SIA.

963. See Helmken (1968). George Dick to S. Galler et al., through Malec, Tillinghast, and Hurlbut, 31 January 1969. Folder "Gamma Ray Balloon Flights 1969–1970." Box 4. RU 262. SIA.

S. Criswell to J. A. Coffey, "Space Requirements for Gamma Ray Flight Experiments Activities," 11 April 1969. Folder "Gamma Ray Astronomy, 1969–1970." Box 4. RU 262. SIA. For an example of continuing efforts, and an indication of how the work has evolved, see Hong et al. (2011).

964. Natick engineers had built a "giant heliostat" as a solar furnace on the shores of Lake Cochituate to "simulate the thermal effects of nuclear explosions," generating temperatures up to 7,000 degrees. The facility was available at night to groups who could pass security background checks. Earls (2005), 73. Night sky work started at Natick in parallel with plans for constructing a suitable astronomical reflector with Hill of MIT. G. Fazio, "History of Air Cerenkov Detector." Acc. 85-215. Box 2. SIA.

965. Fazio et al. (1968a). See also Fazio et al. (1968b), 892–893. The "Gamma Ray Reflector" made it seem to local native Americans that the facility was attracting dangerous radiation.

966. Whipple, in Weekes, ed. (1979), 5.

967. Jacchia (1967), 93; Marra (2000), 88; Horn-d'Arturo (1953), 71–74.

968. Davies and Cotton (1957), 1, 16. Throughout late 1966 and into 1967, the design group considered various options; the issue was not so much mechanical design as optical performance. See folder "Blanton & Co. Architects and Engineers." Box 1. RU 262. SIA. See also "Mount Hopkins Observatory." Box 2. Acc. 85-215. SIA.

969. Burke and Kirchhoff (1968), 284–286.

970. Fazio et al. (1968b), 892–893. A pictorial history of these efforts can be found in Trevor Weekes, "How We Got from There to Here?" delivered at the VERITAS First Light Ceremony in April 2007: see http://veritas.sao.arizona.edu/about-veritas-mainmenu-81/history-aamp-timeline-mainmenu-89 (accessed 21 December 2012).

971. After much negotiation, the Smithsonian submitted an alternative route for the access road to the Montosa Canyon site that did not disturb nearby sweat lodges and talking circle atmospherics. In following years, proposals to greatly expand the facility met with strong resistance. Thomas Stauffer, "Compromise Allows Building Near Sweat Lodge," *Arizona Daily Star*, 2003, http://www.azsj.org/forums/showthread.php?t=6700 (accessed 21 December 2012); see also Mark Muro, "Telescope Plans Blocked by Indian Ritual Site," http://news.sciencemag.org/sciencenow/1999/09/08-01.html (accessed 21 December 2012).

972. Greisen (1971), 61, 71.

973. Fazio (1967).

974. Along the way, there were setbacks, closures, and intervening efforts at photo detector improvement. Weekes formed what was called the Whipple Collaboration, including several dozen physicists at institutions in the United States, Great Britain, Ireland, Hong Kong, Argentina, and France, which provided facilities with improved sensitivity and an imaging camera. Weekes (2012). The Whipple Collaboration was first mentioned explicitly in an annual report from Iowa State University (*BAAS* 18 [1986]: 200–202), consisting of the Smithsonian, University College Dublin, Iowa, and the University of Hawaii.

975. Greisen (1971), 63–64, 69.

976. Weekes et al. (1989). As of December 2012, this paper had garnered some 365 citations according to the SAO/NASA Astrophysics Data System (ADS). http://adsabs.harvard.edu/

977. Whipple OHI (1976), 50. "Preparation of a Master Plan," 27 April 1967. "Blanton & Co. Architects and Engineers." Box 1. RU 262. SIA.

978. Gregory to Bartnik, 1 August 1971. "Program Scientists for the MMT Program." Box 2. RU 7431. SIA.

979. For a general description, see Burke and Kirchhoff (1968), 284–286.

980. John A. Coffey, "Carlton W. Tillinghast, Jr. 1933–1969," *SAO News* 9, no. 9 (September 1969): 1.

981. "The 60-Inch Telescope at Mt. Hopkins Dedicated as 'Tillinghast Reflector,'" *SAO News* 10 no. 6 (September–October 1970): 1.

982. Coffey (1969), 1. Tillinghast, "Joint Government-University Laboratories in the United States," *Smithsonian Year 1969,* 207–217.

983. *Smithsonian Year 1966*, 197.

984. The design and construction of the 60-inch employed a distributed contracting scheme rather than a single-source vendor. At the time, instruments of this class could be procured from commercial firms. But the SAO chose to develop the pieces and act as general contractor to gain intimate experience in the process. See "Smithsonian Astrophysical Observatory, Mt. Hopkins Department, Records," series 1, "Contracts" file. RU 262. SIA. See also, in retrospect, Gregory to Bartnik, "Program Scientists for the MMT Program," 1 August 1971. Box 2. RU 7431. SIA.

985. Ripley, "Statement by the Secretary," 1, in *Smithsonian Year 1969,* 21.

986. *SAO News* 10, no. 3 (March–April 1970): 1.

987. FLWO Event History, 35.

988. Chaffee to Weekes, January 1976. "Summary of 1.5 meter telescope research from Oct 1970 to Jan 1976." "MMT/SAO" files. Box 20. RU 253. SIA.

989. "Land Withdrawal Request," 22 May 1968. "Mt Hopkins Reports." Box 31. RU 468. SIA.

990. On the ERC, see Andrew J. Butrica, "The Electronics Research Center: NASA's Little Known Venture into Aerospace Electronics," https://history.nasa.gov/ercaiaa.html.

991. "Prefatory Remarks and Work Statement for Contract between NASA Electronics Research Center (ERC) and Smithsonian Astrophysical Observatory (SAO)," 17 December 1968. "National Aeronautics and Space Administration: Electronics Research Center—Projects at Mt Hopkins 1968–1970." Box 4. RU 262. See also reports in "Observer Services Division 1969–1970." Box 5. RU 262. SIA.

992. La Count to NASA, 3 July 1969. Folder "Laser Intercomparison Test." "Laser Co-Location" file. Box 4. RU 262. SIA.

993. Weekes to Whipple, 23 January 1970. "Report on scientific activity for C1969." Folder "Monthly progress reports 1966–1970." Box 4. RU 262. SIA.

994. Whipple, introduction, in Trevor C. Weekes, *Mount Hopkins Science Review* (Smithsonian Institution Astrophysical Observatory, October 1970), 2. "Mount Hopkins Reports." Box 31. RU 468. SIA.

995. Meinel (1960), "Design of Reflecting Telescopes," in Kuiper and Middlehurst (1960), 25–42, on 25. McCray (2004a), 61–64, explores Meinel's background and his creation of the Optical Sciences Center. Here we examine in greater detail how his ideas were developed in partnership with the Smithsonian.

996. There were variations on this rough formula. See Learner (1981), 199, and Mertz (1971), 386.

997. "General Notes," *Publications of the Astronomical Society of the Pacific* 79 (1967): 517. Now known as the Bok Telescope, it operates at an f/45 Cassegrain focus.

998. The facility became the College of Optical Sciences. See http://www.optics.arizona.edu/about-osc (accessed 18 February 2013).

999. Dennis (1991).

1000. Robert F. Piper, *History of Space and Missile Systems Organization 1 July 1967–30 June 1969*, vol. 1 (March 1970). SAMSO Chief of Staff History Office. Redacted copy at http://nsarchive.gwu.edu/NSAEBB/NSAEBB235/10.pdf. I thank James David for guidance searching out these records.

1001. Various discussions of MOL are available. See Richelson (1990), 82–86. See also Berger (1970).

1002. Peebles (1987), 251, provides an overview of the life of the program without any details of the optical systems involved.

1003. Berger (1970), 62–65; Richelson (1990), 85.

1004. They turned out to be 72 inches. See redacted copy at http://www.nro.gov/foia/declass/QUILL/25.%20Semi-Annual%20Report%20to%20the%20PFIAB%201965-66.PDF.

1005. Heppenheimer (1998), 204–205. Quotes from John L. McLucas to Melvin Laird, "Taking Stock of the National Reconnaissance Program," 18 December 1972. NRO Release 17 September 2011. "I have seen no adverse effects on the cover programs—and do not expect any of significance," 20. Cited in Perry (1973), accessed at http://www.nro.gov/foia/docs/hosr/hosr-vol1.pdf.

1006. Meinel, 50–52, in MMT Sloan Foundation Videohistory, sess. 6 (11 May 1989). SIA.

1007. Meinel, 4.

1008. John L. McLucas to Melvin Laird, 18 February 1972. "Taking Stock," 21.

1009. *Smithsonian Year 1970*, 49.

1010. Options included a fixed hemisphere (Arecibo type) structure; various types of movable primaries; an "Eyeball," a fully steerable floating sphere containing a stretched, reflecting membrane deflected by a partial vacuum; and a scheme called "CORT," a fixed paraboloid pointed horizontally with a rotating plane mirror feed. There were also several radio-telescope type suggestions. See S,G&H: "Notes for Discussion at first progress meeting on conceptual studies for a major optical reflector," 20 March 1968. Folder "Simpson, Gumpertz & Hager Consulting Engineers 1968–1969." Box 3. RU 262. SIA.

1011. "'Infrared Astronomy' proposal to SI Research Foundation," 4 March 1969. Folder "Infrared Astronomy 1968–1969." Box 4. RU 262. SIA.

1012. Hoffman to Carleton et al., 23 July 1968. "Final Meeting on Initial MOR Study," "MOR 68–70." Box 8. RU 253. SIA.

1013. R. N. Watts "For The Record": MOI Planning Meeting, "Major Optical Instrument," 8 July 1969. "MMT/SAO" files. Box 20. RU 253. SIA.

1014. Mertz, in Dickson, ed. (1970), 507–513. Mertz had been an effective promoter of Fourier techniques. See Mertz (1965, 1969), and Bowen (1964), 816–825, on 825.

1015. L. Mertz (August 1968), "Design for a Giant Telescope," "MOR 68–70." Box 8. RU 253. SIA.

1016. AMB stands for "Astronomy Missions Board"; ASC for "Astronomy Survey Committee." Meinel, "Aperture Synthesis Using Independent Telescopes," 22 May 1970. Box 20. RU 253. "Correspondence 1970–1971." Box 32. RU 468. SIA. Frank Low to "Members of AMB and ASC Infrared panels and Interested Colleagues," 4 June 1970. UA/GK.

1017. McCray (2004a), 62–64. On Low, see Don McCarthy, "Obituary: Frank J. Low (1933–2009)." *BAAS* 43, no. 17 (December 2011). http://adsabs.harvard.edu/abs/2011BAAS...43..017M. The history of the MMT is also detailed in Beckers et al. (1981), which includes lengthy testimony from Ray Weymann and others at Arizona. See http://www.mmto.org/node/442 (accessed 17 March 2016).

1018. Smith (1989), 65. The AMB, chaired by Leo Goldberg, had been convened in 1967 to advise NASA on priorities for support for both long-term and short-term goals For an overview and its relations with the myriad array of boards advising on priorities for astronomy, see Tatarewicz (1990), 100–103. See also D. B. Wood, "Astronomy Missions Board Meetings—Case 710" (12 November 1968). NHO. See http://ntrs.nasa.gov/archive/nasa/casi.ntrs.nasa.gov/19790073768_1979073768.pdf. Goldberg sat on the AMB solar astronomy panel, Whipple was a member of the optical astronomy panel, and Low was a member of the radio astronomy panel. Goldberg and Robert Noyes were also on the Greenstein Committee. See National Research Council (1972).

1019. Meinel, 53–54, in MMT Videohistory, sess. 6 (10 May 1989). RU 9542. SIA. Portions of this exchange appear in McCray (2004a), 65.

1020. Whipple to Low, 29 July 1970. Box 20. RU 253. SIA. See also folder "Corr. 70–71." Box 32. RU 468, which also contains undated notes indicating that Arizona was seeking some $2.5 million from the Air Force.

1021. Whipple, Carleton, and Hoffman, 7–8, 13, 15, in MMT Videohistory, sess. 2 (10 May 1989). RU 9542. SIA. Whipple OHI (1976), 38–39.

1022. Meinel, Aden, 22 May 1970. "Aperture Synthesis Using Independent Telescopes." Box 20. RU 253. SIA.

1023. Frank Low to "Members of AMB and ASC Infrared panels and Interested Colleagues." 4 July 1970. UA/GK. See also "Correspondence 1970–1971." Box 32. RU 468. SIA.

1024. *Smithsonian Year 1970,* 49.

1025. "A Large Astronomical Telescope at Low Cost." SAO submission to OMB, November 1970. RU 7431. SIA.

1026. "Some Questions and Answers to Supplement the Brochure, a Large Astronomical Telescope at Low Cost, concerning SAO's 1972 Request for a Large Six-Element Telescope." Folder "A Large Astronomical Telescope at Low Cost." Box 32. RU 468. SIA.

Chapter 13 – Building the MMT

1027. Lundquist notes "Meeting with U of A on Large Telescope, 15 January 1971. "Multiple Mirror Telescope 1970–1973, folder 1 of 3." Box 8. RU 7431.19710104: Whipple to Kuiper, 4 January 1971. Box 2. RU 7431. SIA.

1028. USRA is a national consortium of more than four dozen universities created to manage large NASA-funded programs. See http://www.usra.edu/about/history/ (accessed 5 March 2013).

1029. Lundquist notes, 15 January 1971. The Mansfield Amendment would be lifted later in 1971, but its dampening effect continued. On the "Mansfield Amendment" to the Military Procurement Authorization Act of 1970 (PL 91-121, Sec. 203) see *National Science Policy* (1970), 84.

1030. Whipple to Seamans, 22 January 1971. Box 8. RU 253. SAO Asst. Director for Science (Lundquist copy). SIA.

1031. Seamans to Whipple, 22 February 1971. Box 8. RU 253. SAO Asst. Director for Science (Lundquist copy). SIA.

1032. McLucas to Laird, "Taking Stock," 21.

1033. Whipple to Kuiper/Low/Meinel/Weymann [cc Goldberg], 25 February 1971. "Six-Element Telescope Project" folder "MMT-FLW." Box 32. RU 468. SIA.

1034. Whipple to Bradley, 16 March 1971. "Congress and the 6-Element Reflector." Folder "Corr. 70–71." Box 32. RU 468. SIA.

1035. "Draft" transcript of testimony between Senator Bible and Whipple, 23 March 1971. Folder "MMT Funding 71–73." Box 31. RU 468. SIA.

1036. Weymann to Morris K. Udall, 12 May 1971. Box 2. RU 7431. SIA.

1037. Whipple to Meinel, 1 July 1971. Folder "Corr. 70–71." Box 32. RU 468. SIA.

1038. Ripley (Bradley) to William D. McElroy, 23 March 1971. Folder "MMT Funding NSF." Box 32. RU 468. SIA.

1039. McElroy to Ripley, 12 April 1971. Folder "MMT Funding NSF." Box 32. RU 468. SIA.

1040. Ripley identical letters to Chet Holifield and John L. McClellan. 5 March 1971. "Dept. of Science and Technology 1971." Box 46. RU 253. SIA.

1041. Weymann to Kuiper et al., 18 February 1971. "Report of a visit to Fleischer at NSF." UA/GK.

1042. On NASA's interests, see Watts to Bartnik and Lundquist, 2 June 1971, and "Lundquist for The Record," 31 July 1971. "Multiple-Mirror Telescope" folder 2." Box 8. RU 253. SIA.

1043. Meinel to Fleischer, 7 February 1972. "MMT Funding NSF." Box 32. RU 468. SIA.

1044. MMT Videohistory, sess. 6 (11 May 1989), 18–19. RU 9542. SIA.

1045. "Some Astronomical, Engineering, and Economic Aspects of a Multiple-Mirror Telescope—a joint project of the University of Arizona and the Smithsonian Astrophysical Observatory, June 1971." Box 2. RU 7431. SIA.

1046. Peter Boyce and Mary Saffell, "Robert Fleischer," http://adsabs.harvard.edu/abs/2003BAAS...35.1460B.

1047. Meeting of "Government Astronomy Administrators Round Table" (GAART). Robert Fleischer to Large Distribution List Attached, 7 April 1969. "NSF 1963–1972." Box 47. RU 253. SIA.

1048. The MMT as a concept for the space telescope is also mentioned in Meinel, "Multiple Mirror Telescopes of the Future," in Weekes, ed. (1979), 8.

1049. Smith (1989), 378.

1050. Gregory to Bartnik, 26 July 1971. Folder "Corr. 70–71." Box 32. RU 468. SIA. See Smith (1989), 69–73, for hints as to the role of the 72 inch mirrors in NASA's plans for space telescopes.

1051. Gregory to Bartnik, 1 August 1971. Folder "Program Scientists for the MMT Program." Box 2. RU 7431. SIA. Gregory "for the Record, MMT Program," 2 August 1971. Folder "MMT Funding 71–73." Box 31. RU 468. Whipple to "All of UA and SAO Interested Scientific Staff." Folder "MMT Program Organization." Box 2. RU 7431. SIA.

1052. Gregory and Hoffman, 16 August 1971, "for the record." Folder "Conferences and Meetings 71–74." Box 32. RU 468. SIA.

1053. Gregory to Attendees, 15 October 1971. "MMT Conf. & Mtgs." Box 32. RU 468. SIA.

1054. Gregory to Distribution, 24 September 1971. Folder "MMT Scientific and Operational Requirements"; Gregory to Bartnik, 28 September 1971. Folder "MMT Conf. and Mtgs." Box 8. RU 253. SIA.

1055. Gregory to Attendees, 15 October 1971. Folder "MMT Conf. & Mtgs." Box 32. RU 468. SIA.

1056. Menzel to Goldberg and Whipple, 9 October 1967; "Minutes of the Meeting of the Committee of Senior Professors," 21 November 1967. Box 19. HUG 4876. HUA.

1057. "Minutes of the Meeting of the Committee of Senior Professors," 16; 20 January 1968. "Harvard College Observatory 1965–1983." Box 19. HUG 4876. HUA.

1058. "Land Withdrawal Request," 22 May 1971. Folder "Mt Hopkins Reports." Box 31. RU 468. SIA.

1059. Whipple to Kuiper/Low/Meinel/Weymann, 25 February 1971. "FLW Correspondence 71–74." Box 32. RU 468. SIA.

1060. Bartnik to Goldberg, 15 June 1971. "MMT/SAO files." Box 20. RU 253. SIA.

1061. Goldberg to Whipple, 21 July 1971. "MMT-FLW." Box 32. RU 468. SIA.

1062. Bartnik "For the Record" to Whipple, 23 July 1971. "MMT-FLW." Box 32. RU 468. SIA.

1063. Goldberg OHI (1978), 108. See also Goldberg OHI (1983), with Owen Gingerich, 14–15; A. Dupree OHI (2007), 49–51.

1064. Goldberg OHI (1978), 106–107.

1065. Robert Fleischer to Weymann, 3 November 1971. "MMT Funding NSF." Box 32. RU 468. SIA.

1066. Weymann to Fleischer, 8 November 1971. "MMT Funding NSF." Box 32. RU 468. SIA.

1067. Whipple to Fleischer, 13 December 1971. "MMT Funding NSF." Box 32. RU 468. SIA.

1068. Fleischer to Whipple, dictated 23 December, 30 December 1971. "MMT Funding NSF." Box 32. RU 468. SIA.

1069. Whipple to Fleischer, draft, 30 December 1971, n.d. "MMT Funding NSF." Box 32. RU 468. SIA.

1070. Whipple to Fleischer, 10 February 1971. "MMT Funding NSF." Box 32. RU 468. SIA.

1071. Weymann to Fleischer, 15 February 1972. "MMT Funding NSF." Box 32. RU 468. SIA.

1072. Fleischer to Weymann, dictated 18 February, 1 March 1972. "MMT Funding NSF." Box 32. RU 468. SIA.

1073. National Research Council (1972), 82–83. Members of the committee were skeptical at first, and, according to Meinel, "very hard to convince." But after Meinel made personal appeals to Jesse Greenstein, the chair, and when Bruce Margon, Weymann recalled, provided particularly effective testimony during the final deliberations, this "swayed the Greenstein Committee." MMT Videohistory, sess. 6 (11 May 1989), 7–8. SIA.

1074. Bartnik to Challinor, 16 June 1972. Folder "MMT Funding 71–73." Box 32. RU 468. SIA.

1075. Meinel to Hon. Harry Davis, 23 December 1971. Folder "MMT Funding NSF." Box 32. RU 468. SIA. Background for Meinel's work can be found in Robert F. Piper "History of Space and Missile Systems Organization 1 July 1967–30 June 1969 Vol. 1" (March 1970), SAMSO Chief of Staff History Office. Redacted copy at http://nsarchive.gwu.edu/NSAEBB/NSAEBB235/10.pdf. See also Meinel and T. S. Byington (1969); Meinel to Hon. Harry Davis, 23 December 1971. "MMT Funding NSF." Box 32. RU 468. SIA.

1076. Weymann to E. Creutz, 13 October; 26 November 1973. "MMT-Funding NSF." Box 32. RU 468. SIA.

1077. John P. Schaefer to H. Guyford Stever, 29 November 1973. "Gen. Corr. 71–74." Box 32. RU 468. SIA.

1078. Ripley to Rhodes, 2 December 1975. Folder "MMT/SAO files." Box 20. RU 253. SIA.

1079. Kuiper to Whipple, 30 December 1970. Folder "Corr. 70–71" and "U of Az/MMT Prof to NSF Corr. 70–71." Box 32. RU 468. SIA.

1080. Weekes to Whipple, 12 May 1970. "MOR File 68–70." Box 8. RU 253. SIA.

1081. David W. Latham, 22 May 1971. "Proposed Site-Testing Program for the Comparison of Mt. Hopkins and Mt. Lemmon." Folder "Corr. 70–71." Box 32. RU 468. SIA.

1082. Whipple "For the Record" conversation with Low, 4 August 1971. Folder "MMT-FLW." RU 468. SIA.

1083. "Multiple Mirror Telescope Facility Agreement," revised MOU, 20 September 1971. Box 2. RU 7431; Gregory to Distribution, 24 September 1971. "MMT Scientific and Operational Requirements." Box 8. RU 253. SIA.

1084. Kuiper to Weaver, 2 November 1971, "U S Forest Service Meeting of November 1 (Mt. Lemmon Policies)." Folder "Gen. Corr. 71–74." Box 32. RU 468. SIA. With clipping, "Catalina Land Use Plan Draws Strong Criticism" [*Tucson Star*]. Kuiper to Doran, 2 November 1971. Folder "Gen. Corr. 71–74." Box 32. RU 468. SIA.

1085. David Latham, 7 January 1972. "Results of the Site-Testing Program for the Comparison of Mt. Hopkins and Mt. Lemmon." "Multiple-Mirror Telescope 1970–1973" folder 3. Box 8. RU 253; Gregory to "MMT Distribution," 12 January 1972. Folder "MMT Steering Committee." Box 32. RU 468. SIA.

1086. The selection was also partly based on economy. Shannon recalled that "Corning had an f2.7 tool available, which saved about $15,000 in the processing." MMT Videohistory, sess. 5 (10 May 1989), 4. SIA.

1087. "Summary of MMT Steering Committee Meeting, 19 January 1972," 19 January; 1 February 1972. Bartnik "for the record," 25 January 1972. "MMT Steering Committee." Box 32. RU 468. Box 32. SIA. See also Weymann statement in Beckers (1981).

1088. "MMT Steering Committee meeting." 9 June 1972. Box 8. RU 253. SIA.

1089. Kassander to Whipple, 7 February 1973. Folder "MMT-FLW." Box 32. RU 468; Kuiper to Whipple, 13 December 1972. Box 2. RU 7431. SIA.

1090. Jesse Greenstein, 20 November 1975. "Report of the Panel to Review the Multiple Mirror Project," B4. Folder "MMT/UA Proposals to NSF." Box 32. RU 468. SIA.

1091. Greenstein, 20 November 1975 draft; 29 January 1976 final, "Report." Jameson to Ripley, Challinor, Field, and Gaynor, 20 February 1976; Challinor to Greenstein, 23 February 1976. "MMT/SAO files." RU 253. Strittmatter and William Hoffmann to Peter Boyce, NSF, 7 April 1976. Formal withdrawal of Arizona's proposal. Folder "U of A MMT prop to NSF." Box 33. RU 468. SIA. On the influence of the Greenstein report, see John F. Jameson to Challinor, Wheeler, and Ullberg, "MMT program funding status." 10 September 1976. "MMT/SAO files." RU 253. SIA.

1092. Pamela Mayhew, "New Telescope to Be Boon or Boondoggle," *Tucson Daily Citizen*, 29 April 1976; Draft of Telegram, n.d., Challinor to McKalip, Editor, "Arizona Citizen." Assistant Secy. for Science 1963–1978. RU 108. Folder 3. Box 31. SIA.

1093. Walter Sullivan, "New Telescope Is Considered a Breakthrough: Third Most Powerful Telescope," *New York Times*, 12 June 1978, B10. Box 2. RU 7431. SIA. On final costs, see Meinel, "Multiple Mirror Telescopes of the Future," in Weekes, ed. (1979), fig. 1, 10.

1094. Beckers (1982), fig. 1. FLWO Event History, 161–163.

Chapter 14 – National Facilities for Astronomy and the Question of Limits

1095. Sullivan (2009), 46–47; 398–403. Whipple and Greenstein (1937).

1096. Sullivan (2009), 412–413; Munns (2002), 126–151. Bart Bok OHI (1978), 39, 51. AIP.

1097. Bok OHI (1978), 67.

1098. Bok and Ewen (1954), 318. On Menzel's organizational efforts, see Munns (2003), 101.

1099. Bok et al. (1957), 8; Heeschen (1956), 588–590; Bok (1956), 141–147.

1100. "Lovell Reports on Latest Jodrell Bank Work," *SAO News* 2, no. 4 (April 1964): 1. On the sense among mainstream astronomers in the 1950s that radio astronomers were not "real" astronomers, see Sullivan (2009), 424–425.

1101. Munns (2003), following England (1982), has detailed the strong resistance in the 1950s to centralized national facilities.

1102. England (1982), 280–285.

1103. See, for instance, Roback (1969), 36–43. On the debate, see McCray (2004a, b).

1104. Lilley (1961).

1105. Barrett and Lilley (1963), 195.

1106. Goldberg OHI (1978), 74.

1107. Background on Lincoln and planetary radar can be found in Butrica (1996), ch. 2. Butrica, ch. 3, examines NEROC, centering on reconciling the technical needs of planetary radar astronomy and mainstream radio astronomy, seeing this as an important ingredient in the eventual demise of the 440-foot dish. Here we offer a complementary interpretation that looks at competition to create national facilities.

1108. "Haystack Radar-Radio" (1964).

1109. Meeks and Lilley (1963); Barath et al. (1964).

1110. Lilley to File, 24 March 1963 (n.d.), "Joint Radio and Radar Facility; Log of relevant Meetings, Actions and Correspondence." "Lilley Log." Acc. 83-121. SIA. Hereinafter AELL/SIA.

1111. Findlay (1963), 73.

1112. A "key" meeting was noted in Butrica (1996), 67, ref. 36, "Lilley to File."

1113. National Research Council (1964).

1114. National Research Council (1964), 24. The report was highlighted in Philip Abelson. "Ground Based Astronomy" Science 146 (4 December 1964), 1255. See also Lubkin (1975), courtesy Patrick McCray and Gloria Lubkin.

1115. National Research Council (1964), 57.

1116. Lilley log entry for 13 November 1963.

1117. Lilley log entries, December 1963.

1118. Hawkins et al. (1956), 179.

1119. Whipple, "Meteor Work at Harvard College Observatory and Smithsonian Astrophysical Observatory in 1957, 1958, and 1959," in Whipple (1972), 1:422–424; Whipple and Hawkins (1959); Hawkins (1963), 53–62; Gilbert (1977, repr. 2013).

1120. Lilley log entries from 2 December 1963 through 20 February 1964.

1121. On early evaluations, see Meeks et al. (1968), 746–751; Meeks and Ruze (1971), 723–728.

1122. Ripley to Haworth, 26 February 1965. "NSF 1972." Box 14. RU 108. SIA.

1123. Bradley to Ripley w/copy to Galler, 1 December 1965. "Trip to the Smithsonian Astrophysical Observatory with George Evans." Box 7. RU 137. SIA. Sugar Grove was a Navy project to build a surveillance system capable of monitoring radio communications in Soviet-held territories, using the Moon as a radio reflector. The Navy attempted to give it an astronomical cover. On Sugar Grove, see Greenberg (1962), 327. See also van Keuren (2001) and DeVorkin (2000b).

1124. Bradley to Ripley, 1 December 1965.

1125. Lilley log entry for 26 August 1965.

1126. On CAMROC and NEROC, see Butrica (1996), 63, 73–85. NEROC comprised nine northeastern institutions. See also Lilley log entries for 27 May 1966; 6 July 1966. Joel Orlen to Wiesner, Purcell, Ruina, and Whipple, 21 June 1967; Jerome B. Wiesner to Howard W. Johnson, 21 June 1967 and 22 June 1967: "Certification of Incorporation of Northeast Radio Observatory Corporation." Folder "NEROC 1967–69 and undated." Box 6. RU 256. SIA.

1127. Brooks was dean of the Division of Engineering and Applied Sciences at Harvard.

1128. Lilley log entry for 14 August 1967. On Brooks's role and on the Dicke Panel, see Lubkin (1975), 5–6; and see McCray (2004a, b) for similar discussions regarding priorities for optical astronomy.

1129. McElheny (1967), 907.

1130. "Ad Hoc Panel" to Haworth, 21 August 1967, 14. "NEROC 1967–69 and undated." Box 6. RU 256. SIA. McElheny (1967), 909–910. Butrica (1996), 75, notes that in terms of competitiveness for planetary radar, "the Dicke Panel report devastated NEROC plans, not to mention planetary radar astronomy at Lincoln Laboratory." Wiesner's reply is from: Orlen to CAMROC members, 11 October 1967, attaching Wiesner to Haworth. Folder 9 "CAMROC, 1964–69 and undated." Box 1. RU 256. SIA.

1131. Hurlburt call to Lilley, 23 May 1968. Lilley log entry that date.

1132. Ripley to Whipple, 22 November 1967. "Astrophysical Observatory Smithsonian (1.AOS) 1963–1973." Box 1. RU 108. SIA.

1133. Smithsonian Institution appropriations hearings, 95th Cong., 2d sess., pt. 1: Senate (1979), 1076. http://library.si.edu/digital-library/book/smithsonianin95211979smit#sthash.vz4imabv.dpuf (accessed 20 October 2014).

1134. Ripley to Haworth, 6 March 1968. "NSF-SI Relations." Box 250. RU 99. SIA. On Albert Thomas and his penchant for "equitable distribution" of NSF funds, see "Albert Thomas" (1966), 1064–1065.

1135. Ripley to Haworth. "Godmother" and "universal institution" from Dupree (1986), 330.

1136. "Memo," 8 February 1968, and "A Statement on Sponsored Research and the National Science Foundation," 1 March 1968. Folder "NSF," box 341; and folder "NSF-SI Relations," 6. Box 250. RU 99. SIA.

1137. Lilley log entries for 8 August and 11 August 1968. Bradley to Ripley, "Proposal for Radio Astronomy at SAO," 22 August 1968; Lilley to Wiesner, 22 August 1968. "NEROC 1967–69 and undated." Box 6. RU 256. SIA. See also Lubkin (1975), 12.

1138. Lilley to Wiesner, "NEROC Status," 22 August 1968. Folder "NEROC 1967–1969." Box 6. RU 256. Bradley, AEL to FLW, 27 August 1968. Bradley to Ripley, 27 August 1968, reporting on a meeting at MIT with Wiesner, Purcell, Shapiro, Whipple, Lilley, Lundquist, Tillinghast, and Bradley, clipped to Bradley to Fleischer, 17 January 1969. "NSF." Box 341. RU 99. SIA.

1139. Lilley log entries for 13, 15, and 21 August 1968.

1140. Bradley to Ripley, 27 August 1968. Folder "NSF" regarding MIT meeting report. Box 341. RU 99. SIA.

1141. Loweth's role at the Office of Management and Budget (OMB) and related activities regarding the organization of the science budget during the Nixon administration can be inferred from Greenberg (2003), ch. 5.

1142. Even though the term "Big Dish" had certain negative connotations, in the wake of the Sugar Grove fiasco, the term was used by SI staff informally for the radio-radar telescope. See Bradley to Ripley, 12 September 1969 [RU 99, box 341, folder "NSF"] and "Congress Gets 'Big Dish' Bill," *SAO News* 9, no. 4 (March 1969): 1.

1143. Bradley to Ripley, 11 September 1968, clipped to Bradley to Fleischer, 17 January 1969. Folder "NSF." Box 341. RU 99. SIA.

1144. Bradley to Ripley, 11 September 1968.

1145. Inferences from Roback (1969).

1146. Lilley to Bradley. Lilley log entry for 20 September 1968.

1147. See "Smithsonian Science Information Exchange, Records, 1946–1983." Box 1. RU 482. SIA.

1148. Bradley to Ripley, 27 August 1968, regarding "meeting with Weisner on radio telescope." See also entry for 1 December 1968, and items in the file dating back to the previous June. Box 1. RU 108. SI Asst. Secy. for Science 1963–1973. SIA.

1149. Victor Cohn, "$3-Million Request to Increase Range of Big Telescope Denied," *Washington Post*, 19 March 1968. Copy in "NEROC 1967–1969." Box 6. RU 256. SIA.

1150. Lilley log entries in August, especially 27 August 1968 and 4 October 1968.

1151. Peter Gwynne, "U.S. Falls behind Europe in Radio Astronomy," *Boston Herald Traveler*, 6 October 1968, n.p. Copy clipped to Lilley to NEROC Distribution, 7 October 1968. Folder "NEROC 1967–69." Box 6. RU 256. SIA.

1152. Findlay, "Summary of a meeting," 1 December 1968. Folder "Astrophysical Observatory Smithsonian (1.AOS)." Box 1. RU 108. SI Asst. Secy. for Science 1963–1973. SIA.

1153. Lubkin (1975), 5; J. W. Findlay, April 1967. "A Synthetic Antenna Which May Meet the LFST Requirements," manuscript FPST/9. Folder 9, "CAMROC 1964–1969." Box 1. RU 256. SIA.

1154. Lilley log entry for 11 December 1968.

1155. Ripley to Regents, 25 November 1968. See also "Statement on Radio and Radar Astronomy," January 1969; and "A Bill to Authorize the Smithsonian Institution to acquire lands and to design a radio-radar telescope for the Smithsonian Astrophysical Observatory." 15 January 1969 draft letter to Regents. "Astrophysical Observatory, Smithsonian." Box 1. RU 108. SIA.

1156. Ripley to Regents, 25 November 1968.

1157. Senators Anderson, Fulbright, and Scott, authorization bill S. 705 introduced 28 January 1969. Clipped to Ripley to Haworth, 17 March 1969. Folder "NSF." Box 341. RU 99. SIA.

1158. As Lilley recorded, "Lilley receives call from Orlen at Lincoln Laboratory. He is extremely concerned about Whipple's assertion that STAG reports to Smithsonian." Lilley log entry for 6 January 1969.

1159. Lilley log entries for 3, 6, 14, 27, 28 February 1969.

1160. Lilley log entry for 27 February 1969. On Shapiro's work at Millstone and Haystack, see Butrica (1996) throughout, as well as Benjamin Wilson and David Kaiser, "Calculating Times: Radar, Ballistic Missiles, and Einstein's Relativity," in Oreskes and Krige, eds. (2014), 273–316.

1161. Edmondson (1997), 63.

1162. Edmondson (1997), 59–61.

1163. Indications of Air Force intentions are in Lilley log entry for 17 October 1968, recording call to Lilley from Clauser. On NEROC's fate, see Wilson and Kaiser in Oreskes and Krige, eds. (2014), 301–303.

1164. Lilley log entry for 10 March 1969.

1165. Ripley to Haworth, 17 March 1969. Follow-up to letter to BoB of January 20. Folder "NSF." Box 341. RU 99. SIA. Lubkin (1975), 8–10.

1166. Weisner to Lilley, 7 March 1969; Bradley to Whipple, 10 March 1969, regarding "Committee of Cooperating Users"; and Lilley to Whipple, 12 March 1969. Folder "Management" in "Radio Radar Astronomical Telescope '65–'71." Box 21. RU 7431. SIA.

1167. Bradley to Whipple, 10 March 1969. Folder "Management" in "Radio Radar Astronomical Telescope '65–'71." Box 21. RU 7431. SIA.

1168. Galler to Ripley, 27 March 1969, "Personal and Confidential." "APO" folder 3. "Correspondence." Box 1. RU 108. SIA.

1169. Lubkin (1975), 10.

1170. Lilley log entry 23 July 1969, recording phone call from Whipple. H.R. 13024, counterpart to S. 705, introduced into the House by Rep. Thompson of New Jersey, 22 July 1969.

1171. Bradley, 3 September 1960, n.d. "Briefing Notes for Hearing before the Subcommittee." Folder "Conference." Box 21. RU 7431. SIA. The date of this document is not clear, but it appears that it was prepared well before September at a time when McElroy was just beginning his tenure.

1172. McElroy wholeheartedly endorsed the full and final transfer of authority and responsibility for the Science Information Exchange to the Smithsonian.

1173. James Frey to Pell, 2 September 1969. "APO" folder 1 of 2. Box 8. RU 108. SIA. There were actually several delays. The hearings began in mid-July and continued in the Senate through 29 July, with the House hearings on 30 July. See Ripley to Regents, 17 June 1970. "Correspondence to Regents re Thompson Cong. Hearings." Box 64. RU 137. SIA.

1174. Lubkin (1975), 8–10.

1175. Greenstein to "Members of the American Astronomical Society." 6 January 1970. Folder "NSF." Box 341. RU 99. SIA.

1176. Lubkin (1975), 15–18.

1177. Lubkin (1975), 15–18, on 15.

1178. The National Science Board, "A History in Highlights 1950–2000." http://www.nsf.gov/nsb/documents/2000/nsb00215/nsb50/1970/mansfield.html. David H. Guston and Kenneth Keniston, "Introduction: The Social Contract for Science," in Guston and Keniston, eds. (1994), 7, 33, note 2.

1179. Wilson and Kaiser in Oreskes and Krige, eds. (2014), 302; Butrica (1996), 79–82. MIT was willing to retain Lincoln but drop its association with Draper Labs. V. McElheny, "M.I.T. Administration Makes Public Its Intentions on Disposition of Draper and Lincoln Laboratories," *Science* 168 (29 May 1970): 1074–1075.

1180. Joan Hoff, "The Presidency, Congress, and the Deceleration of the U.S. Space Program in the 1970s," in Launius and McCurdy, eds. (1997), 101.

1181. Richard D. Lyons, "Nixon Reorganizing Vast Federal Science Complex," *New York Times*, 1 November 1970.

1182. Lyons, "Nixon."

1183. Maxine Cheshire, "Smithsonian's Fund Use Criticized," *Boston Globe*, 26 January 1970, 6; Cheshire, "Smithsonian Probe Asked," *Washington Post*, 16 April 1970, D1; Jack Anderson, "Mall Memorial to Hirshhorn Probed," *Washington Post*, 11 April 1970, C11. Jieun Shin and Pam Henson kindly provided insights into Frank Thompson.

1184. "Report by the Committee on House Administration," 91st Cong., 2d sess. House Report No. 91-1801 (30 December 1970), 1. Union Calendar No. 865. Box 64. RU 137. SIA.

1185. H.R. 13024, in 91st Cong., 2d sess. House Report No. 91-1801. "The Smithsonian Institution," 30 December 1970. 48-006. http://archive.org/stream/smithsonianit9121970smit/smithsonianit9121970smit_djvu.txt (accessed 20 October 2014).

1186. Thompson, "General Hearings Before the Subcommittee on Library and Memorials of the Committee on House Administration of the House of Representatives, 981st Congress, Second Session, July 1970." Quoted in Bradley OHI, no. 12 (10 March 1976), 225–226

1187. Statements by Ripley and Whipple, 29 July 1970 to "Subcommittee on Library and Memorials on H.R. 13024. Conference." Box 21. RU 7431. SIA.

1188. "Statements," 29 July 1970, 554.

1189. "Statements," 29 July 1970, 554–555.

1190. "Statements," 29 July 1970, 563–565.

1191. "Statements," 29 July 1970, 566.

1192. "Statements," 29 July 1970, 575.

1193. "Statements," 29 July 1970, 595.

1194. "Statements," 29 July 1970, 595–596.

1195. "Statements," 29 July 1970, 596. Thompson also recognized Connecticut Senator Emilio Daddario's success in securing a small but substantial funding increase for basic research.

1196. Ripley, "Announcement," 4 August 1970. "Thompson Hearings House Res 910." Box 64. RU 137. SIA. There are many factors, both internal and external, not discussed here that prompted the Thompson Hearings and led to profound changes in Ripley's administration in the 1970s. I am indebted to Pamela Henson for many helpful discussions about this watershed in Smithsonian history, which deserves deeper attention than can be given here. Thompson was appointed a regent in 1979.

1197. Lundquist notes—July 1970. Folder "Multiple Mirror Telescope 1970–1973," 1 of 3." Box 8. RU 253. SIA.

1198. Bradley to Challinor, 7 April 1970; Bradley to Ripley, 9 January 1970. Folder "APO." Box 2. RU 108. SIA. On BoB's suggestion, see Jameson to Bradley and Galler, 24 December 1969. "Admin Confidential—SAO and the Federal Budget." "APO" folder 1 of 2. Box 8. RU 108. SIA.

1199. Jameson to Ripley et al., 24 August; 1 September 1970. Folder "OMB General Correspondence." Box 11. Acc. 82-018. SIA. The "SAO Issue Paper" was an edited copy, Loweth to Siscoe. Routing slip, 26 August 1970. Courtesy Dan Fabricant, SAO.

1200. "SAO Issue Paper" attached to Anderson to Jameson, 23 February 1971, and Jameson to Raymond N. Watts, 1 March 1971.

1201. "SAO Issue Paper," 2.

1202. "SAO Issue Paper," 6.

1203. "SAO Issue Paper," 9.

1204. "SAO Issue Paper," 10.

1205. "SAO Issue Paper," 11.

1206. John Jameson to Raymond N. Watts, 1 March 1971. "OMB Reports on SAO" clipped to Dean Anderson to John Jameson "OMB Route Slip," 23 February 1971. "Office of Programming and Budget." SAO Folder. Box 13. SIA. T90068.

1207. Dean Anderson statement, clipped to Jameson and Watts, 1 March 1971. "APO" folder 1. Box 8. RU 108. SIA. There were repeated suggestions in the 1970s for closing the SAO and transferring it to NSF or to NASA. See, for instance, "Argument against transferring Mt. Hopkins to Kitt Peak," 16 March 1973. "APO" folder 4. Box 17. RU 108. SIA.

1208. Ripley to Frank Thompson, 5 January 1971. "Thompson Hearings." Box 64. RU 137. 19710107: John H. Dobkin to The Record (cc Bradley, Blitzer and Powers), 7 January 1971. "Corr. re Thompson Hearings." Box 64. RU 137. SIA.

1209. Challinor's focus was, moreover, on relations between the National Museum of Natural History's research and NSF funding. 19710204: Challinor to Ripley, 4 February 1971. "SI and the NSF." Box 11. RU 108. SIA.

1210. Senate Bill 1421 was identical to the original S. 705 the previous year and HR 13024. See Ripley to Lucien Nedzi, 2 April 1971. NSF "Nedzi" folder. Box 47. RU 137; Bradley to. Whipple, 2 April 1971. Folder "Congress and Bills" in "Radio Radar Astronomical Telescope '65–'71." Box 21. RU 7431. SIA.

1211. Call from B. F. Burke, Lilley log entry for 20 April 19761. See also Lubkin (1975), 21.

1212. During this time, there was also some talk of Navy support, but the Smithsonian rejected it, feeling that it was contrary to its mandate for basic research rather than "mission-oriented" research. Telephone call to Lundquist from Huxley, Lilley log entry for 26 April 1971. Woody Price to Ripley (through Powers and Bradley), 26 April 1971 on proposed DOD Amendment to radar telescope legislation. Folder "Management" in "Radio Radar Astronomical Telescope '65–'71." Box 21. RU 7431. See also Price to Ripley, 28 April 1971. "DoD Amendment to Radar Telescope Bill." "APO" folder 1 of 2. Box 8. RU 108. SIA.

1213. Bradley to Ripley, 11 May 1971. "Confidential." NSF "Nedzi" folder. Box 47. RU 137. SIA.

1214. Harvey Brooks to Whipple, 1 June 1971. Folder "Congress and Bills" in "Radio Radar Astronomical Telescope '65–'71." Box 21. RU 7431. SIA.

1215. Ripley to Thompson, 23 June 1971; Ripley to Anderson, 23 June 1971. "Thompson Committee Hearings." "APO" folder 1 of 2. Box 8. RU 108. SIA.

1216. Lilley log entry for 21 May 1971. On continued NEROC planning, see Butrica (1996), 82–83.

1217. Whipple to all NEROC Participants, 21 May 1971. Folder "Congress and Bills" in "Radio Radar Astronomical Telescope '65–'71." Box 21. RU 7431. SIA.

1218. Lubkin (1975), 23; Richard Lyons, "U.S. Proposes Mammoth Radio Telescope at New Mexico Site," *New York Times*, 16 March 1972, 24.

1219. Butrica remarks on how shifts in desired wavelength regimes changed priorities in radio astronomy that were distinct from the needs of planetary radar. Butrica (1996), 82–85.

1220. Whipple OHI (1976), 16.

1221. Bradley OHI, no. 12 (1976), 235–236, 242.

Chapter 15 – Growing Fault Lines in Cambridge

1222. Marsden OHI session II (2005); Field OHI (2007). https://www.aip.org/history-programs/niels-bohr-library/oral-histories.

1223. Whipple, "A Word from the Director," *SAO News* 10 (March–April 1970): 1.

1224. Each situation had its peculiarities. Stanford was similar, but primarily it was an issue of scaling up, not heterogeneity. Galison et al. (1992), 46–77.

1225. Laura Whipple OHI (2007), 30. On strained relations, see also Babette Whipple OHI (2007), 40.

1226. McCray (2000), 14; (2004a).

1227. DeVorkin (2000b).

1228. Greenstein to Goldberg, 7 June 1960. Folder 9. Box 12. Greenstein Papers. Caltech.

1229. Goldberg to Greenstein, 14 June 1960. Folder 9. Box 12; also folder 8. Box 43. Greenstein Papers. Caltech.

1230. Goldberg to Whipple, 6 October 1960. Folder "Goldberg, Leo." Box 2. RU 7431. Ser. 1. SIA.

1231. Whipple to Greenstein, 11 April 1961. Folder 8. Box 43. Greenstein Papers. Caltech.

1232. Remington Kellog to John Macy, 1 March 1962. "APO Personnel 55–64." Box 21. RU 50. SIA.

1233. See articles on the decision to build an observatory on Mauna Kea instead of Haleakala, in "STADAD—Hawaii, Visits and Visitors, 1965–66." Box 47. RU 263. SIA.

1234. John Hsia to Lundquist, 17 August 1965, on the University of Michigan activity on Maui. "STADAD—Hawaii—Public Info, 1965." Box 47. RU 263. SIA.

1235. Goldberg OHI (1978), 119–120. Goldberg et al. (1965), 676. Goldberg's shock tube laboratory and staff building are noted in Gingerich OHI (2005), 65.

1236. Goldberg OHI (1978), 123–124. On OSO, see Bester (1966), 17, and throughout the volume.

1237. Aller (1997), 122. http://www.nasonline.org/publications/biographical-memoirs/memoir-pdfs/goldberg-leo.pdf (accessed 31 August 2013).

1238. Leo Goldberg, "Harvard College Observatory," *Bulletin of the American Astronomical Society* 1 (1969): 20–25.

1239. Goldberg to Dunlop, 5 November 1970. Box 2. Goldberg Papers. HUG FP 83.6. HUA. "Zero tolerance" from Aller, "Leo Goldberg," 122. See also comments by Avrett OHI (2005), 84; Gingerich OHI (2005), 79; and Dupree OHI (2007), 40. By the late 1960s, Smithsonian staff were responsible for over half the Harvard teaching load, with only "token salaries from Harvard." Smithsonian was definitely second class. Carl Sagan had a split appointment, but was denied promotion and tenure at Harvard. Whipple was ready to offer him a full-time Smithsonian appointment, but Sagan wanted an academic professorship. Lundquist OHI (2011), 192. Sagan's tenure problems may have resulted from Harvard faculty concerns but definitely from highly critical external reviews. See Davidson (1999), 202–204.

1240. Andrea Dupree OHI, 49.

1241. Goldberg OHI (1978), 105.

1242. K. Drummond to Whipple, 15 March 1960. "Laying of Cornerstone April 15." Box 2. RU 7431. SIA.

1243. Whitney OHI (2005), 56.

1244. Avrett OHI (2005), 84.

1245. When SAO staff were interviewed as part of this history, they were shown spreadsheets based on the *Annual Report of the Director* that identified at least three levels of administrative structure, including group leaders and subgroups. Respondents were consistently skeptical that such groupings were real from an administrative standpoint. They were real enough for the specific problem at hand, but groups constantly reformed and floated.

1246. Whitney OHI (2005), 53–54. Gingerich OHI (2005), 79–80.

1247. Lundquist OHI (2011), 190.

1248. Latham OHI (2005), 111.

1249. Avrett OHI (2005), 90.

1250. Goldberg OHI (1983).

1251. Whipple to John D. Buchanan, 11 March 1970. Found in Whipple's desk after his death by Donna Coletti. Copy in author's files.

1252. Greenstein liner notes to: Goldberg "Report to the Visiting Committee," 10 April 1970. Greenstein Papers 63.3. Caltech. See also "HCO-SAO related" folder. Box 2. Goldberg Papers. HUG FP 83.6. HUA. For x-ray astronomy, see Hirsh (1983).

1253. Goldberg to "Visiting Committee," 10 April 1970. Goldberg Papers. Box 2. HUG FP 83.6. Box 2. HUA.

1254. Goldberg OHI (1983), 22–35.

1255. Ripley to Whipple, 5 May 1970. "Common Problems Shared by Harvard and SAO." Box 64. RU 137. SIA.

1256. Whipple to Ripley, 25 May 1970. "Astrophysical Observatory Smithsonian." Box 2. Goldberg Papers. HUG FP 83.6. HUA; also box 1. RU 108. SI Asst. Secy. for Science 1963–1973. SIA.

1257. Goldberg to Whipple, 28 May 1970. Box 2. Goldberg Papers. HUG FP 83.6. HUA.

1258. Goldberg to Whipple, 28 May 1970. See also "Report on Smithsonian's Science Activities for the House Subcommittee on Library and Memorials." Folder "Conference." Box 21. RU 7431. SIA.

1259. Goldberg OHI (1983), 35.

1260. Anderson, 39 June 1970. "SAO Field Trip Report." Folder 1 "APO." Box 8. RU 108. SIA. Report attached to Jameson to Watts, 1 March 1971; Anderson to Jameson, 23 February 1971. Whipple announced that he would be away from the SAO for three weeks in June, in Whipple to Ripley, 25 May 1970. "APO." Box 1. RU 108. SIA.

1261. Whipple to Dean Bundy, 6 June 1970, in Goldberg to Dunlop, 5 November 1970. Box 2. Goldberg Papers. HUG FP 83.6. HUA.

1262. Anderson, "SAO Field Trip Report," 2.

1263. "SAO Field Trip Report."

1264. "SAO Field Trip Report," 3.

1265. "SAO Field Trip Report," 4.

1266. Goldberg to Dunlop, 5 November 1970. Box 2. Goldberg Papers. Folder "HCO-SAO." HUG FP 83.6. HUA. No comparable report has been found in Whipple's papers.

1267. Goldberg to Dunlop, 5 November 1970.

1268. Goldberg to Pusey, 28 October 1970. Box 2. Folder "HCO-SAO." Goldberg Papers. HUG FP 83.6. HUA. On Goldberg's influence, see McCray (2004a), ch. 1.

1269. As implied in Goldberg OHI (1978), sess. II.

1270. R. G. Reed to Goldberg, 29 October 1970; Goldberg to Dunlop, 5 November 1970. Folder "HCO-SAO." Box 2. Goldberg Papers. HUG FP 83.6. HUA.

1271. Goldberg to Dunlop, 11 November 1970. Folder "HCO-SAO." Box 2. Goldberg Papers. HUG FP 83.6. HUA.

1272. Goldberg to Dunlop, 11 November 1970. He made similar criticisms about the telescope builders at Kitt Peak before his tenure. McCray (2004a), 48.

1273. Emphasis in original. Ripley to Greenstein Committee, 3 December 1970; Ripley to Pusey, 4 December 1970; Galler to Ripley, 9 December 1970. Box 64. RU 137. SIA.

1274. David Layzer to John T. Dunlop, 30 December 1970. Folder "HCO-SAO." Box 2. Goldberg Papers. HUG FP 83.6. HUA.

1275. Layzer to Dunlop, 30 December 1970. Years later, rereading his letter, he felt that the problems could be summed up by the evident "demoralization of students, [and] the frustration of having an institution with this huge budget and no scientific stature except the one person that Leo and I had brought." He was referring to Dalgarno. Layzer OHI (2007), 64.

1276. Greenstein to Alan Pifer, 30 April 1971. Greenstein Papers, 64.2. Caltech.

1277. Draft letter, Pusey and Ripley to "Dear Mr." 3 December 1970. "APO." Box 1. RU 108. SIA.

1278. Greenstein to Ripley and Pusey, with draft report of the "Ad Hoc Committee on the Relations Between . . . ," 22 January 1971. Final report dated 1 March 1971. "APO." Box 1. RU 108. SIA.

1279. Greenstein to Ripley and Pusey, 22 January 1971.

1280. Layzer OHI (2007), AIP.

1281. Greenstein "draft report."

1282. Greenstein's notes. "HCO, Dept-SAO Relations." 22 January 1971. 13 pages, handwritten. Greenstein Papers 63.7. Caltech.

1283. Dalgarno OHI (2007), 42–45.

1284. Greenstein's notes.

1285. Greenstein's notes.

1286. Greenstein's notes.

1287. Greenstein's notes.

1288. Greenstein's notes.

1289. Greenstein's notes.

1290. Greenstein's notes.

1291. Greenstein to Ripley and Pusey, n.d. "Draft Report of 1 January 1971 meeting of "Ad Hoc Committee." "APO." Folder 1. Box 1. RU 108. SIA.

1292. Greenstein's notes.

1293. See, for instance, Merritt (2002) and Bix (1993).

1294. Greenstein to Ripley and Pusey, 22 January 1971.

1295. Greenstein to Ripley and Pusey, 22 January 1971. In fact, Bradley was thinking of retiring at the time, owing to the stresses caused preparing for, and in the aftermath of, the Thompson hearings. He soon did so, returning as special consultant to Ripley for several years. Pam Henson, private communication.

1296. Spitzer to Greenstein, 10 February 1971. Greenstein Papers 63.7. Caltech.

1297. Kuiper to Greenstein, 19 February 1971. Greenstein Papers 63.7. Caltech. On Kuiper's ALSOS work, see DeVorkin (1992).

1298. Greenstein to members of the ad hoc committee, 23 February 1971; Spitzer to Greenstein, 24 February 1971, and, after phoning him, sent telegrams to Ripley and Pusey; Greenstein to Dunlop, 26 February 1971. Greenstein Papers 63.7. Caltech.

1299. Goldberg to Greenstein, 26 February 1971. Greenstein Papers 63.4. Caltech.

1300. Greenstein to Dean Dunlop, 1 March 1971. Greenstein Papers 63.4. Caltech.

1301. Whipple to Kuiper, February 1971. "Summary of Programs and Achievements of the SAO." Appendix 2 to 1 March 1971. Greenstein, "Ad Hoc Committee on Harvard-Smithsonian Observatory Relations," "Personal and Confidential—Not for Publication—Recommendations to Harvard." Greenstein Papers 62.2. Caltech.

1302. Whipple to Kuiper, February 1971.

1303. Greenstein, "Ad Hoc Committee," 2. Greenstein 62.2. Caltech.

1304. Whipple to Greenstein, 7 May 1971. Greenstein Papers 63.6. Caltech.

1305. Whipple to Greenstein, 7 May 1971.

1306. Whipple even alleged that Goldberg never visited Lilley's laboratory, just "a mile up the street." Whipple to Greenstein, 5 May 1971.

1307. Whipple to Greenstein, 7 May 1971.

1308. Whipple to Greenstein, 7 May 1971.

1309. Greenstein to Whipple, 10 June 1971. Provided by Donna Coletti. Author's files.

1310. Greenstein to Whipple, 10 June 1971.

1311. Menzel to Dunlop, 1 March 1971. Greenstein Papers 63.4. Caltech.

1312. Menzel to Dunlop.

1313. Greenstein to Goldberg, 2 March 1971. Greenstein Papers 12.9. Caltech.

Chapter 16 – Whipple Steps Down

1314. Greenstein to Ripley and Pusey, 22 January 1971.

1315. Layzer to Dunlop, 30 December 1970; Reeves to Goldberg, 15 February 1970; and Goldberg (at Kitt Peak) to Dunlop, 30 November 1971. Goldberg Papers. HUG FP 83.6. Box 7. HUA.

1316. Challinor for file "Smithsonian Astrophysical Observatory Briefing," 14 April 1971. "APO" folder 1 of 2. Box 8. RU 108. SIA.

1317. Goldberg, "Report to the Visiting Committee," 9 April 1971, in Ripley file. "APO" folder 1 of 2. Box 8. RU 108. SIA.

1318. Goldberg, 9 April 1971. "Report to the Visiting Committee."

1319. Greenstein notes, 23 July 1971, after conversation with Ed Purcell. Greenstein 64.2. Caltech.

1320. Field (1956).

1321. Field OHI, sess. I (1980), by Richard Hirsh.

1322. Field OHI (2007), 1–5.

1323. Field OHI (2007), 9.

1324. Greenstein to Ripley, 29 November 1971; 14 January 1972. "APO" folder 1 of 2. Box 8. RU 108. SIA.

1325. Goldberg, 30 November 1971. Folder "Harvard College Observatory." Box 7. HUG FP 83.6. HUA. The MOA was dated October 1971 and was signed by Ripley and Derek Bok.

1326. Dalgarno, April 1972. "Report to the Visiting Committee." Greenstein Papers 104.6. Caltech.

1327. "George Field Named New HCO Head," *SAO News* 12, no. 1 (Winter 1971–1972): 1.

1328. Ripley to Whipple, 5 May 1972. "APO" folder 2. Box 8. RU 108. SIA.

1329. Ripley added: "You have pioneered this entire relationship so successfully that it appear[s] to me (as I believe it appears equally to all the authorities at Harvard) that your help in the evolution of this new, more integrated arrangement is absolutely crucial." Ripley to Whipple, 5 May 1972. "APO" folder 2. Box 8. RU 108. SIA.

1330. Ripley to Whipple, 5 May 1972.

1331. Babette Whipple OHI (2007), n.p. https://www.aip.org/history-programs/niels-bohr-library/oral-histories/37898.

1332. Ripley to Participants, NRAO Users Committee. "Memorandum," "APO" folder 2. Box 8. RU 108. SIA.

1333. Bradley to Ripley, 4 May 1972. "SAO Director." Box 58. RU 137. SIA.

1334. Dunlop to Goldberg, 8 May 1972. "Harvard College Observatory." Box 7. Goldberg Papers. HUG FP 83.6. HUA.

1335. Whipple to Ripley, 8 May; 11 May 1972, clipped to Ripley to Dunlop, n.d. "APO" folder 2. Box 8. RU 108. SIA.

1336. Whipple to Ripley.

1337. Whipple to Ripley.

1338. Challinor to Ripley, 21 March 1972. "Stever visit to SAO." Folder "NSF 1963–1972." Box 14. SIA.

1339. Challinor to Ripley, 21 March 1972. Bradley to Ripley, 23 March 1972. Folder "NSF 1963–1972." Box 14. RU 108. SIA.

1340. Bradley to Ripley, 10 May 1972. Folder "NSF 1963–1972." Box 14. RU 108. SIA.

1341. Ripley to Stever, 10 May 1972. Folder "NSF 1972." Box 14. RU 108. SIA.

1342. Ripley to Whipple, 22 May 1972. Folder "APO 2." Box 8. RU 108. SIA.

1343. Ripley to Dunlop, 11 May 1972. Folder "APO 2." Box 8. RU 108. SIA.

1344. Challinor to Bradley, 15 May 1972. Folder "SAO Director." Box 58. RU 137. SIA.

1345. Bassi (2015), 96, 160–162.

1346. Branscomb (1963), 660–662. Dalgarno OHI (2007), 59.

1347. Field to Challinor. 5 August 1972. "SAO" folder 2. Box 8. RU 108. SIA. On JILA, see Bassi (2015).

1348. Branscomb to Field, 15 August 1972, clipped to Field to Challinor, 5 August 1972. "SAO" folder 2. Box 8. RU 108. SIA.

1349. Bradley to Ripley, 15 August 1972. "APO" folder 2. Box 8. RU 108. SIA.

1350. Osterbrock (1992); Sandage (2004), 555.

1351. According to Allan Sandage, the observational cosmologist who spent his career there. Sandage (2004), 556. The joint observatories were dissolved in 1980. Schmidt (1980), 589–593.

1352. Bradley to Ripley, 15 August 1972; Challinor to Ripley, 15 August 1972. "APO" folder 2. Box 8. RU 108. SIA. Copy in Bradley OHI, no. 8 app.

1353. Bradley, through Brooks to Ripley, 13 September 1972. "Further conversation with George Field: Staff Analysis needed on SAO and Harvard Observatory." "APO" folder 2. Box 8. RU 108. SIA.

1354. Challinor to Ripley, 30 October 1972. "APO" folder 2. Box 8. RU 108. SIA.

1355. George Field, 6 November 1972. "A Proposal for a Harvard-Smithsonian Astrophysics Center, for Discussion Purposes Only." Box 7. Goldberg Papers. HUG FP 83.6. HUA.

1356. Lundquist to Field, 29 September 1972. Box 1. RU 437.1. Lundquist Papers. SIA.

1357. Lundquist to Field. 29 September 1972.

1358. Lundquist to Field, 29 September 1972.

1359. Lundquist to Field, 29 September 1972. On the nature and boundaries of space science regarding disciplinarity, see Newell (1980), Hirsh (1983), DeVorkin (1992), Mack and DeVorkin (1982).

1360. Bradley to Ripley, 25 September 1972. "SAO NASA Geodesy." Box 83. RU 137. SIA. At the same time, Lundquist alerted Ripley, Brooks, and Bradley to how NASA's new "Earth and Ocean Physics Application Program" might result in the SAO becoming one of four major centers, which would double the present satellite-tracking grant. Dorothy Rosenberg to "For the Record," 26 September 1972. Copies to Ripley, Brooks, Bradley, and Challinor. "SAO NASA Geodesy." Box 83. RU 137. SIA.

1361. Bradley to Ripley, 25 September 1972.

1362. Bradley to Ripley, 11 September; 25 September 1972. Soon after he announced he was stepping down, Whipple terminated Robert V. Bartnik, replacing him with John Gregory. At the time, there was considerable upheaval in SAO management. Mark Malec was also terminated but was soon reinstated on a probationary basis. "APO" folder 2. Box 8. RU 108. SIA.

1363. Bradley to Ripley, 25 September 1972.

1364. Lundquist to Distribution, "Anticipated roles of Al Cameron," 7 February 1973. Lundquist folder. Box 1. RU 437.1 SIA.

1365. Field to Challinor, 4 December 1972. "SAO1972." Box 14. RU 108. SIA. Hirsh (1983), 106–107; Field OHI (2007), 12–13.

1366. Field OHI (2007), 12–13. Greenstein to Dalgarno, 22 January 1973. Endorses Giacconi's joint appointment. Greenstein Papers, 8.5. Caltech.

1367. Lundquist to Distribution, 7 February 1973. "Anticipated roles of Riccardo Giacconi." "Lundquist" folder. Box 1. RU 437.1 SIA.

1368. Lundquist to Field, 22 February 1973. "Lundquist" folder. Box 1. RU 437.1. SIA.

1369. Alex Dalgarno, April 1973. "Report to the Visiting Committee." "HCO" folder 2. Box 19. HUG 4876. HUA.

1370. Dalgarno to Greenstein, 4 April 1973. 19730416: Greenstein to Dalgarno, 16 April 1973. Greenstein Papers. 8.5. Caltech.

1371. "Harvard Corporation approves establishment of CfA," 7 May 1973. Folder "SAO." RU 613. Office of the Secy. [1972–1984]. SIA. On the double overhead problem, see Gregory to Ripley, 26 April 1973, and subsequent records lasting through the 1970s dealing with the establishment of a hybrid corporate entity to manage all federal contracting. "SAO." Box 17.3. RU 108. SIA.

1372. Carl Larsen to David Challinor, 23 May 1973. Folder 7.1 "SAO." Box 1. RU 108.1. SIA.

1373. "Smithsonian and Harvard to Establish Center for Astrophysics July 1," 1 June 1973. Box 7. Goldberg Papers. HUG FP 83.6. HUA.

1374. Field to Challinor, 16 February 1973. Folder 17.1. RU 108. SIA.

1375. "HCO, Dept-SAO Relations," 22 January 1971. Greenstein's notes. Greenstein Papers 63.7. Caltech.

1376. Lundquist OHI (2011), 227.

1377. Lundquist to Challinor, 18 July 1973. Box 17.3. RU 108. SIA. Lundquist also suggested that Challinor visit Huntsville to see their space museum to help him visualize what might be possible for the Smithsonian's planned Air and Space Museum.

Chapter 17 – Whipple's Legacy

1378. Cornell OHI (2010), 188–189, reading excerpts from working notes prepared for Irwin Shapiro's remarks honoring Whipple. Veracity was less a concern than was spiritual flavor.

1379. Foreword to *The Collected Contributions of Fred L. Whipple*, vol. 3, pt. 1 (Cambridge, MA: Smithsonian Astrophysical Observatory, 1972a), xi.

1380. Greenberg (1965). Also quoted in Alexis Doster, "Smithsonian's Progress," in letters to the editor, *Washington Post*, 11 May 1966, A24, wherein he commented on the original article by Carl Bode, "The New School of Old Smithsonian," from 28 April.

1381. Whipple to Carmichael. 7 April 1965. "C." Box 8. RU 7431. SIA. Whipple wished to keep his criticisms private in his letter to Carmichael, whereas Ripley wrote publically to the *Post*. Ripley, letter to the editor, *Washington Post*, 11 May 1966, A24.

1382. Carmichael to Whipple, 12 April 1965. "C." Box 8. RU 7431. SIA.

1383. "Forging Ties between Harvard and the Smithsonian—Cooperating for Mutual Benefit," *Harvard Crimson*, 19 April 1989.

1384. "Forging Ties."

1385. "Forging Ties."

1386. See tables in Malec/Field (1977), exhibit 3, graphics on pp. 8 and 9, in "Report by the Director to the Smithsonian Astrophysical Observatory Visiting Committee, April 1975–April 1977," exhibit 3 in folder "Congressional Investigation of SAO." Box 22. RU 468. SIA.

1387. Babette Whipple OHI (2007), 40–42.

1388. Whipple OHI (1976), 85.

1389. Whipple OHI (1977). http://www.aip.org/history/ohilist/5403.html (accessed 25 December 2013).

1390. Whipple OHI (1977).

1391. Whipple OHI (1977).

1392. Ortolano (2009), 10–11.

1393. Whipple OHI 1977. AIP, http://www.aip.org/history/ohilist/5403.html (accessed 25 December 2013).

1394. Whipple to McCrosky, 13 January 1958. Folder "McCrosky, Richard." Box 3. RU 7431. SIA. McCrosky OHI (2005).

1395. Hacker (2000), 313.

1396. Fred L. Whipple, "Knowledge and Understanding of the Physical Universe as Determinants of Man's Progress," in *Knowledge Among Men*, ed. P. Oehser (New York: Simon & Schuster, 1966), 173.

1397. Needell (2000), 366.

BIBLIOGRAPHY

Archival Source Abbreviations

AELL	A. Edward Lilley, "Joint Radio and Radar Research Facility—Log of Relevant Meetings, Actions and Correspondence." Accession 83-121. SIA. AELL/SIA.
AIP	American Institute of Physics, Center for History of Physics Archives: Papers of the American Astronomical Society; numerous oral histories.
BANC	Bancroft Library, U. of California, Berkeley: R. T. Birge Papers (BANC/RTB); A. O. Leuschner Papers (BANC/AOL); Department of Astronomy Records CU-25 (BANC/DOA); President's Files CU-5 (BANC/UCP).
CIT	California Institute of Technology Archives: George Ellery Hale Papers, microfilm edition (AIP/GEH, also cited as AIP/CIT/GEH); Greenstein Papers; Gell-Mann Papers.
HL	Huntington Library, Manuscripts Division: Walter Sydney Adams Papers (HL/WSA); George Ellery Hale Papers (HL/GEH); Ira S. Bowen Papers.
HUA	Harvard U. Archives: Harlow Shapley Director's Correspondence, UA V.630.xx (HUA/HSDC); Personal correspondence HUG 42 4773.xx (HUG/HSPC); Whipple Papers (HUG 4876); Goldberg Papers (HUG FP 83.6).
LAUC	Mary Lea Shane Archives of the Lick Observatory, McHenry Library, Santa Cruz: W. W. Campbell Papers (LAUC/WWC), R. G. Aitken Papers (LAUC/RGA), W. H. Wright Papers (LAUC/WHW); C. D. Shane Papers (LAUC/CDS); Gerald E. Kron Papers (LAUC/GEC).
LC	Library of Congress Manuscripts Division: Simon Newcomb Papers (LC/SN). Library of Congress: Vannevar Bush Papers, Library of Congress: correspondence with Conant, Carmichael, Hunsaker, Menzel, Oppenheimer, Roberts, Shapley, the Smithsonian Institution, and Struve.
NAS	National Academy of Sciences, Space Science Board (SSB) papers, also IGY papers (SSB/NAS, IGY/NAS).
NASA	National Aerospace S Agency History Office: NHO; OAO Records
NASM	National Air and Space Museum, Archives
NSF	National Science Foundation

OHI	Oral history interview	
PUL	Princeton University Library. Henry Norris Russell Papers (PUL/HNR)	
RU	Record Unit (see SIA)	
SIA	Smithsonian Institution Archives, abstracted records from the following RUs:	
	50	SI Secretary
	108	Asst. Secy. for Science
	168	SAO Contracts
	188	SAO Contracts
	253	SAO Asst. Director
	260	Celescope
	263	Satellite Tracking
	437	George Field
	522	Celescope Contracts
	7091	Science Service
	7431	Fred Whipple
SL	M. Louis Salmon Library, University of Alabama, Huntsville. Lundquist Papers.	
SLRC	Schlesinger Library, Radcliffe College. Dorrit Hoffleit Papers (SLRC/DH).	
UA	University of Arizona Archives, Gerard P. Kuiper Papers.	

Interviews

Unless otherwise specified, each OHI was conducted by the author. Most AIP sources are keyword addressable if not paginated, and are available at https://www.aip.org/history-programs/oral-histories/search?search_api_views_fulltext=&=Apply.

Avrett OHI (2005). Eugene Avrett, 14 October 2005. NASA/NASM/AIP.

Baker OHI (1980). James Baker, 9 June 1980. NSF/AIP.

Baum OHI (1983); (1997). William A. Baum, 12 January 1983 NASM; 14 January 1997. AAS/AIP.

Bok OHI (1978). Bart J. Bok, 15 May 1978–14 June 1978. NSF/AIP.

Bowen OHI (1969). Ira S, Bowen, interviewed by Charles Weiner. 26 August 1969. AIP.

Bradley OHI (1975, 1976). James C. Bradley interviewed by Pamela Henson and Miriam Freilicher, 1974–1978. RU009515 SIA.

Code OHI (1982). Arthur D. Code, 30 September 1982–1 October 1982. NASM.

Cornell OHI (2010). James C. Cornell, 9 June 2010. NASA/NASM/AIP.

Dalgarno, OHI (2007). Alexander Dalgarno, 6 December 2007. NASA/NASM/AIP.

Davis OHI (2005). Robert J. and Ruth Davis, 15 October 2005. NASA/NASM/AIP.

Dupree OHI (2007). Andrea Dupree, 29 October 2007. NASA/NASM/AIP.

Field OHI (1980), George Field interviewed by Richard Hirsh, 14–15 July 1980. NASA/AIP.

Field OHI (2007). George Field, 5 December 2007. NASA/NASM/AIP.

Gingerich OHI (2005). 18 October 2005. NASA/NASM/AIP.

Goldberg OHI (1978). Leo Goldberg interviewed by Spencer Weart, 16–17 May 1978. NSF/AIP.

Goldberg OHI (1983). Leo Goldberg interviewed by Owen Gingerich, 9 August–10 October 1983). AIP.

Goldberg OHI (1983). Leo Goldberg, 22 February 1983. NASM.

Haramundanis OHI (2005). Katherine Haramundanis, 20 October 2005. NASA/NASM/AIP.

Hoffleit (1979). E. Dorrit Hoffleit, 4 August 1979. AIP.

Keller OHI (1983) Sam Keller 15 December 1983. AIP.

Kozai OHI (1997). Yoshihide Kozai, 2 September 1997. NASM/AIP.

Latham OHI (2005). David Latham, 8 October 2005. NASA/NASM/AIP.

Layzer OHI (2007). David Layzer, 5 December 2007. NASA/NASM/AIP.

Lundquist OHI (2011). Charles Lundquist, 16–18 November 2011. NASA/NASM/AIP.

Marsden OHI (2005). Brian Marsden, 17 October 2005. NASA/NASM/AIP.

McCrosky OHI (2005). Richard McCrosky, 9 October 2005. NASA/NASM/AIP.

McGrath OHI (2005). Leo McGrath, 10 October 2005. NASA/NASM/AIP.

Multiple-Mirror Telescope VI, Videohistory Collection, Smithsonian Institution (1989). Group interviews with Nathaniel Carleton, Frederic H. Chaffee, Craig Foltz, Thomas E. Hoffman, Aden Meinel, Michael Reed, Robert R. Shannon, Ray J. Weymann, Joseph T. Williams, and Fred Lawrence Whipple. RU 9542. SIA.

Noyes OHI (2005). Robert W. Noyes, 5 December 2007. NASA/NASM/AIP.

Schwarzschild OHI (1977). Martin Schwarzschild interviewed by Spencer Weart and David DeVorkin, 16 December 1977. NSF/AIP.

Singer OHI (1991). S. Fred Singer interviewed by Allan A. Needell and David DeVorkin, 23 April 1991. NASM/AIP.

Whipple, B. OHI (2007). Babette Whipple, 29 October 2007. NASA/NASM/AIP.

Whipple, F. L. OHI (1976). Fred L. Whipple interview by Pamela M. Henson, 24–25 June 1976. SIA.

Whipple, F. L. OHI (1977). Fred L. Whipple, 29 April 1977. NSF/AIP.

Whipple, L. OHI (2007). Laura Whipple, 26 April 2007. NASM/AIP.

Whitney OHI (2005). Charles Whitney, 13 October 2005. NASA/NASM/AIP.

Publications

Abbot, C. G. "The Astrophysical Observatory of the Smithsonian Institution." *Leaflets of the Astronomical Society of the Pacific*, February 1947.

———. *Adventures in the World of Science.* Washington, DC: Public Affairs Press, 1958a.

———. "On Sterne and Dieter's Paper, 'The Constancy of the Solar Constant.'" *Smithsonian Contributions to Astrophysics* 3 (1958b): 1–21.

Aikens, R., J. A. Hynek, W. A. Baum, and J. Kimmel. "The Use of Telescope-Image Orthicon Systems with Hypersentization at Readout." Abstract. *Astronomical Journal* 65 (August 1960): 339.

"Albert Thomas: Late Congressman Who Supervised NSF Budget Had Witty Views on Science and Politics." *Science* 151, no. 3714 (1966): 1064–1065.

Aller, Lawrence. "Leo Goldberg." *Biographical Memoirs of the NAS* 72 (1997): 115–134.

American Astronomical Society. "Abstracts of Papers Presented at the One Hundred-Fifth Meeting of the American Astronomical Society." *Astronomical Journal* 65 (August 1, 1960): 339.

American Geophysical Union. "Abstracts of the Papers Presented at the Forty-Third Annual Meeting, American Geophysical Union Washington, DC, April 25–28, 1962." *Journal of Geophysical Research* 67, no. 9 (1962): 3537–3612.

Anderson, P. H., C. G. Lehr, and L. A. Maestre. *Use of a Laser for Satellite-Range Measurements.* SAO Special Report 190R. Washington, DC: Smithsonian Institution, Astrophysical Observatory, 1965. https://ui.adsabs.harvard.edu/#abs/1965SAOSR.190.....A/abstract.

Arm, David L., ed. *Science in the Sixties: The Tenth Anniversary AFOSR Scientific Seminar. June 1965.* Albuquerque: University of New Mexico, Office of Publications, 1965.

Arnold, K. "The Use of Satellites for Geodetic Studies." *Space Science Reviews* 7, no. 1 (1967): 4–68.

Astronomical Society of the Pacific. "From the University of Arizona." *Publications of the Astronomical Society of the Pacific* 79 (October 1967): 517.

Atkins, Charles M. *NASA and the Space Science Board of the National Academy of Sciences.* HHN-62. Washington, DC: NASA Office of Policy Analysis, 1966.

Bailey, Solon I. *The History and Work of Harvard Observatory, 1839 to 1927; an Outline of the Origin, Development, and Researches of the Astronomical Observatory of Harvard College Together with Brief Biographies of Its Leading Members.* New York: McGraw-Hill Book, 1931.

Baker, James G. "A Family of Flat-Field Cameras, Equivalent in Performance to the Schmidt Camera." *Publications of the American Astronomical Society* 9 (1939): 255. http://adsabs.harvard.edu/abs/1939PAAS....9..255B.

Baldwin, Ralph B. "The Moon." *Annual Review of Astronomy and Astrophysics* 2, no. 1 (1964): 73–94.

Barath, F. T., A. H. Barrett, D. E. Jones, J. Copeland, and A. E. Lilley. "Symposium on Radar and Radiometric Observations of Venus during the 1962 Conjunction: Mariner 2 Microwave Radiometer Experiment and Results." *Astronomical Journal* 69 (1964): 49–58.

Barrett, Alan H., and Edward Lilley. "Mariner-2 Microwave Observations of Venus." *Sky & Telescope* 25 (April 1963): 192–195.

Bartky, Ian R. *Selling the True Time: Nineteenth-Century Timekeeping in America.* Stanford, CA: Stanford University Press, 2000.

Bassi, Joseph P. *A Scientific Peak: How Boulder Became a World Center for Space and Atmospheric Science.* Boston: American Meteorological Society Books, 2015.

Baum, W. A., W. A. Hiltner, H. L. Johnson, and A. R. Sandage. "The Main Sequence of the Globular Cluster M13." *Astrophysical Journal* 130 (1959).

Baxter, James Phinney. *Scientists against Time.* Cambridge, MA: MIT Press, 1968.

Beardsley, Wallace R., and J. Richard Hansen. "Use of an SEC Camera Tube for Recording Faint Objects." *Astronomical Journal* 73 (1968): S4–S5.

Becker, Barbara J. *Unraveling Starlight: William and Margaret Huggins and the Rise of the New Astronomy.* Cambridge: Cambridge University Press, 2011.

Beckers, Jacques M, Bobby Ulrich, Robert Shannon, Nathaniel P. Carleton, John C. Geary, David W. Latham, J. Roger P. Angel, William F. Hoffmann, Frank J. Low, Ray J. Weymann, and Neville J. Woolf. "The Multiple Mirror Telescope," in *Telescopes for the 1980s.* Annual Reviews of Astronomy and Astrophysics, 1981, 63-128.

Beckers, Jacques M. "Observatory Reports: Multiple Mirror Telescope Observatory." *Bulletin of the American Astronomical Society* 14 (1 January 1982): 322–325.

Berendzen, Richard, ed. *International Conference on Education in and History of Modern Astronomy.* New York: New York Academy of Sciences, 1972.

Bergaust, Erik, and William S. Beller. *Satellite!* Garden City, NY: Hanover House, 1956.

Berger, Carl. "History of the Manned Orbiting Laboratory Program (MOL)." (MOL Program Office, February 1970). NRO-approved release, 1 July 2015.

Bergin, Thomas. *50 Years of Army Computing, from ENIAC to MSRC: A Record of a Symposium and Celebration, November 13 and 14, 1996, Aberdeen Proving Ground.* Aberdeen, MD: U.S. Army Research Laboratory, 2000. http://www.worldcat.org/title/50-years-of-army-computing-from-eniac-to-msrc-a-record-of-a-symposium-and-celebration-november-13-and-14-1996-aberdeen-proving-ground/oclc/51538455&referer=brief_results.

Bester, Alfred. *The Life and Death of a Satellite.* Boston: Little, Brown, 1966.

Bix, Amy Sue. "'Backing into Sponsored Research': Physics and Engineering at Princeton University." *History of Higher Education Annual* 13 (1993): 9–52.

Bok, Bart J. "21-Centimeter Research." *Smithsonian Contributions to Astrophysics* 1 (1956). https://ui.adsabs.harvard.edu/#abs/1956SCoA....1..141B/abstract.

Bok, Bart J., and Harold I. Ewen. "Research of the 21-cm radiation at Agassiz Station." *Astronomical Journal* 59 (1954): 318.

Bok, Bart J., Harold I. Ewen, and David S. Heeschen. "The George R. Agassiz Radio Telescope of Harvard Observatory." *Astronomical Journal* 62 (1957): 8.

Bowen, I. S. "Telescopes." *Astronomical Journal* 69 (1964): 816–825.

Bower, Ernest Clare, and Fred L. Whipple. "Elements and Ephemeris of the Lowell Observatory Object: Second Paper." *Lick Observatory Bulletin* 421 (1930).

Boyd, Sylvia L. *Portrait of a Binary: The Lives of Cecilia Payne and Sergei Gaposchkin.* Rockland, ME: Penobscot Press, 2014.

Branscomb, Lewis M. "Joint Institute for Laboratory Astrophysics of the National Bureau of Standards and University of Colorado, Boulder, Colorado Report." *Astronomical Journal* 68 (November 1963): 660–662.

Brooks, Courtney G., James M. Grimwood, and Loyd S. Swenson. *Chariots for Apollo: A History of Manned Lunar Spacecraft.* NASA SP-4205. Washington, DC: Scientific and Technical Information Branch, National Aeronautics and Space Administration, 1979.

Brouwer, Dirk. "Solution of the Problem of Artificial Satellite Theory without Drag." *Astronomical Journal* 64 (1959): 378–397.

———. "Review of Celestial Mechanics." *Annual Review of Astronomy and Astrophysics* 1, no. 1 (1963): 219–234.

Burgers, J. M. "Third Symposium on Cosmical Gas Dynamics." *Physics Today* 10, no. 9 (1957): 50.

Burgers, Johannes Martinus, and Richard Nelson Thomas. *Cosmical Gas Dynamics, Proceedings from IAU Symposium No. 8* (1958). http://adsabs.harvard.edu/abs/1958IAUS....8.....B.

Burggraaf, Pieter. *Harqua Hala Letters: The Story of Arizona's Forgotten 1920's Smithsonian Institution Observatory.* Phoenix: Arizona State Office of the Bureau of Land Management, 1996.

Burke, John J., and Werner Kirchhoff. "Mount Hopkins Observatory." *Sky & Telescope* 36 (November 1968): 284–286.

Burnham, D. "Mobile Explorers & Beasts of Burden: A History of NASA's Prospector and Lunar Logistics Vehicle Projects." *Journal of the British Interplanetary Society* 48, no. 5 (1995): 213.

Bushnell, David. *Project Cat Eye: A History of Light Amplification Research at the Aeronautical Research Laboratory 1952–1960.* Pub. no. AD 274028. Arlington, VA: Armed Services Technical Information Agency, March 1962.

Butler, Orville Roderick. "The Birth of American Astrophysics: The Development of a Science in Its Cultural Context." PhD diss., Iowa State University, 1993.

Butrica, Andrew J. *To See the Unseen: A History of Planetary Radar Astronomy.* NASA SP-4218. Washington, DC: National Aeronautics and Space Administration, 1996.

Cannon, Annie J. "Report of the Astronomical Fellowship Committee." *Annual Report of the Maria Mitchell Association* 20 (1922): 14–15.

Carroll, Philip, Richard McCrosky, Robert Wells, and Fred L. Whipple. "Film Molding for the Baker Super-Schmidt Meteor Cameras." *Astronomical Journal* 56 (1951).

Cassidy, David C. *J. Robert Oppenheimer and the American Century*. New York, N.Y.: Pi Press, 2005.

Ceplecha, Z., and R. E. McCrosky. "Fireball End Heights: A Diagnostic for the Structure of Meteoric Material." *Journal of Geophysical Research* 81, no. 35 (1976): 6257–6275.

Cherniack, J. R., and E. M. Gaposchkin. *Smithsonian Astrophysical Observatory Program Writeup (SCROGE)*. SAO Special Report AOSR 121. Cambridge, MA: Smithsonian Astrophysical Observatory, 1963.

Clarke, Arthur C. *The Exploration of Space*. New York: Harper, 1951.

Clauser, F. H. *Preliminary Design of an Experimental World-Circling Spaceship*. RAND Report No. SM-11827. Santa Monica, CA: Douglas Aircraft Co. Engineering Division, 1946.

Cloud, John. "Imaging the World in a Barrel: Corona and the Clandestine Convergence of the Earth Sciences." *Social Studies of Science* 31 (2001): 231–251.

Clymer, Adam. "Smithsonian Will Move Observatory Here Soon: College Astronomers Plan Joint Research with Whipple Holding Dual Chairmanship." *Harvard Crimson*, 10 May 1955. http://www.thecrimson.com/article/1955/5/10/smithsonian-will-move-observatory-here-soon/.

Code, Arthur D., ed. *The Scientific Results from the Orbiting Astronomical Observatory (OAO-2)*. NASA SP-310. Washington, DC: Scientific and Technical Information Office, National Aeronautics and Space Administration, 1972.

Collins, Martin J. *Cold War Laboratory: RAND, the Air Force, and the American State, 1945–1950*. Washington, DC: Smithsonian Institution Press, 2002.

Committee on Science and Public Policy (NAS). "Ground-Based Astronomy: A Ten-Year Program: A Report Prepared by the Panel on Astronomical Facilities for the Committee on Science and Public Policy of the National Academy of Sciences." *Science* 146, no. 3652 (1964): 1641–1648.

Compton, William David. *Where No Man Has Gone Before. A History of Apollo Lunar Exploration Missions*. NASA SP-4214. Washington, DC: National Aeronautics and Space Administration, 1989. http://catalog.hathitrust.org/api/volumes/oclc/21926772.html.

Compton, William David, and Charles D. Benson. *Living and Working in Space: A History of Skylab*. NASA SP-4214. Washington, DC: Scientific and Technical Information Branch, National Aeronautics and Space Administration, 1983.

Computer Sciences Corporation. *NASA Directory of Observation Station Locations*. Washington, DC: U.S. National Aeronautics and Space Administration, 1970.

"The Construction and Use of Star Catalogues Conference." *Astronomical Journal* 72, no. 5 (1967): 551–552.

Corliss, William R. (Goddard Space Flight Center, United States, and National Aeronautics and Space Administration). *The Evolution of the Satellite Tracking and Data Acquisition Network (STADAN)*. Greenbelt, MD: Goddard Space Flight Center, 1967.

Cortright, Edgar M. *Apollo Expeditions to the Moon*. NASA SP-350. Washington, DC: Scientific and Technical Information Office, National Aeronautics and Space Administration, 1975.

David, James E. *Spies and Shuttles: NASA's Secret Relationship with the DoD and CIA*. Smithsonian: University Press of Florida, 2015.

Davidson, Keay. *Carl Sagan: A Life*. New York: Wiley, 1999.

Davies, John M., and Eugene S. Cotton. "Design of the Quartermaster Solar Furnace." *Solar Energy* 1, no. 2–3 (April 1957): 16–22.

Davis, Robert J. "A 21-Centimeter Survey of the Milky Way between Galactic Longitudes 166 Degrees and 250 Degrees." PhD diss., Harvard University, 1959.

———. "The Use of the Uvicon-Celescope Television System for Ultra-Violet Astronomical Photometry." *Advances in Electronics and Electron Physics* 22 (1966): 875–884.

Davis, Robert J., William A. Deutschman, and Katherine Haramundanis. *Celescope Catalog of Ultraviolet Stellar Observations: 5068 Objects Measured by the Smithsonian Experiment Aboard the Orbiting Astronomical Observatory (OAO-2)*. Washington, DC: Smithsonian Institution, 1973.

Day, Dwayne A., John M. Logsdon, and Brian Latell, eds. *Eye in the Sky: The Story of the Corona Spy Satellites*. Washington, DC: Smithsonian Institution Press, 1998.

DeFelice, J., G. G. Fazio, and E. L. Fireman. "Cosmic-Ray Exposure Age of the Farmington Meteorite from Radioactive Isotopes." *Science* 142, no. 3593 (1963): 673–674.

Dennis, Michael A. "A Change of State: The Political Cultures of Technical Practice at the MIT Instrumentation Laboratory and the Johns Hopkins University Applied Physics Laboratory, 1930–1945." PhD diss., Johns Hopkins University, 1991.

Deutschman, W. A., R. J. Davis, and R. E. Schild. "The Galactic Distribution of Interstellar Absorption as Determined from the Celescope Catalog of Ultraviolet Stellar Observations and a New Catalog of UBV, H-Beta Photoelectric Observations." *Astrophysical Journal Supplement Series* 30 (February 1976): 97.

DeVorkin, David H. "The Harvard Summer School in Astronomy." *Physics Today* 37, no. 7 (1984): 48.

———. "Organizing for Space Research: The V-2 Rocket Panel." *Historical Studies in the Physical and Biological Sciences* 18 (1987): 1–24.

———. "Defending a Dream—Charles Greeley Abbot's Years at the Smithsonian." *Journal for the History of Astronomy* 21 (February 1990): 121–136.

———. *Science with a Vengeance: How the Military Created the US Space Sciences after World War II*. New York: Springer-Verlag, 1992.

———. "Charles Greeley Abbot, 1872–1973." In *Biographical Memoirs of the National Academy of Sciences*, 73. Washington, DC: National Academies Press, 1998. http://

www.nasonline.org/publications/biographical-memoirs/memoir-pdfs/abbot_charles.pdf.

———. *Henry Norris Russell: Dean of American Astronomers.* Princeton, NJ: Princeton University Press, 2000a.

———. "Who Speaks for Astronomy? How Astronomers Responded to Government Funding after World War II." *Historical Studies in the Physical and Biological Sciences* 31 (2000b): 55–92.

———. "Menzel at Princeton." *Journal for the History of Astronomy* 33 (May 2002): 119–131.

———. *SAO during the Whipple Years: The Origins of Project Celescope.* Astrophysics and Space Science Library, 229–250. New York: Springer, 2005.

———. "A Referendum on the State of Astronomy at Harvard: Choosing Harlow Shapley's Successor." *American Astronomical Society Meeting Abstracts* 209 (December 2006a): 964.

———. "Serving the Muse in Shapley's Wake." In *The Hoffleit Centennial: A Year of Celebration*, edited by W. Van Altena and R. Koopmann, 11–20. Schenectady, NY: L. Davis Press, 2006b.

———. "The Space Age and Disciplinary Change in Astronomy." In *NASA's First 50 Years: Historical Perspectives*, edited by Steven J. Dick. NASA SP 2010-4704.Washington, DC: NASA, 2010.

Dick, Steven J. *The Biological Universe.* Cambridge: Cambridge University Press, 1996.

———. *Sky and Ocean Joined: The U.S. Naval Observatory 1830–2000.* Cambridge: Cambridge University Press, 2003.

———. *NASA's First 50 Years: Historical Perspectives.* NASA SP 2010-4704. Washington, DC: National Aeronautics and Space Administration, Office of Communications, History Division, 2010.

———. *Discovery and Classification in Astronomy Controversy and Consensus.* Cambridge: Cambridge University Press, 2013. http://public.eblib.com/choice/publicfullrecord.aspx?p=1303700.

Dick, Steven J., and LeRoy E. Doggett, eds. *Sky with Ocean Joined: Proceedings of the Sesquicentennial Symposia of the U.S. Naval Observatory, December 5 and 8, 1980.* Washington, DC: U.S. Naval Observatory, 1983.

Dickson, J. Home, ed. *Optical Instruments and Techniques 1969: Proceedings of the Conference Held at the University of Reading during 14th–19th July, 1969, under the Auspices of the International Commission for Optics and Arranged by the Optical Sub-Committee of the British National Committee for Physics.* Newcastle-upon-Tyne: Oriel Press, 1970.

Doel, Ronald E. "Redefining a Mission: The Smithsonian Astrophysical Observatory on the Move." *Journal for the History of Astronomy* 21 (February 1990): 137–153.

———. *Solar System Astronomy in America: Communities, Patronage, and Interdisciplinary Science, 1920–1960.* Cambridge: Cambridge University Press, 1996.

———. "Constituting the Postwar Earth Sciences: The Military's Influence on the Environmental Sciences in the USA after 1945." *Social Studies of Science* 33, no. 5 (2003): 635–666.

Duncombe, R. L., and J. S. Duncombe. "The Star Catalogue Project. A Progress Report." *Astronomical Journal* 72 (1967): 582–583.

Dupree, A. Hunter. "Central Scientific Organisation in the United States Government." *Minerva* 3, no. 4 (Summer 1963): 453–469.

———. *Science in the Federal Government: A History of Policies and Activities.* Baltimore: Johns Hopkins University Press, 1986.

———. *Asa Gray: American Botanist, Friend of Darwin.* Cambridge, MA: Harvard University Press, 1959. Reprint, Baltimore: Johns Hopkins University Press, 1988.

Earls, Alan R. *U.S. Army Natick Laboratories: The Science behind the Soldier.* Portsmouth, NH: Arcadia, 2005.

Edge, David O., and M. J. Mulkay. *Astronomy Transformed: The Emergence of Radio Astronomy in Britain.* New York: Wiley, 1976.

Edmondson, Frank K. *AURA and Its US National Observatories.* Cambridge: Cambridge University Press, 1997.

Elbers, Astrid. "Early Dutch Radio Astronomy (1940–1970): The People and the Politics." PhD diss., Leiden University, 2015.

Elliott, Clark A., and Margaret W. Rossiter. *Science at Harvard University: Historical Perspectives.* Bethlehem, PA: Lehigh University Press, 1992.

England, J. Merton. *A Patron for Pure Science.* Washington, DC: National Science Foundation, 1982.

Evans, David S., and J. Derral Mulholland. *Big and Bright: A History of the McDonald Observatory.* Austin: University of Texas Press, 1986.

Evans, J. W. "Princeton University Observatory Report." *Astronomical Journal* 72 (1967): 1128.

Fazio, G. "Gamma Radiation from Celestial Objects." *Annual Reviews of Astronomy and Astrophysics* 5 (1967), 481–524.

Fazio, G., H. F. Helmken, G. H. Rieke, and T. C. Weekes. "A Search for Discrete Sources of Cosmic Gamma Rays of Energies near 2×10^{12} eV." *The Astrophysical Journal* 154 (1968a).

———. "Upper Limits to Gamma Ray Fluxes from Three Pulsating Radio Sources." *Nature* 220, no. 5170 (1968b): 892–893.

Field, George B. "Radiation by Plasma Oscillations." *Astrophysical Journal* 124 (1956): 555–570.

Findlay, John W. "The 300-Foot Transit Telescope at the National Radio Astronomy Observatory." *Astronomical Journal* 68 (1963): 73.

Fireman, E. L. "Program for Collecting Meteorites." *Science* 131, no. 3416 (1960): 1824–1831.

———. "Freshly Fallen Meteorites from Portugal and Mexico." *Sky & Telescope* 37 (May 1969). https://ui.adsabs.harvard.edu/#abs/1969S&T....37..272F/abstract.

Florence, Ronald. *The Perfect Machine: Building the Palomar Telescope.* New York: Harper-Collins, 1994.

Flügge, Siegfried, ed. *Handbuch der Physik.* Vol. 46, no. 1 "Cosmic Radiation." Berlin: Springer, 1961.

Forman, Paul. "Behind Quantum Electronics: National Security as Basis for Physical Research in the United States, 1940–1960." *Historical Studies in the Physical and Biological Sciences* 18 (1987): 149–229.

Foukal, P. V., P. E. Mack, and J. E. Vernazza. "The Effect of Sunspots and Faculae on the Solar Constant." *Astrophysical Journal* 215 (1977).

(FLWO) *Fred L. Whipple Observatory Event History* by Don Hogan. http://www.sao.arizona.edu/FLWO/flwohis.html

Fries, Sylvia D. "The Ideology of Science during the Nixon Years, 1970–76." *Social Studies of Science* 14 (1984): 323–341.

Galison, Peter. *How Experiments End.* Chicago: University of Chicago Press, 1987.

Galison, Peter, and Bruce Hevly, eds. *Big Science: The Growth of Large-Scale Research.* Stanford, CA: Stanford University Press, 1992.

Galison, Peter, Bruce Hevly, and Rebecca Lowen. "Controlling the Monster: Stanford and the Growth of Physics Research, 1935–1962." In *Big Science: The Growth of Large-Scale Research*, edited by Peter Galison and Bruce Hevly, 46–77. Stanford, CA: Stanford University Press, 1992.

Gaposchkin, E. M. *Differential Orbit Improvement (DOI-3).* SAO Special Report 161. Cambridge, MA: Smithsonian Astrophysical Observatory, 1964.

———. *1973 Smithsonian Standard Earth (III).* Cambridge, MA: Smithsonian Astrophysical Observatory, 1973.

Gaposchkin, E. M., and Kurt Lambeck. *1969 Smithsonian Standard Earth (II).* SAO Special Report 315. Cambridge, MA: Smithsonian Astrophysical Observatory, 1970.

Gebel, R. K. H., and Lloyd R. Wylie. "The Cateye." *Sky & Telescope* 18 (December 1958): 84–85.

Geiger, Roger L. *Research and Relevant Knowledge: American Research Universities since World War II.* New York: Oxford University Press, 1993.

Gilbert, G. Nigel. "Growth and Decline of a Scientific Specialty: The Case of Radar Meteor Research." *Eos, Transactions, American Geophysical Union*, 58 (May 1977), 273–277. Repr. *History of Geophysics*, 2 (2013): 95–99.

Gingerich, O., R. W. Noyes, W. Kalkofen, and Y. Cuny. "The Harvard-Smithsonian Reference Atmosphere." *Solar Physics* 18, no. 3 (1971): 347–365.

Gingerich, Owen. "The Central Bureau for Astronomical Telegrams." *Physics Today* 21, no. 12 (1968): 36.

———. “Shapley’s Impact.” In *The Harlow Shapley Symposium on Globular Cluster Systems in Galaxies*, edited by Jonathan E. Grindlay and A. G. Philip, 23–33. Proceedings of IAU Symposium No. 126. Dordrect: Kluwer, 1988. http://www.worldcat.org/title/shapleys-impact/oclc/6681992379&referer=brief_results.

Goetze, G. W. “Transmission Secondary Emission from Low Density Deposits of Insulators.” In *Image Intensifier Symposium*. NASA-SP-2. Washington, DC: NASA, 1962.

Gold, T. “The Lunar Surface.” *Monthly Notices of the Royal Astronomical Society* 115 (1955): 585–603.

Goldberg, Leo, Martin Huber, W. H. Parkinson, E. M. Reeves, and F. L. Tobey. “Simultaneous Shock Tube Measurements of Fe and CR F Values of Astronomical Interest.” Abstract. *Astronomical Journal* 70 (1965): 676.

Green, Constance McLaughlin, and Milton Lomask. *Vanguard: A History*. Washington, DC: Scientific and Technical Information Division, National Aeronautics and Space Administration, 1970.

Green, Martin. “Sensitivity and Image Radius Characteristics of an SEC Camera Used to Record Faint Stars.” Abstract. *Astronomical Journal* 72 (1967): S14.

———. “Observations with an Ultrasensitive TV Camera.” *Sky & Telescope* 35, no. 3 (March 1968): 140–145.

Greenberg, Daniel S. “Navy’s Big Dish: Zooming Costs, Reduced Need, Bring End to Plans for Biggest Radio Telescope.” *Science* 137 (1962), 327.

———. “Smithsonian: Under New Secretary It Is Seeking to Regain Place as Center for Scientific Research.” *Science* 147 (March 12, 1965): 1266–1269.

———. *Science, Money and Politics: Political Triumph and Ethical Erosion*. Chicago: University of Chicago Press, 2003.

Greisen, Kenneth. *The Physics of Cosmic X-Ray, [Gamma]-Ray, and Particle Sources*. New York: Gordon and Breach, 1971.

“Ground-Based Astronomy: A Ten-Year Program: A Report Prepared by the Panel on Astronomical Facilities for the Committee on Science and Public Policy of the National Academy of Sciences.” *Science* 146, no. 3652 (1964): 1641–1648.

Guston, David H., and Kenneth Keniston, eds. *The Fragile Contract: University Science and the Federal Government*. Cambridge, MA: MIT Press, 1994.

Hacker, Barton C. “Military Patronage and the Geophysical Sciences in the United States.” *Historical Studies in the Physical and Biological Sciences* 30 (2000): 309–313.

Hall, R. Cargill. *Lunar Impact: A History of Project Ranger*. NASA SP-4210. Washington, DC: NASA, Scientific and Technical Information Office, 1977.

Haramundanis, K. “Experience of the Smithsonian Astrophysical Observatory in the Construction and Use of Star Catalogues.” *Astronomical Journal* 72, no. 5 (1967): 588–597.

Haramundanis, K., and Cecilia Payne-Gaposchkin. “Intrinsic Ultraviolet Colors from OAO-2 Celescope Observations for Stars on the Main Sequence.” *Astronomical Journal* 78 (1973): 395–400.

———. *Cecilia Payne-Gaposchkin: An Autobiography and Other Recollections*. Cambridge: Cambridge University Press, 1996.

Hagihara, Yusuke. *Celestial Mechanics: Dynamical Principles and Transformation Theory*. Foundations of Celestial Mechanics, Vol. 1. Cambridge: MIT Press, 1970.

———. *Celestial Mechanics: Perturbation Theory*. Foundations of Celestial Mechanics, Vol. 2. Cambridge: MIT Press, 1972.

———. *Celestial Mechanics: Differential Equations in Celestial Mechanics*. Foundations of Celestial Mechanics, Vol. 3. Tokyo: Japan Society for the Promotion of Science, 1975.

———. *Celestial Mechanics: Periodic and Quasi-Periodic Solutions*. Foundations of Celestial Mechanics, Vol. 4. Tokyo: Japan Society for the Promotion of Science, 1976.

———. *Celestial Mechanics: Topology of the Three-Body Problem*. Foundations of Celestial Mechanics, Vol. 5. Tokyo: Japan Society for the Promotion of Science, 1977.

Harwit, Martin. *Cosmic Discovery: The Search, Scope, and Heritage of Astronomy*. New York: Basic Books, 1981.

Hawkins, G. S. "The Harvard Radio Meteor Project." *Smithsonian Contributions to Astrophysics* 7 (1963): 53–62.

Hawkins, Gerald S., Curtis L. Hemenway, and Fred L. Whipple. "The Harvard Radio-Meteor Program." *Astronomical Journal* 61 (1956).

Hayes, E. Nelson. "The Smithsonian's Satellite-Tracking Program: Its History and Organization," part 1. In *Annual Report of the Board of Regents of the Smithsonian Institution [for] 1961*. Publication 4478. Washington, DC: Smithsonian Institution, 1962.

———. *Trackers of the Skies*. Washington, DC: Smithsonian Institution, 1967.

"Haystack Radar-Radio Telescope Dedicated." *Sky & Telescope* 28 (December 1964): 349–351.

Heeschen, David S. "Harvard's New Radio Telescope." *Sky & Telescope* 15 (July 1956): 588–590.

Hellman, Geoffrey. *The Smithsonian: Octopus on the Mall*. Westport, CT: Greenwood Press, 1978.

Helmken, H. "Study of the Low-Energy Solar and Celestial Gamma Radiation. Final Report, March 18–September 17, 1969." Smithsonian Astrophysical Observatory, 1968.

Heppenheimer, T. A. *The Space Shuttle Decision: NASA's Search for a Reusable Space Vehicle*. NASA SP-4221. Washington, DC: National Aeronautics and Space Administration, NASA History Office, Office of Policy and Plans, 1999.

Herget, Paul. "Armin Otto Leuschner, January 16, 1868–April 22, 1953." *Biographical Memoirs of the National Academy of Sciences* 49 (1978): 129–147.

Hershberg, James G. *James B. Conant: Harvard to Hiroshima and the Making of the Nuclear Age*. New York: Knopf, 1993.

Hirsh, Richard F. *Glimpsing an Invisible Universe: The Emergence of X-Ray Astronomy*. Cambridge: Cambridge University Press, 1983.

Hoffleit, Dorrit. "The Milton Bureau Revisited." *Journal of the American Association of Variable Star Observers (JAAVSO)* 28 (1 June 2000): 127–148.

———. *Misfortunes as Blessings in Disguise: The Story of My Life*. Cambridge, MA: American Association of Variable Star Observers, 2002. http://www.worldcat.org/title/misfortunes-as-blessings-in-disguise-the-story-of-my-life/oclc/51864161&referer=brief_results.

Hoffleit, Dorrit, A. G. Davis Philip, William F. Van Altena, and Arthur R. Upgren, eds. *Anni Mirabiles: A Symposium Celebrating the 90th Birthday of Dorrit Hoffleit: Held at Yale University, New Haven, Connecticut, March 7–8, 1997*. Schenectady, NY: L. Davis Press, 1997.

Holton, Gerald James, and Erik H Erikson. *The Twentieth-Century Sciences: Studies in the Biography of Ideas*. New York: Norton, 1970.

Hong, J., B. Allen, J. Grindlay, S. Barthelemy, R. Baker, A. Garson, H. Krawczynski, et al. "Flight Performance of an Advanced CZT Imaging Detector in a Balloon-Borne Wide-Field Hard X-Ray Telescope—ProtoEXIST1." *Nuclear Instruments and Methods in Physics Research, A* 654, no. 1 (2011): 361–372.

Hoppe, J. "Zur physikalischen Theorie des Ganges der Sternschnuppenhäufigkeit." *Astronomische Nachrichten* 264, no. 14 (1937): 233–240.

Horak, Henry G., and Charles A. Lundquist. "The Transfer of Radiation by an Emitting Atmosphere. II." *Astrophysical Journal* 119 (January 1954): 42–50.

Horn-d'Arturo, Guido. "The Tessellated Mirror." *Journal of the British Astronomical Association* 63 (1953): 71–74.

Hounshell, David A. "Rethinking the Cold War; Rethinking Science and Technology in the Cold War; Rethinking the Social Study of Science and Technology." *Social Studies of Science*. 31 (2001): 289–297.

Hoyt, Douglas V. "The Smithsonian Astrophysical Observatory Solar Constant Program." *Reviews of Geophysics* 17, no. 3 (1979): 427–458.

Hufbauer, Karl. *Exploring the Sun: Solar Science since Galileo*. Baltimore: Johns Hopkins University Press, 1991.

Hynek, J. Allen. *Astrophysics; a Topical Symposium Commemorating the Fiftieth Anniversary of the Yerkes Observatory and a Half Century of Progress in Astrophysics*. New York: McGraw-Hill, 1951.

IGY, ed. IGY World Data Center A: Rockets and Satellites. *Corpuscular Radiation, Magnetic Field, and Micrometeorite Observations with Satellites and Space Probes*. Washington, DC: National Academy of Sciences, 1961.

Izsák, Imre G. *On the Critical Inclination in Satellite Theory*. SAO Special Report 90. Cambridge, MA: Smithsonian Astrophysical Observatory, 1962.

Jacchia, Luigi Giuseppe. *The Earth's Gravitational Potential as Derived from Satellites 1957 Beta One and 1958 Beta Two*. SAO Special Report 19. Cambridge, MA: Smithsonian Astrophysical Observatory, 1958. http://adsabs.harvard.edu/abs/1958SAOSR.19....1J.

———. "Corpuscular Radiation and the Secular Acceleration of Satellites." *Astronomical Journal* 64 (1959).

———. "An Italian Astronomer." *Sky & Telescope* 34 (August 1967): 93.

Jacchia, Luigi G., and Jack Slowey. *Accurate Drag Determinations for Eight Artificial Satellites: Atmospheric Densities and Temperatures.* SAO Special Report 100. Cambridge, MA: Smithsonian Astrophysical Observatory, 1962.

Jacchia, Luigi G., and Fred L. Whipple. "Precision Orbits of 413 Photographic Meteors." *Smithsonian Contributions to Astrophysics* 4, no. 4 (1961): 97–129.

Jacobs, David Michael. *The UFO Controversy in America.* Bloomington: Indiana University Press, 1975.

Jeffreys, Harold. "The Determination of the Earth's Gravitational Field." *Geophysical Journal International* 5 (1941): 1–22.

Jones, Bessie Zaban. *Lighthouse of the Skies: The Smithsonian Astrophysical Observatory: Background and History, 1846–1955.* Washington, DC: Smithsonian Institution, 1965.

Jones, Bessie Zaban, and Lyle Gifford Boyd. *The Harvard College Observatory: The First Four Directorships, 1839–1919.* Cambridge, MA: Belknap Press of Harvard University Press, 1971.

Jones, R. C. "On the Detective Quantum Efficiency of Television Camera Tubes." *SMPTE Motion Imaging Journal* 68, no. 7 (July 1959): 462–466.

Jones, R. V. *Most Secret War.* London: Hamilton, 1978.

Kargon, Robert, and Stuart Leslie. "Imagined Geographies: Princeton, Stanford and the Boundaries of Useful Knowledge in Postwar America." *Minerva* 32, no. 2 (1994): 121–143.

Kaula, W. M. "Statistical and Harmonic Analysis of Gravity." *Journal of Geophysical Research* 64, no. 12 (1959): 2401–2421.

Keller, Morton, and Phyllis Keller. *Making Harvard Modern: The Rise of America's University.* New York: Oxford University Press, 2001. http://www.dawsonera.com/depp/reader/protected/external/AbstractView/S9780198033011.

Kevles, Daniel J. *The Physicists: The History of a Scientific Community in Modern America.* New York: Knopf, 1978.

Kidwell, P. "Harvard Astronomers in the Second World War." *Journal for the History of Astronomy* 21, no. 1 (1990): 105–106.

———. "Harvard Astronomers and World War II—Disruption and Opportunity." In *Science at Harvard University: Historical Perspectives*, edited by M. W. Rossiter and Clark A. Elliott, 285–299. Cranbury, NJ: Associated University Presses, 1992.

King, E. A., E. Schonfeld, K. A. Richardson, and J. S. Eldridge. "Meteorite Fall at Pueblito de Allende, Chihuahua, Mexico: Preliminary Information." *Science* 163, no. 3870 (1969): 928–929.

Klett, K. K. "Jet Propulsion Laboratory/Table Mountain Observatory, Wrightwood, California 92397. Report for the Period Jul 1990–Jul 1991." *Bulletin of the American Astronomical Society* 24 (January 1992): 119–120.

Kopal, Zdeněk. "Topography of the Moon." *Space Science Reviews* 4, no. 5–6 (1965): 737–855.

Koppes, Clayton R. *JPL and the American Space Program: A History of the Jet Propulsion Laboratory*. New Haven, CT: Yale University Press, 1982.

Kozai, Yoshihide. "The Motion of a Close Earth Satellite." *Astronomical Journal* 64 (1959): 367–377.

Krige, John. "Felix Bloch and the Creation of a 'Scientific Spirit' at CERN." *Historical Studies in the Physical and Biological Sciences* 32 (2001): 57–69.

Kuiper, Gerard P., and Barbara M. Middlehurst. *Stars and Stellar Systems*. Vol. 1, *Telescopes*. Chicago: University of Chicago Press, 1960.

Kurucz, Robert L. *ATLAS: A Computer Program for Calculating Model Stellar Atmospheres*. Cambridge, MA: Smithsonian Institution, 1970.

Kurucz, Robert L., Eric Peytremann, and Eugene H. Avrett. *Blanketed Model Atmospheres for Early-Type Stars*. Washington, DC: Smithsonian Institution, 1974.

Langley, S. P. "Solar Physics—Part I." *The Observatory* 3 (1880): 501–506.

———. *The New Astronomy*. Boston: Ticknor & Co., 1888.

———. "Smithsonian Standards for Physical Apparatus." *Nature* 45 (1 December 1891): 197.

Langway, Chester C., and Ursula B. Marvin. "Some Characteristics of Black Spherules." *Annals of the New York Academy of Sciences* 119, no. 1 (1964): 205–223.

Lankford, John, and Ricky L. Slavings. *American Astronomy: Community, Careers, and Power, 1859–1940*. Chicago: University of Chicago Press, 1997.

Launius, Roger D. *Exploring the Solar System: The History and Science of Planetary Exploration*. New York: Palgrave, 2013. http://www.worldcat.org/title/exploring-the-solar-system-the-history-and-science-of-planetary-exploration/oclc/793573175&referer=brief_results.

Launius, Roger D., and Howard E. McCurdy, eds. *Spaceflight and the Myth of Presidential Leadership*. Urbana: University of Illinois Press, 1997.

LeCompte, G. P. *Project Cat Eye: A History of Light Amplification Research at the Aeronautical Research Laboratory 1952–1960*. Pub. no. AD 274028. Washington, DC: OAR-4, Historical Division, U.S. Air Force Office of Aerospace Research, 1962.

Learner, Richard. *Astronomy through the Telescope*. New York: Van Nostrand Reinhold, 1981.

Lenoir, Timothy. *Instituting Science: The Cultural Production of Scientific Disciplines*. Stanford, CA: Stanford University Press, 1997.

Leslie, Stuart W. *The Cold War and American Science: The Military-Industrial-Academic Complex at MIT and Stanford*. New York: Columbia University Press, 1993.

Leslie, Stuart W., and Bruce Hevly. "Steeple Building at Stanford: Electrical Engineering, Physics, and Microwave Research." *Proceedings of the IEEE* 73, no. 7 (n.d.): 1169–1180.

Levine, Arnold S. *Managing NASA in the Apollo Era*. NASA SP 4102. Washington, DC: NASA, 1982.

Levinson, A. A., ed. *Proceedings of the Apollo 11 Lunar Science Conference, Houston, Texas, January 5–8, 1970*. Vol. 1. New York: Pergamon, 1970.

Levy, David H. *The Man Who Sold the Milky Way: A Biography of Bart Bok*. Tucson: University of Arizona Press, 1993.

———. "Harvard Twilight: The Exit of Shapley and Bok." *Sky & Telescope* 87 (February 1994), 41–44.

Lilley, A. E. "The Temperature of Venus." *Astronomical Journal* 66 (March 1961): 290.

Lindemann, F. A., and G. M. B. Dobson. "A Theory of Meteors, and the Density and Temperature of the Outer Atmosphere to Which It Leads." *Proceedings of the Royal Society of London. Series A, Containing Papers of a Mathematical and Physical Character* 102, no. 717 (1923): 411–437.

Lindsay, E. M. "The ADH (Armagh-Dunsink-Harvard) Telescope." *Irish Astronomical Journal* 2 (1953): 140–146.

Logsdon, John M. ed. w/ Amy Paige Snyder, Roger D. Launius, Stephen J. Garber, and Regan Anne Newport. *Exploring the Unknown: Selected Documents in the History of the U.S. Civilian Space Program*. Vol. 5. NASA SP-2001-4407. Washington DC: NASA, 2001.

Logsdon, John M., ed. w/ Linda J. Lear, J. Warren Findley, R. A. Williamson, and D. Day. *Exploring the Unknown: Selected Documents in the History of the U.S. Civil Space Program*. Vol. 1. NASA SP-4407. Washington, DC: NASA, 1995.

Lomask, Milton. *A Minor Miracle: An Informal History of the National Science Foundation*. Washington, DC: National Science Foundation, 1976.

Lowen, Rebecca S. *Creating the Cold War University: The Transformation of Stanford*. Berkeley: University of California Press, 1997. http://search.ebscohost.com/login.aspx?direct=true&scope=site&db=nlebk&db=nlabk&AN=8674.

Lubkin, Gloria. "The Decision to Build the Very Large Array." Unpublished manuscript, 1975.

Lubowe, Anthony G. "How Critical Is the Critical Inclination?" *Celestial Mechanics* 1, no. 1 (1969): 6–10.

Lundquist, C. A. *Geodetic Satellite Results during 1967*. SAO Special Report 264. Cambridge, MA: Smithsonian Astrophysical Observatory, 1967. https://ui.adsabs.harvard.edu/#abs/1967SAOSR.264.....L/abstract.

Lundquist, Charles A., and Henry D. Friedman, eds. *Scientific Horizons from Satellite Tracking*. Cambridge, MA: Smithsonian Astrophysical Observatory, 1966.

Lynds, Beverly T., and Leo Goldberg. *Employment Problems in Astronomy: Report by the Astronomy Manpower Committee*. Washington, DC: National Academy of Sciences, 1975.

Mack, Pamela E., and David H. DeVorkin. "Pro-Seminar on Space-History." *Technology and Culture* 23 (1982): 202–206.

Mack, Pamela Etter. *Viewing the Earth: The Social Construction of the Landsat Satellite System*. Cambridge, MA: MIT Press, 1990.

MacKenzie, Donald A. *Inventing Accuracy: An Historical Sociology of Nuclear Missile Guidance*. Cambridge, MA: MIT Press, 1993.

Marché, Jordan D., and Adam J. Walsh. "The Wisconsin Experiment Package (WEP) aboard the Orbiting Astronomical Observatory (OAO-2)." *Journal of Astronomical History and Heritage* 9 (2006). https://ui.adsabs.harvard.edu/#abs/2006JAHH....9..185M/abstract.

Marra, Monica. "New Astronomy Library in Bologna Is Named After Guido Horn D'Arturo: A Forefather of Modern Telescopes." *Journal of the British Astronomical Association* 110 (April 2000): 88.

Marsden, Brian G. "Amateur Astronomers and the IAU Central Bureau for Astronomical Telegrams and Minor Planet Center." *International Astronomical Union Colloquium* 98 (1988): 64–76.

Marvin, Ursula B. "Edward Leonard Fireman, 1922–1990." *Bulletin of the American Astronomical Society* (1992): 1322–1323.

———. "Oral Histories in Meteoritics and Planetary Science: XIII. Fred L. Whipple." *Meteoritics and Planetary Science,* supplement 39 (2004): A199–A214.

Marvin, Ursula B., and Cornelis Klein. "Meteoritic Zircon." *Science* 146, no. 3646 (1964): 919–920.

Marvin, Ursula B., J. A. Wood, and J. S. Dickey. "Ca-Al Rich Phases in the Allende Meteorite." *Earth and Planetary Science Letters* 7, no. 4 (1970): 346–350.

Mayall, Margaret W., ed. *Centennial Symposia, December 1946. Contributions on Interstellar Matter, Electronic and Computational Devices, Eclipsing Binaries [and] the Gaseous Envelope of the Earth.* Harvard Observatory Monograph No. 7. Cambridge, MA: Harvard College Observatory, 1948.

McArthur, Charles W. *Operations Analysis in the U.S. Army Eighth Air Force in World War II.* Providence, RI: American Mathematical Society, 1990.

McCall, Gerald Joseph Home, A. J. Bowden, and Richard John Howarth, eds. *The History of Meteoritics and Key Meteorite Collections: Fireballs, Falls and Finds.* London: The Geological Society, 2006.

McCorkell, R., E. L. Fireman, C. C. Langway, and U. B. Marvin. "Aluminum-26 and Beryllium-10 in Greenland Ice." *Science* 158, no. 3809 (1967): 1690–1692.

McCray, W. Patrick. "Large Telescopes and the Moral Economy of Recent Astronomy." *Social Studies of Science* 30, no. 5 (2000): 685–711.

———. *Giant Telescopes: Astronomical Ambition and the Promise of Technology*. Cambridge, MA: Harvard University Press, 2004a.

———. "National Observatories: Contention Continues." *Physics Today* 57, no. 4 (2004b): 14.

———. "Amateur Scientists, the International Geophysical Year, and the Ambitions of Fred Whipple." *Isis* 97 (2006): 634–658.

———. *Keep Watching the Skies! The Story of Operation Moonwatch & the Dawn of the Space Age.* Princeton, NJ: Princeton University Press, 2008.

McCrosky, Richard E. "Fragmentation of Faint Meteors." *Astronomical Journal* 60 (1955a).

———. "Some Physical and Statistical Studies of Meteor Fragmentation." PhD diss., Harvard University, 1955b.

McCrosky, Richard E., and H. Boeschenstein Jr. "The Prairie Meteorite Network." *SAO Special Report* 173 (May 1, 1965): 173.

McDougall, Walter A. *The Heavens and the Earth: A Political History of the Space Age*. New York: Basic Books, 1985.

McElheny, Victor K. "Radio Astronomy: Dicke Panel Reaches Its Conclusions." *Science* 157, no. 3791 (1967): 907–910.

McGee, James Dwyer, and L. Mandel. *Advances in Electronics and Electron Physics Proceedings of the Second Symposium held at Imperial College, London, September 5–8, 1961*. Vol. 16. New York: Academic Press, 1962. http://site.ebrary.com/id/10235549.

McGee, James Dwyer, and William Leslie Wilcock, eds. *1st Symposium on Photo-Electronic Image Devices*. London: Academic Press, 1960.

Meadows, Arthur Jack. *Science and Controversy: A Biography of Sir Norman Lockyer*. Basingstoke: Macmillan, 2008.

———. *Big Science: Fifty Years of the Rutherford Appleton Laboratory (1957–2007)*. Morrisville, NC: Lulu, 2009.

Meeks, M. L., J. Ball, and A. Hull. "The Pointing Calibration of the Haystack Antenna." *IEEE Transactions on Antennas and Propagation* 16, no. 6 (1968): 746–751.

Meeks, M. L., and A. E. Lilley. "The Microwave Spectrum of Oxygen in the Earth's Atmosphere." *Journal of Geophysical Research* 68, no. 6 (1963): 1683–1703.

Meeks, M. L., and J. Ruze. "Evaluation of the Haystack Antenna and Radome." *IEEE Transactions on Antennas and Propagation* 19, no. 6 (1971): 723–728.

Meeus, J. "Smithsonian Astrophysical Observatory. Star Catalog." *Ciel et Terre* 83 (1967): 268.

Meinel, A. B. "Astronomical Observations from Space Vehicles." *Publications of the Astronomical Society of the Pacific* 71 (October 1959): 369.

Meinel, Aden B., and T. S. Byington. *Program for Electro-Optical Aerospace Surveillance Supported by Work at Cloudcroft Electro-Optical Experimental Station*. DTIC 1969. Tucson: Arizona University Tucson Optical Sciences Center. Ft. Belvoir Defense Technical Information Center, 1969. http://oai.dtic.mil/oai/oai?&verb=getRecord&metadataPrefix=html&identifier=AD0863697.

Menzel, Donald H. "The Harvard-M. I. T. Eclipse Expedition." *Publications of the Astronomical Society of the Pacific* 48 (1936): 164.

———. *Donald Menzel Autobiography*. College Park, MD: American Institute of Physics Center for History of Physics, 1974.

Menzel, Donald H., Joseph Boyce, Henry Hemmindinger, Robert d'E. Atkinson, and Wallace Brode. *The Design and Operation of the Harvard-M.I.T. 1936 Eclipse Equipment*. Cambridge, MA: Harvard College Observatory, 1937.

Merritt, J. I. *Princeton's James Forrestal Campus: Fifty Years of Sponsored Research*. Princeton, NJ: Trustees of Princeton University, 2002.

Mertz, Lawrence. "Astronomical Infrared Spectrometer." *Astronomical Journal* 70 (1965): 548–551.

———. "Astronomical Fourier Spectrometry." PhD diss., Harvard University, 1969.

———. "Analysis of Telescope Costs." *Bulletin of the American Astronomical Society* 3 (June 1971): 386.

Mihalas, D. "Summary/Prognosis of the Workshop." *Astronomical Society of the Pacific Conference Series* 288 (2003): 677–694.

Millman, P., and D. W. R. McKinley. "Meteors." In *The Moon, Meteorites and Comets*, edited by G. Kuiper and B. Middlehurst. Chicago: University of Chicago Press, 1963.

Morgenthaler, George W., and Howard D. Greyber, eds. *Astronomy from a Space Platform; Proceedings.* Science and Technology Series, American Astronautical Society. Vol. 28. Tarzana, CA: American Astronautical Society, 1972.

Morrison, P. "On Gamma-Ray Astronomy." *Il Nuovo Cimento* 7, no. 6 (1958): 858–865.

Morrison-Low, Alison D., Sven Dupré, Giorgio Strano, and Stephen Johnston, eds. *From Earth-Bound to Satellite Telescopes, Skills and Networks.* Leiden: Brill, 2012.

Morrow, C. T., ed. *Ballistic Missile and Aerospace Technology: Proceedings of the 6th Symposium on Ballistic Missile and Aerospace Technology, Held in Los Angeles, California, August 1961.* Vol. 3, *Propulsion, Space Science and Space Exploration.* New York: Academic Press, 1961.

Mudgway, Douglas J. *Uplink-Downlink: A History of the NASA Deep Space Network, 1957–1997.* NASA SP-2001-4227. Washington, DC: National Aeronautics and Space Administration, 2001.

Muir-Harmony, Teasel. "Tracking Diplomacy: The International Geophysical Year and American Scientific and Technical Exchange with East Asia." Master's thesis, Notre Dame University, 2009.

Muir-Harmony, Teasel, Peter Abrahams, and David H. DeVorkin. "Wide-Field Photographic Telescopes: The Yale, Harvard and Harvard/Smithsonian Meteor and Satellite Camera Network." In *From Earth-Bound to Satellite Telescopes, Skills and Networks*, edited by Alison D. Morrison-Low, Sven Dupré, Giorgio Strano, and Stephen Johnston. Leiden: Brill, 2012.

Mukerji, Chandra. *A Fragile Power: Scientists and the State.* Princeton, NJ: Princeton University Press, 1989. http://public.eblib.com/choice/publicfullrecord.aspx?p=1700551.

Multhauf, Robert, and Gregory Good. *A Brief History of Geomagnetism and a Catalogue of the Collection of the National Museum of American History.* Smithsonian Studies in History and Technology 48. Washington, DC: Smithsonian Institution Press, 1987.

Munns, David P. D. "'Wizards of the Micro-Waves': A History of the Radio Astronomy Community." PhD diss., Johns Hopkins University, 2002.

———. "If We Build It, Who Will Come? Radio Astronomy and the Limitations of 'National' Laboratories in Cold War America." *Historical Studies in the Physical and Biological Sciences* 34 (2003).

NASA. *Apollo 11 Lunar Landing Mission.* Press Kit NASA Release 69-83K. Washington, DC: NASA, 1969.

———. ed. *Legislative History on H.R. 12575: The Space Act of 1958: Providing for Research into Problems of Flight within and outside the Earth's Atmosphere.* Washington, DC: NASA Headquarters, Law Library, 1960.

Nasim, Omar. *Observing by Hand: Sketching the Nebulae in the Nineteenth Century.* Chicago: University of Chicago Press, 2013.

National Research Council (U.S.). *Ground-Based Astronomy: A Ten-Year Program.* Washington, DC: NAS/NRC, 1964. Reprint, 1966.

———. *Astronomy and Astrophysics for the 1970's: Report of the Astronomy Survey Committee.* Vol. 1. Washington, DC: National Academy of Sciences, 1972.

———. ed. *Space Research: Directions for the Future; Report of a Study by the Space Science Board.* Washington, DC: National Academy of Sciences-National Research Council, 1966.

———. ed. *Funding a Revolution: Government Support for Computing Research.* Washington, DC: National Academy Press, 1999. http://site.ebrary.com/id/10041036.

National Science Policy, H. Con. Res. 666. Hearings before the Subcommittee on Science, Research, and Development of the Committee on Science and Astronautics U.S. House of Representatives, Ninety-First Congress, Second Session, July 7, 8, 21, 22, 23, 28, 29; August 4, 5, 11, 12, 13; September 15, 16, and 17, 1970. Washington, DC: U.S. Government Printing Office, 1970.

Needell, Allan A. "Lloyd Berkner, Merle Tuve, and the Federal Role in Radio Astronomy." *Osiris* 3 (1987): 261–288.

———. *Science, Cold War and the American State: Lloyd V. Berkner and the Balance of Professional Ideals.* Amsterdam: Harwood Academic, 2000.

Neufeld, Michael J. "Orbiter, Overflight, and the First Satellite: New Light on the Vanguard Decision." In *Reconsidering Sputnik: Forty Years since the Soviet Satellite*, edited by Roger D. Launius, John M. Logsdon, and Robert W. Smith, 231–257. Amsterdam: Harwood Academic, 2000.

———. "The End of the Army Space Program: Interservice Rivalry and the Transfer of the von Braun Group to NASA, 1958–1959." *Journal of Military History* 69 (2005): 737–757.

———. *Von Braun: Dreamer of Space, Engineer of War.* New York: Knopf, 2007.

Newell, Homer E. *Beyond the Atmosphere: Early Years of Space Science.* Washington, DC: NASA, 1980.

Newkirk, G. "Variations in Solar Luminosity." *Annual Review of Astronomy and Astrophysics* 21, no. 1 (1983): 429–464.

Odishaw, Hugh, ed. *The Earth in Space.* New York: Basic Books, 1967.

Oehser, Paul H. *The Smithsonian Institution.* New York: Praeger, 1970.

Ordway, Frederick I., Randy Liebermann, and Ben Bova. *Blueprint for Space: Science Fiction to Science Fact.* Washington, DC: Smithsonian Institution Press, 1992.

Oreskes, Naomi, and John Krige, eds. *Science and Technology in the Global Cold War*. Cambridge, MA: MIT Press, 2014. http://search.ebscohost.com/login.aspx?direct=true&scope=site&db=nlebk&db=nlabk&AN=884429.

Ortolano, Guy. *The Two Cultures Controversy: Science, Literature and Cultural Politics in Postwar Britain*. Cambridge: Cambridge University Press, 2009.

Osterbrock, Donald E. "The California-Wisconsin Axis in American Astronomy." *Sky & Telescope* 51 (1976): 9–14, 91–97.

———. "The Appointment of a Physicist as Director of the Astronomical Center of the World." *Journal for the History of Astronomy* 23 (August 1992): 155–165.

———. Yerkes *Observatory, 1892–1950: The Birth, Near Death, and Resurrection of a Scientific Research Institution*. Chicago: University of Chicago Press, 1997.

Osterbrock, Donald E., John R. Gustafson, and W. J. Shiloh Unruh. *Eye on the Sky: Lick Observatory's First Century*. Berkeley: University of California Press, 1988.

Parsons, S. B., J. D. Wray, G. F. Benedict, Y. Kondo, and K. G. Henize. "SKYLAB Ultraviolet Stellar Spectra—Cool Stars with Hot Secondaries." *Astrophysical Journal* 203 (January 1976): 435–437.

Payne-Gaposchkin, Cecilia, A. G. Davis Philip, and Rebecca A. Koopmann, eds. *The Starry Universe: The Cecilia Payne-Gaposchkin Centenary; Proceedings of a Symposium Held at the Harvard-Smithsonian Center for Astrophysics, Cambridge, Massachusetts, October 26–27, 2000*. Schenectady, NY: L. Davis Press, 2001.

Peebles, Curtis. *Guardians: Strategic Reconnaissance Satellites*. Novato, CA: Presidio Press, 1987.

Perry, Robert. *A History of Satellite Reconnaissance*. Vol. 1, *Corona*. Chantilly, VA: National Reconnaissance Office. http://www.nro.gov/foia/docs/hosr/hosr-vol1.pdf.

Peytremann, E., and R. J. Davis. "Ultraviolet Television Data from the Orbiting Astronomical Observatory. II—Stellar Ultraviolet Colors and Interstellar Extinction." *Astrophysical Journal Supplement Series* 28 (August 1974): 211–245.

Philip, A. G. Davis, Rebecca Koopmann, and William F. Van Altena, eds. *The Hoffleit Centennial: A Year of Celebration*. Institute for Space Observations 21. Schenectady, NY: L. Davis Press, 2006.

Physical Society of London, ed. *The Emission Spectra of the Night Sky and Aurorae; Papers Read at an International Conference Held under the Auspices of the Gassiot Committee of the Royal Society in London in July 1947*. London: Physical Society, 1948.

Plotkin, H. "The Henderson Network versus the Prairie Network: The Dispute between the Smithsonian's National Museum and the Smithsonian Astrophysical Observatory over the Acquisition and Control of Meteorites, 1960–1970." *Journal of the Royal Astronomical Society of Canada* 91, no. 1 (1997): 32–38.

Pollack, J. B., and G. G. Fazio. "Nuclear Interactions of Cosmic Rays as the Source of the Synchrotron Radiating Particles of Our Galaxy." *Astrophysical Journal* 141 (1965): 730–744.

Portolano, Marlana. "John Quincy Adams's Rhetorical Crusade for Astronomy." *Isis* 91, no. 3 (2000): 480–503.

Reece, J. S., and J. G. Marsh. "Simultaneous Observation Solutions for NASA-MOTS and SPEOPT Station Positions on the North American Datum." *Bulletin Géodésique (1946–1975)* 111, no. 1 (1974): 53–71.

Reed, B. Cameron. "Catalog of Galactic OB Stars." *Astronomical Journal* 125, no. 5 (May 2003): 2531–2533.

Richelson, Jeffrey. *America's Secret Eyes in Space: The U.S. Keyhole Spy Satellite Program.* New York: Harper & Row, 1990.

Rinehart, John S. *Adventures of a Physicist: From Peddling News to Making It.* Santa Fe, NM: Sunstone Press, 1999.

Roback, H. "Do We Need a Department of Science and Technology?" *Science* 165, no. 3888 (4 July 1969): 36–43.

Roemer, Elizabeth. "Discovery and Observation of Close-Approach Asteroids." *Proceedings of International Astronomical Union Colloquium, NASA SP-267* 12 (1971): 643–648.

Roland, Alex. *Model Research: The National Advisory Committee for Aeronautics, 1915–1958.* Washington, DC: NASA: Scientific and Technical Information Branch, 1985.

Rudney, Robert S. *A Preliminary History of the OAO Program (1966–1968).* HHN-115. Washington, DC: NASA History Office, 1971.

Sandage, Allan. *Centennial History of the Carnegie Institution of Washington.* Vol. 1, *The Mount Wilson Observatory.* Cambridge: Cambridge University Press, 2004. https://ui.adsabs.harvard.edu/#abs/2004mwo..book.....S/abstract.

Savage, B. D. "Ultraviolet Photometry from the Orbiting Astronomical Observatory. XX. The Ultraviolet Extinction Bump." *The Astrophysical Journal* 199 (July 1975): 92–109.

Schaffer, Simon. "The Leviathan of Parsonstown: Literary Technology and Scientific Representation." In *Inscribing Science: Scientific Texts and the Materiality of Communication*, edited by Timothy Lenoir. Palo Alto, CA: Stanford University Press, 1998.

Schmidt, M. "Hale Observatories. Annual Report of the Director 1979–1980." *Carnegie Institution of Washington Year Book* 79 (1980): 589–593.

Sears, Derek W. G. "Oral Histories in Meteoritics and Planetary Science—XV: John Wood." *Meteoritics & Planetary Science* 47, no. 5 (2012): 903–915.

Shapley, Harlow. "General Notes." *Popular Astronomy* 42 (1934): 113.

Shapley, Harlow. "The South African Station of the Harvard Observatory." *Science* 67, no. 1733 (1928): 297–298.

———. ed. *Climatic Change: Evidence, Causes, and Effects.* Cambridge, MA: Harvard University Press, 1953. Reprint 1960.

Shayler, David, and Colin Burgess. *NASA'S Scientist-Astronauts.* Berlin: Springer, 2007. http://public.eblib.com/choice/publicfullrecord.aspx?p=336711.

Shindell, Matthew Benjamin. "Domesticating the Planets: Instruments and Practices in the Development of Planetary Geology." *Spontaneous Generations: A Journal for the History and Philosophy of Science* 4, no. 1 (2010): 191–230.

———. "From the End of the World to the Age of the Earth: The Cold War Development of Isotope Geochemistry at the University of Chicago and Caltech." In *Science and Technology in the Global Cold War*, edited by Naomi Oreskes and John Krige, 107–139. Cambridge, MA: MIT Press, 2014.

Slayton, Rebecca. "From a 'Dead Albatross' to Lincoln Labs: Applied Research and the Making of a Normal Cold War University." *Historical Studies in the Natural Sciences* 42, no. 4 (2012): 255–282.

Smith, Harlan J., and Richard E. McCrosky. "Night Cloud Coverage in the Southwest with Reference to Astronomical Observing Conditions." *Astronomical Journal* 59 (1954): 156–162.

Smith, Robert W. *The Space Telescope: A Study of NASA, Science, Technology, and Politics.* Cambridge: Cambridge University Press, 1989.

Smith, R. W., and J. N. Tatarewicz. "Replacing a Technology: The Large Space Telescope and CCDs." *Proceedings of the IEEE* 73, no. 7 (1985): 1221–1235.

Smithsonian Astrophysical Observatory (SAO). *A Catalog of Positions and Proper Motions of 258 997 Stars for the Epoch and Equinox of 1950.0.* "SAOSR 151." Cambridge, MA: Smithsonian Institution, Astrophysical Observatory, 1964.

———. *Star Catalog: Positions and Proper Motions of 258,997 Stars for the Epoch and Equinox of 1950.0.* Washington, DC: Smithsonian Institution, 1966.

———. *Smithsonian Astrophysical Observatory Star Atlas of Reference Stars and Nonstellar Objects.* Cambridge, MA: MIT Press, 1969.

Smithsonian Institution. *Annual Report of the Board of Regents of the Smithsonian Institution. 1950.* Washington, DC, 1950.

———. *Annual Report of the Board of Regents of the Smithsonian Institution, 1959.* Washington, DC, 1959.

———. *Annual Report of the Board of Regents of the Smithsonian Institution, 1964.* Washington, DC, 1964.

———. *Annual Report of the Board of Regents of the Smithsonian Institution, 1966.* Washington, DC, 1966.

———. *Report of the Secretary of the Smithsonian Institution and financial report of the Executive Committee of the Board of Regents for the Year Ending June 30 1963.* Washington, DC, 1963.

———. *Smithsonian Year: Annual Report of the Smithsonian Institution for the Year Ended 30 June 1965.* Washington, DC, 1965.

———. *Smithsonian Year: Annual Report of the Smithsonian Institution for the Year Ended 30 June 1966.* Washington, DC, 1966.

———. *Smithsonian Year: Annual Report of the Smithsonian Institution for the Year Ended 30 June 1968.* Washington: DC, 1968.

———. *Smithsonian Year: Annual Report of the Smithsonian Institution for the Year Ended 30 June 1969*. Washington: DC, 1969.

Snyder, Gordon Lewis. "S-66 Satellite Laser Tracking Experiment." Master's thesis, University of Pennsylvania, 1967.

Solomon, Leonard H. *Some Results at Baker-Nunn Tracking Stations.* SAO Special Report 244. Cambridge, MA: Smithsonian Astrophysical Observatory, 1967.

Sterne, T. E. "On Periodicities in Measures of the Solar Constant." *Proceedings of the National Academy of Sciences* 25, no. 11 (1939): 559–564.

———. "Formula for Inferring Atmospheric Density from the Motion of Artificial Earth Satellites." *Science* 127, no. 3308 (1958): 1245.

Sterne, T. E., K. F. Guthe, and W. O. Roberts. "On Possible Changes in the Solar 'Constant.'" *Proceedings of the National Academy of Sciences* 26, no. 6 (1940): 399–406.

Sterne, Theodore E., and Nannielou Dieter. "The Constancy of the Solar 'Constant.'" *Astronomical Journal* 62 (1957): 147.

———. "The Constancy of the Solar 'Constant.'" *Smithsonian Contributions to Astrophysics* 3 (1959): 9–12.

Stewart, Irvin. *Organizing Scientific Research for War: The Administrative History of the Office of Scientific Research and Development.* Boston: Little, Brown, 1948.

Strand, K. Aa., ed. *Stars and Stellar Systems III. Basic Astronomical Data.* Chicago: University of Chicago Press, 1963.

Strauss, David. *Percival Lowell: The Culture and Science of a Boston Brahmin.* Cambridge, MA: Harvard University Press, 2001.

Strömgren, B. "McDonald Observatory." *Astronomical Journal* 61 (October 1956): 354–355.

Sturdevant, Rick W. "From Satellite Tracking to Space Situational Awareness: The USAF and Space Surveillance, 1957–2007." *Air Power History* 55 (2008): 6–22.

Sullivan, Woodruff Turner. *Cosmic Noise: A History of Early Radio Astronomy.* Cambridge: Cambridge University Press, 2009.

Tatarewicz, Joseph N. *Space Technology and Planetary Astronomy, 1958–1981.* Bloomington: Indiana University Press, 1990.

Thomas, Shirley. *Satellite Tracking Facilities: Their History and Operation.* New York: Holt, Rinehart and Winston, 1963.

Tombaugh, Clyde William, J. C. Robinson, B. A. Smith, and A. S. Murrell. *The Search for Small Natural Earth Satellites: Final Technical Report.* University Park: New Mexico State University, Physical Science Laboratory, 1959. http://catalog.hathitrust.org/api/volumes/oclc/68160008.html.

True, Webster Prentiss. *The Smithsonian Institution.* Vol. 1. New York: Smithsonian Institution Scientific Series, 1929. http://siris-libraries.si.edu/ipac20/ipac.jsp?profile=liball&term=939691&index=BIB.

Trumpbour, John. *How Harvard Rules: Reason in the Service of Empire.* Boston: South End Press, 1989.

Tuve, M. A., W. Kent Ford, J. S. Hall, and W. A. Baum. "Results of Preliminary Tests of Cascaded Image Converters." *Publications of the Astronomical Society of the Pacific* 70, no. 417 (1958): 592–594.

Ulivi, Paolo, and David M. Harland. *Lunar Exploration: Human Pioneers and Robotic Surveyors.* London: Springer-Verlag, 2004.

Underhill, Anne Barbara. "Some Methods for Computing Model Stellar Atmospheres." *Quarterly Journal of the Royal Astronomical Society* 3 (1962): 7–20.

Van Allen, James A. *Scientific Uses of Earth Satellites.* Ann Arbor: University of Michigan Press, 1956.

———. *Origins of Magnetospheric Physics.* Iowa City: University of Iowa Press, 2004. http://public.eblib.com/choice/publicfullrecord.aspx?p=843261.

Vanguard Staff. "Project Vanguard Report No. 1." *NRL Report 4700* (13 January 1956).

Van Helden, Albert. "Building Large Telescopes: 1900–1950." In *The General History of Astronomy*, edited by M. A. Hoskin, vol. 4, 134–152. Cambridge: Cambridge University Press, 1984.

Van Keuren, David K. "Cold War Science in Black and White: U.S. Intelligence Gathering and Its Scientific Cover at the Naval Research Laboratory 1948–62." *Social Studies of Science* 31 (2001): 207–229.

Varsavsky, C. M. "Dynamical Limits on a Lunar Origin for Tektites." *Geochimica et Cosmochimica Acta* 14, no. 4 (1958): 291–303.

Veis, George. "Optical Tracking of Artificial Satellites." *Space Science Reviews* 2, no. 2 (1963): 250–296.

Visher, Stephen Sargent. *Scientists Starred, 1903–1943, in "American Men of Science"; a Study of Collegiate and Doctoral Training, Birthplace, Distribution, Backgrounds, and Developmental Influences.* Baltimore: Johns Hopkins Press, 1947.

Wang, Zuoyue. *In Sputnik's Shadow: The President's Science Advisory Committee and Cold War America.* New Brunswick, NJ: Rutgers University Press, 2008. http://public.eblib.com/choice/publicfullrecord.aspx?p=361661_0.

Warner, Deborah Jean. "From Tallahassee to Timbuktu: Cold War Efforts to Measure Intercontinental Distances." *Historical Studies in the Physical and Biological Sciences* 30, no. 2 (2000): 393–415.

———. "Political Geodesy: The Army, the Air Force, and the World Geodetic System of 1960." *Annals of Science* 59 (2002): 363–389.

Watson, Alan A. "The Discovery of Cherenkov Radiation and Its Use in the Detection of Extensive Air Showers." *Nuclear Physics B (Proceedings Supplements)* 212 (2011): 13–19.

Weekes, T. C., M. F. Cawley, D. J. Fegan, K. G. Gibbs, A. M. Hillas, P. W. Kowk, R. C. Lamb, et al. "Observation of TeV Gamma Rays from the Crab Nebula Using the Atmospheric Cerenkov Imaging Technique." *Astrophysical Journal* 342 (July 1989): 379–395.

Weekes, Trevor C., ed. *The MMT and the Future of Ground-Based Astronomy: Proceedings of a Symposium Held to Mark the Dedication of the Multiple Mirror Telescope at the*

Mount Hopkins Observatory, Arizona, on May 9, 1979. Cambridge, MA: Smithsonian Astrophysical Observatory, 1979.

———. "The Crab Nebula as a Standard Candle for TeV Gamma-Ray Astronomy." Abstract J7.003, American Physical Society April Meeting, 2012.

Weitekamp, Margaret A. *Right Stuff, Wrong Sex: America's First Women in Space Program*. Baltimore: Johns Hopkins University Press, 2004.

Wesselink, A. J. "Heat Conductivity and Nature of the Lunar Surface Material." *Bulletin of the Astronomical Institutes of the Netherlands* 10 (1948): 352–363.

Westwick, Peter J. *The National Labs: Science in an American System, 1947–1974*. Cambridge, MA: Harvard University Press, 2003.

Whipple, Fred L. "A Spectrophotometric Study of the Cepheid Variables [Eta] Aquilae and [Delta] Cephei." *Lick Observatory Bulletin* 16 (1932): 1.

———. "Photographic Meteor Studies, I." *Proceedings of the American Philosophical Society* 79, no. 4 (1938): 499–548.

———. "Upper Atmosphere Densities and Temperatures from Meteor Observations." *Popular Astronomy* 47 (1939): 419–425.

———. "Meteors and the Earth's Upper Atmosphere." *Reviews of Modern Physics* 15, no. 4 (1943): 246–264.

———. *Earth, Moon and Planets*. Philadelphia: Blakiston Company, 1941.

———. "A Comet Model I. The Acceleration of Comet Encke." *Astrophysical Journal* 111 (1950a): 375–394.

———. "On Tests of the Icy Conglomerate Model for Comets." *Astronomical Journal* 55 (1950b): 83.

———. "The Baker Super-Schmidt Meteor Cameras." *Astronomical Journal* 56 (1951): 144–145.

———. "Astronomy from the Space Station." *Sky & Telescope* 12 (April 1953): 151.

———. "Why Conquer Space?" Typescript (1953). Box 19. RU 7431. SIA. Published in *Astronautics* 1 no. 1 (1954): 7.

———. *New Horizons in Astronomy*. Washington, DC: Smithsonian Institution, 1956a.

———. "Smithsonian Contributions to Astrophysics." *Smithsonian Contributions to Astrophysics* 1, no. 1 (1956b): iii–iv.

———. "The Coming Exploration of Space." *Saturday Evening Post* 231 (16 August 1958a): 32–33.

———. "Optical and Visual Tracking of Artificial Satellites." In *Proceedings of the VIIIth International Astronautical Congress*, 429–435. Vienna: Springer-Verlag, 1958b.

———. "On the Lunar Dust Layer." *Vistas in Astronautics* 2 (1959): 267–272.

———. "Knowledge and Understanding of the Physical Universe as Determinants of Man's Progress." In *Knowledge among Men*, edited by P. Oehser, 173–191. New York: Simon & Schuster, 1966.

———. "Smithsonian Astrophysical Observatory." *Smithsonian Year 1966: Annual Report of the Smithsonian Institution*. Washington, DC, 1966.

———. "Stochastic Painting." *Leonardo* 1 (1968): 81–83.

———. *The Collected Contributions of Fred L. Whipple.* Vols. 1, 2. Cambridge, MA: Smithsonian Astrophysical Observatory, 1972a.

———. "The Incentive of a Bold Hypothesis: Hyperbolic Meteors and Comets." *Annals of the New York Academy of Sciences* 198 (1972b): 219–224.

Whipple, Fred L., and Wernher von Braun. "The Exploration." *Collier's* (25 October 1952): 36–47.

Whipple, Fred L., and Robert J. Davis. "Proposed Stellar and Interstellar Survey." *Astronomical Journal* 65 (1960): 287.

Whipple, Fred L., and Jesse L. Greenstein. "On the Origin of Interstellar Radio Disturbances." *Proceedings of the National Academy of Sciences* 23, no. 3 (1937): 177–181.

Whipple, Fred L., and Gerald S. Hawkins. "Meteors." In *Handbuch der Physik.* Vol. 52, 519–564. Berlin: Springer-Verlag, 1959.

Whipple, Fred L., and J. Allen Hynek. "Observations of Satellite I." *Scientific American* 197, no. 6 (1957): 37–43.

Whitten, C. A., and Kenneth H. Drummond, eds. *Contemporary Geodesy; Proceedings of a Conference Held at the Harvard College Observatory—Smithsonian Astrophysical Observatory, Cambridge, Massachusetts, December 1–2, 1958.* Vol. 4. AGU Geophysical Monograph. Washington, DC: American Geophysical Union, 1959.

Wilkes, Owen, and Nils Petter Gleditsch. "Optical Satellite Tracking: A Case Study in University Participation in Preparation for Space Warfare." *Journal of Peace Research* 15, no. 3 (1978): 205–225.

Williams, Thomas R., and Michael Saladyga. *Advancing Variable Star Astronomy: The Centennial History of the American Association of Variable Star Observers.* New York: Cambridge University Press, 2011.

Wilson, R., and A. Boksenberg. "Ultraviolet Astronomy." *Annual Review of Astronomy and Astrophysics* 7 (1969): 421–472.

Wood, John A. *Meteorites and the Origin of Planets.* New York: McGraw-Hill, 1968.

———. "Fragments of Terra Rock in the Apollo 12 Soil Samples and a Structural Model of the Moon." *Icarus* 16, no. 3 (1972): 462–501.

Wright, Helen. *Explorer of the Universe: A Biography of George Ellery Hale.* New York: Dutton, 1966.

Wright, Helen, Joan Warnow-Blewett, and Charles Weiner. *The Legacy of George Ellery Hale; Evolution of Astronomy and Scientific Institutions, in Pictures and Documents.* Cambridge, MA: MIT Press, 1972.

INDEX

Page numbers in italic refer to figures.